AF388981

Synthesis of Marine Natural Products and Molecules Inspired by Marine Substances

Synthesis of Marine Natural Products and Molecules Inspired by Marine Substances

Editor

Emiliano Manzo

MDPI • Basel • Beijing • Wuhan • Barcelona • Belgrade • Manchester • Tokyo • Cluj • Tianjin

Editor
Emiliano Manzo
Institute of Biomolecular
Chemistry
National Council of Research
Pozzuoli (Naples)
Italy

Editorial Office
MDPI
St. Alban-Anlage 66
4052 Basel, Switzerland

This is a reprint of articles from the Special Issue published online in the open access journal *Marine Drugs* (ISSN 1660-3397) (available at: www.mdpi.com/journal/marinedrugs/special_issues/ Synthesis_Inspired).

For citation purposes, cite each article independently as indicated on the article page online and as indicated below:

LastName, A.A.; LastName, B.B.; LastName, C.C. Article Title. *Journal Name* **Year**, *Volume Number*, Page Range.

ISBN 978-3-0365-1768-1 (Hbk)
ISBN 978-3-0365-1767-4 (PDF)

Contents

marine drugs

MDPI

Editorial

Synthesis of Marine Natural Products and Molecules Inspired by Marine Substances

Emiliano Manzo

Bio-Organic Chemistry Unit, Institute of Biomolecular Chemistry of the National Research Council (CNR), Via Campi Flegrei 34, 80078 Naples, Italy; emanzo@icb.cnr.it

Citation: Manzo, E. Synthesis of Marine Natural Products and Molecules Inspired by Marine Substances. *Mar. Drugs* **2021**, *19*, 208. https://doi.org/10.3390/md19040208

Received: 1 April 2021
Accepted: 7 April 2021
Published: 8 April 2021

Publisher's Note: MDPI stays neutral with regard to jurisdictional claims in published maps and institutional affiliations.

The sea covers more than 70% of Earth's surface and contains more than 300,000 organisms with huge biodiversity. These organisms represent an enormous tank of substances whose chemical structures are the result of the enormous ecological pressure for survival. Given their high chemical diversity, biological activity, biochemical specificity and other crucial molecular properties, marine natural products have always played a crucial role in the search for and discovery of novel bioactive molecules potentially useful for pharmacological applications. In fact the development of new "lead" compounds from the sea has always been—and is one of the main purposes of drug discovery [1].

The study and advanced progress of these products cannot be separated from the development of chemical synthesis and synthetic strategies aimed at the preparation and optimization of these substances and/or analogs, opening the way to new classes of biologically active compounds with pharmacological potential.

This Special Issue, comprising eight articles and one review, describes the synthetic methodologies and biological activity of different classes of bioactive marine metabolites and analogs crucial to favor pharmacological applications of these molecules.

Esposito et al. [2], with the aim to optimize the antimicrobial activity of marine derived imminosugars, efficient glycomimetics, reported a synthetic strategy for the preparation of their lipophilic analogs characterized by promising antimicrobial activity due to improved internalization within the bacterial cell. This behavior was in fact favored by major lipophilicity of these molecules. The one-pot strategy involved the conjugation of iminosugars with lipophilic moieties, such as cholesterol, through a cleavable succinic acid linker that would positively contribute to substance transfer and release within the bacterial cell. The synthetic procedure, favored by combined use of the polymer-supported triphenylphosphine, molecular iodine and imidazole [3] in a one-pot strategy with cholesterol, succinic acid and unprotected iminosugars, led to the final compounds in high yields.

Mazzotta et al. developed a synthetic strategy for the preparation of different new marine-derived labdane diterpenes analogs, characterized by the homodrimane backbone bearing flexible tails with various chemical properties [4]. They start from the commercial (+)-sclareolide. In particular, different modifications, such as amide, ester and carboxylic acid functions, were introduced on the sclaerolide. The subsequent derivatization reaction led to a series of homodrimane analogs characterized by important biological activities involving the activation of the transient receptor potential channel subfamily V member 4 and 1 (TRPV4 and TRPV1) channels [5–7] that have recently emerged as a pharmacological target for several respiratory diseases, including the severe acute respiratory syndrome coronavirus 2 (SARS-CoV-2) infection. Chemical determinants crucial for TRPV4 and TRPV1 antagonism were identified by structure–activity relationships. This study represents the first report of semisynthetic homodrimane TRPV4 antagonists, selective over TRPV1, and potentially useful as pharmacological tools for the development of novel TRPV4 channel modulators.

Martinez et al. showed a concise strategy for the preparation of sponge derived tetracyclic meroterpenoids characterized by a sesquiterpene skeleton linked to a phenolic or quinone part [8]. Aureol, strongylin A, cyclosmenospongine and smenoqualone are representative examples of this class of bioactive compounds. The preparation of aureol was made possible by using a short and efficient synthetic route relying on a C–C bond-forming reaction between albicanal and an aryllithium-derivative and a sequence of 1,2-hydride and 1,2-methyl shifts mediated by $BF_3 \bullet Et_2O$ as activator and water as initiator. Aureol and 5-epi-aureol obtained by this strategy are key intermediates opening the way for the synthesis of a large number of natural and synthetic tetracyclic meroterpenoids with potential antitumor and antiviral activity.

Ouyang et al. described the first eight steps of the total synthesis of 5′-*O*-α-D-Glucopyranosyl Tubercidin, a disaccharide 7-deazapurine nucleoside characterized by fungicidal activity [9]. The chemical approach, based on trichloroacetimidate strategy, consisted in one-pot Vorbrüggen glycosylation of protected ribose with 6-chloro-7-bromo-7-deazapurine and stereoselective α-*O*-glycosylation of 7-deazapurine nucleoside derivative with 2,3,4,6-tetra-*O*-benzyl-glucopyranosyl trichloroacetimidate.

Zhang et al. showed how the substitution on the defined position on the chemical structure of bioactive molecules analogs could be a determinant for biological activity [10]. In particular, the authors experienced that fluorination at the C18 position of the marine-derived largazole showed good tolerance towards inhibitory activity and selectivity of histone deacetylases (HDACs). Further substitution on valine residue in the fluoro-largazole's macrocyclic moiety with S-Me L-Cysteine or Glycine moieties was performed, leading to derivatives with totally different biological activity. In particular the S-Me L-Cysteine-modified analog displayed enhanced inhibition of all the tested HDACs. Furthermore, a molecular modeling analysis provided a rational explanation and structural evidence for the enhanced inhibitory activity. This new finding will aid the design of novel, potent HDAC inhibitors.

Eldehna et al. designed and synthetized different anticancer marine inspired indoles and bis-indoles, as Topsentin and Nortopsentin analogs [11]. This synthesis was based on replacing the heterocyclic spacer in the natural leads with a more flexible hydrazide linker between indole rings. In this approach, in fact, the rigid heterocyclic spacer in the marine natural products Topsentin and Nortopsentin was replaced by the flexible linker and this change resulted in the development of bis-indole scaffold with promising in vitro antitumor activity toward breast cancer cell lines. Different analogs were prepared and all the synthetic bis-indoles were characterized for antiproliferative action against human breast cancer (MCF-7 and MDA-MB-231) cell lines. These results suggested that the reported bis-indoles are good lead compounds for further optimization and development as potential efficient anti breast cancer drugs.

Pemha et al., starting from recinoleic acid, synthetized a series of 1-*O*-alkylglycerols containing different functional groups (methoxy, amide, fluoro, azide and hydroxy) on the alkyl chain and antimicrobial activity was evaluated for all compounds [12]. The hydroxy derivative displayed more promising and significant activity. It was evident that synthetic non-natural 1-*O*-alkylglycerols can be further explored as a new source of drugs.

Imperatore et al. reported the chemical synthesis of two prenyl-quinones and their corresponding dioxothiazine fused quinones [13]. The synthesis of the prenylated compounds was performed through an efficient and versatile strategy designed and developed in order to easily reproduce the chemodiversity within the thiazinoquinones library. These molecules were inspired to an antitumor marine compound, a geranylquinone with 1,1-dioxo-1,4-thiazine ring isolated from the ascidian *Aplidium conicum* and named aplidinone A [14–16]. The synthetic molecules showed a comparable toxicity against three different human cancer cell lines, breast adenocarcinoma (MCF-7), pancreas adenocarcinoma (Bx-PC3) and bone osteosarcoma (MG-63).

Vessella et al. in their review have collected different and representative reports regarding the total synthesis of fucosylated chondroitin sulfate (fCS) oligosaccharides and semisynthetic strategies to obtain fCS oligosaccharides and low molecular weight polysaccharides [17]. The importance of this review was based on the various potential biomedical applications of marine derived fCS, among them its anticoagulant activity. This is the first report on (semi)-syntheses of (macro)molecules resembling the structure of natural fCS polysaccharides. Total synthetic and/or semi-synthetic strategies have been discussed, underlining the advantages and drawbacks for each approach and also reporting the main results on the structure–bioactivity relationships. The authors are sure that the research targeting the (semi)-synthesis of fCS oligo- and polysaccharides and analogs will attract a growing interest in the next few years.

As guest editor, I am grateful to all the authors who contributed their excellent results to this Special Issue, all the reviewers who carefully evaluated the submitted manuscripts, and *Marine Drugs* for their support and kind help.

Funding: This research received no external funding.

Conflicts of Interest: The author declare no conflict of interest.

References

1. Carroll, A.R.; Copp, B.R.; Davis, R.A.; Keyzers, R.A.; Prinsep, M.R. Marine natural products. *Nat. Prod. Rep.* **2020**, *37*, 175–223. [CrossRef] [PubMed]
2. Esposito, A.; D'Alonzo, D.; D'Errico, S.; De Gregorio, E.; Guaragna, A. Toward the Identification of Novel Antimicrobial Agents: One-Pot Synthesis of Lipophilic Conjugates of N-Alkyl D- and L-Iminosugars. *Mar. Drugs* **2020**, *18*, 572. [CrossRef] [PubMed]
3. De Fenza, M.; D'Alonzo, D.; Esposito, A.; Munari, S.; Loberto, N.; Santangelo, A.; Lampronti, I.; Tamanini, A.; Rossi, A.; Ranucci, S.; et al. Exploring the effect of chirality on the therapeutic potential of N-alkyl-deoxyiminosugars: Anti-inflammatory response to Pseudomonas aeruginosa infections for application in CF lung disease. *Eur. J. Med. Chem.* **2019**, *175*, 63–71. [CrossRef]
4. Mazzotta, S.; Carullo, G.; Moriello, A.S.; Amodeo, P.; Di Marzo, V.; Vega-Holm, M.; Vitale, R.M.; Aiello, F.; Brizzi, A.; De Petrocellis, L. Design, Synthesis and In Vitro Experimental Validation of Novel TRPV4 Antagonists Inspired by Labdane Diterpenes. *Mar. Drugs* **2020**, *18*, 519. [CrossRef]
5. Vitale, R.M.; Moriello, A.S.; De Petrocellis, L. Natural compounds and synthetic drugs targeting the ionotropic cannabinoid members of transient receptor potential (TRP) channels. In *New Tools to Interrogate Endocannabinoid Signalling From Natural Compounds to Synthetic Drugs*; Maccarrone, M., Ed.; RSC: London, UK, 2020; in press.
6. Liedtke, W.; Choe, Y.; Martí-Renom, M.A.; Bell, A.M.; Denis, C.S.; Šali, A.; Hudspeth, A.J.; Friedman, J.M.; Heller, S. Vanilloid receptor-related osmotically activated channel (VR-OAC), a candidate vertebrate osmoreceptor. *Cell* **2000**, *103*, 525–535. [CrossRef]
7. Strotmann, R.; Harteneck, C.; Nunnenmacher, K.; Schultz, G.; Plant, T.D. OTRPC4, a nonselective cation channel that confers sensivity to extracellular osmolarity. *Nat. Cell Biol.* **2000**, *2*, 695–702. [CrossRef] [PubMed]
8. Martínez, A.R.; Enríquez, L.; Jaraíz, M.; Morales, L.P.; Rodríguez-García, I.; Ojeda, E.D. A Concise Route for the Synthesis of Tetracyclic Meroterpenoids: (±)-Aureol Preparation and Mechanistic Interpretation. *Mar. Drugs* **2020**, *18*, 441. [CrossRef]
9. Ouyang, W.; Huang, H.; Yang, R.; Ding, H.; Xiao, Q. First Total Synthesis of 5'-O-α-D-Glucopyranosyl Tubercidin. *Mar. Drugs* **2020**, *18*, 398. [CrossRef] [PubMed]
10. Zhang, B.; Ruan, Z.-W.; Luo, D.; Zhu, Y.; Ding, T.; Sui, Q.; Lei, X. Unexpected Enhancement of HDACs Inhibition by MeS Substitution at C-2 Position of Fluoro Largazole. *Mar. Drugs* **2020**, *18*, 344. [CrossRef]
11. Eldehna, W.M.; Hassan, G.S.; Al-Rashood, S.T.; Alkahtani, H.M.; Almehizia, A.A.; Al-Ansary, G.H. Marine-Inspired Bis-indoles Possessing Antiproliferative Activity against Breast Cancer; Design, Synthesis, and Biological Evaluation. *Mar. Drugs* **2020**, *18*, 190. [CrossRef]
12. Pemha, R.; Kuete, V.; Pagès, J.-M.; Pegnyemb, D.E.; Mosset, P. Synthesis and Biological Evaluation of Four New Ricinoleic Acid-Derived 1-O-alkylglycerols. *Mar. Drugs* **2020**, *18*, 113. [CrossRef] [PubMed]
13. Imperatore, C.; Della Sala, G.; Casertano, M.; Luciano, P.; Aiello, A.; Laurenzana, I.; Piccoli, C.; Menna, M. In Vitro Antiproliferative Evaluation of Synthetic Meroterpenes Inspired by Marine Natural Products. *Mar. Drugs* **2019**, *17*, 684. [CrossRef] [PubMed]
14. Menna, M.; Aiello, A.; D'Aniello, F.; Imperatore, C.; Luciano, P.; Vitalone, R.; Irace, C.; Santamaria, R. Conithiaquinones A and B, Tetracyclic Cytotoxic Meroterpenes from the Mediterranean Ascidian Aplidium conicum. *Eur. J. Org. Chem.* **2013**, *2013*, 3241–3246. [CrossRef]
15. Aiello, A.; Fattorusso, E.; Luciano, P.; Macho, A.; Menna, M.; Munoz, E. Antitumor Effects of Two Novel Naturally Occurring Terpene Quinones Isolated from the Mediterranean Ascidian Aplidium conicum. *J. Med. Chem.* **2005**, *48*, 3410–3416. [CrossRef] [PubMed]
16. Aiello, A.; Fattorusso, E.; Luciano, P.; Mangoni, A.; Menna, M. Isolation and structure determination of aplidinones A-C from the Mediterranean ascidian Aplidium conicum: A successful regiochemistry assignment by quantum mechanical 13C NMR chemical shift calculations. *Eur. J. Org. Chem.* **2005**, *2005*, 5024–5030. [CrossRef]
17. Vessella, G.; Traboni, S.; Laezza, A.; Iadonisi, A.; Bedini, E. (Semi)-Synthetic Fucosylated Chondroitin Sulfate Oligo- and Polysaccharides. *Mar. Drugs* **2020**, *18*, 293. [CrossRef]

 *marine drugs* 

Article

Toward the Identification of Novel Antimicrobial Agents: One-Pot Synthesis of Lipophilic Conjugates of *N*-Alkyl D- and L-Iminosugars

Anna Esposito [1], Daniele D'Alonzo [1], Stefano D'Errico [2], Eliana De Gregorio [3] and Annalisa Guaragna [4,*]

1   Department of Chemical Sciences, University of Naples Federico II, Via Cintia, 80126 Naples, Italy; anna.esposito5@unina.it (A.E.); daniele.dalonzo@unina.it (D.D.)
2   Department of Pharmacy, University of Naples Federico II, Via Domenico Montesano, 49, 80131 Napoli, Italy; stefano.derrico@unina.it
3   Department of Molecular Medicine and Medical Biotechnology, University of Naples Federico II, Via S. Pansini 5, 80131 Naples, Italy; eliana.degregorio@unina.it
4   Department of Chemical, Materials and Production Engineering, University of Naples Federico II, Piazzale V. Tecchio 80, 80125 Naples, Italy
*   Correspondence: annalisa.guaragna@unina.it

Received: 1 November 2020; Accepted: 16 November 2020; Published: 19 November 2020 

Abstract: In the effort to improve the antimicrobial activity of iminosugars, we report the synthesis of lipophilic iminosugars **10a–b** and **11a–b** based on the one-pot conjugation of both enantiomeric forms of *N*-butyldeoxynojirimycin (NBDNJ) and *N*-nonyloxypentyldeoxynojirimycin (NPDNJ) with cholesterol and a succinic acid model linker. The conjugation reaction was tuned using the established PS-TPP/I$_2$/ImH activating system, which provided the desired compounds in high yields (94–96%) by a one-pot procedure. The substantial increase in the lipophilicity of **10a–b** and **11a–b** is supposed to improve internalization within the bacterial cell, thereby potentially leading to enhanced antimicrobial properties. However, assays are currently hampered by solubility problems; therefore, alternative administration strategies will need to be devised.

Keywords: lipophilic iminosugars; polymer-supported triphenyl phosphine; cholesterol; antibacterial iminosugars

---

1. Introduction

Iminosugars are naturally occurring or synthetic glycomimetics with an amino function replacing the endocyclic oxygen of the corresponding carbohydrates [1]. Thanks to their excellent ability to act as modulators (inhibitors/enhancers) of the activity of carbohydrate processing enzymes, including glycosyl hydrolases [2,3], glycosyltransferases [4] or glycogen phosphorylases [5], iminosugars can be considered as the main and best-settled class of sugar mimetics described so far. Over the last few decades, interesting broad-spectrum therapeutic applications have been found that have led to the development of three iminosugar drugs (Figure 1), namely Glyset (*N*-hydroxyethyl-deoxynojirimycin, Miglitol, **1**), which is used for the therapeutic treatment of type 2 diabetes [6], Zavesca (*N*-butyldeoxynojirimycin, NBDNJ, **2**), which is used for lysosomal storage disorders including Gaucher and Niemann–Pick type C diseases [7–9], and Galafold (deoxygalactonojirimycin, DGJ, **3**), which was recently approved for the treatment of Fabry disease [10,11]. A variety of other iminosugars have also been identified as therapeutic candidates against malignancies [12], viral infections [13,14] and other genetic disorders, including cystic fibrosis [15,16].

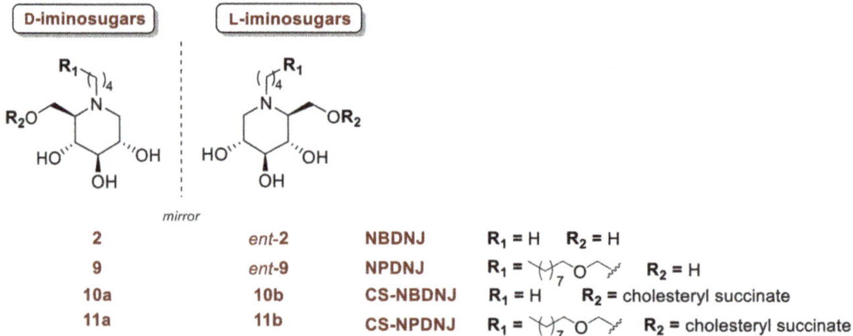

Figure 1. Marketed iminosugar drugs.

However, only rarely has this class of glycomimetics been considered for its antimicrobial activity. The antibacterial effect of iminosugars was first observed to be exhibited by the iminosugar progenitor nojirimycin (NJ, **4** Figure 2) against *Xanthomanos oryzae*, *Shigella flexeneri* and *Mycobacterium* 607 [17]. Similarly, deoxynojirimycin (DNJ, **5**) was found to inhibit biofilm formation in *Streptococcus mutans* [18,19]. The antimicrobial activity against *Staphylococcus epidermidis* was also highlighted by iminosugars extracted from a marine organism, such as batzellaside A–C (**6–8**, Figure 2), which were isolated from a Madagascar *Batzella* sponge [20,21] (Figure 2).

Figure 2. Naturally occurring iminosugars with antimicrobial properties.

Recently, as part of our longstanding research program that focused on the analysis of the effect of sugar chirality in the biological properties of iminosugars [15,22–24] and other enantiomeric bioactive compounds [25–28], we evaluated the antimicrobial potential of D- and L-DNJ, as well as that of a small library of their *N*-alkyl derivatives, against *Staphylococcus aureus* ATCC 29,213 [29]. Our data recognize a role of the lipophilicity of iminosugars in contributing to their antibacterial activity against *S. aureus* more evidently for the L-enantiomers. This led to the identification of *N*-nonyloxypentyl-L-DNJ (L-NPDNJ, *ent*-**9** Figure 3) as an interesting candidate, because of its capacity to affect growth, biofilm formation and virulence factor expression of *S. aureus* [29].

Figure 3. *N*-alkyl D- and L-iminosugars as antimicrobial agents against *S. aureus.*

Based on the assumption that the role of lipophilicity in the antimicrobial activity of *N*-alkyl iminosugars is mainly due to the favorable internalization of the molecules increasing along with the *N*-alkylation degree, we suggested that the conjugation of iminosugars with lipophilic moieties (even at different positions than the amino group) through a cleavable linker would positively contribute to iminosugar delivery and release within the bacterial cell. Accordingly, we conceived the preparation of iminosugars **10** and **11** bearing a cholesterol unit as the lipophilic group jointed by a short succinic bis-ester, as model linker [30,31], to the iminosugar moiety. In these early synthetic efforts, we selected as iminosugars *N*-butyl D- and L-DNJ (D- and L-NBDN, **2** and *ent*-**2** Figure 3) and *N*-nonyloxypentyl D- and L-DNJ (D- and L-NPDNJ, **9** and *ent*-**9** Figure 3). Contrary to the aforementioned activity of the latter, the former did not display antimicrobial properties against *S. aureus* ATCC 29,213 [29]. In line with our assumption, we studied whether conjugation could confer antibacterial activity to otherwise ineffective *N*-alkyl iminosugars while improving the properties of those already bearing an antimicrobial potential.

2. Results and Discussion

Chemistry

Our synthetic strategy was aimed to afford iminosugar conjugates **10a–b** and **11a–b** from the corresponding starting materials in a single reaction vessel, using a method that was able to activate the carboxylic moieties of succinic acid while leaving the remaining functionalities of cholesterol and iminosugars unreacted. To this end, we conceived as an activating agent the system based on the combined use of triphenyl phosphine (TPP) or its polymer-supported variant (PS-TPP), molecular iodine (I_2) and imidazole (ImH). The PS-TPP/I_2/ImH system has long been at the core of our synthetic studies aimed at many transformations, including the synthesis of β-amino acids [32] and of deuterated fatty acids [33,34], the formation of glycosyl iodides [35], the phosphorylation of nucleosides [36], the acetalization of sugars [37] and steroids [38] and the one-pot alkylation of iminosugars [15]. In all cases, the activation of alcohol or carboxylic acid functions to the corresponding iodides was achieved by the reaction of the last ones with pre-formed triphenyl-di-imidazolyl phosphorane [39] and congeners thereof, ultimately releasing the iodinating product and triphenylphosphine oxide. The use of TPP/I_2/ImH in one-pot procedures has already been successfully demonstrated [15]. In this case, a still unanswered question regards the selectivity of carboxylic acids vs. alcohols (Scheme 1), if both nucleophiles occur concurrently when performing the reaction in a single vessel.

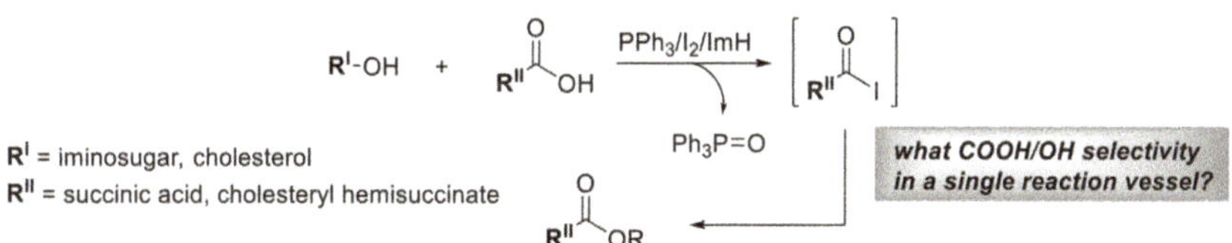

Scheme 1. Selectivity (COOH vs. OH) in the TPP/I_2/ImH reaction.

A stepwise procedure was initially explored to optimize the single reaction conditions. First, we studied the coupling reaction of succinic acid with cholesterol (Scheme 2). Succinic acid (**12**) was first treated with the pre-formed TPP/I_2/ImH complex in refluxing dioxane for 1.5 h (dioxane has been identified as the most suitable solvent among all those tested, mostly considering the best solubility of all reagents in view of the one step procedure). Then, upon cholesterol addition, the reaction mixture was left while stirring under reflux for 8 h, yielding cholesteryl hemisuccinate (**13**) in 82% (Scheme 2). The reaction gave roughly the same results when cholesterol was reacted, using a previously described procedure [40] with succinic anhydride and triethylamine in hot toluene (60 °C) for 8 h.

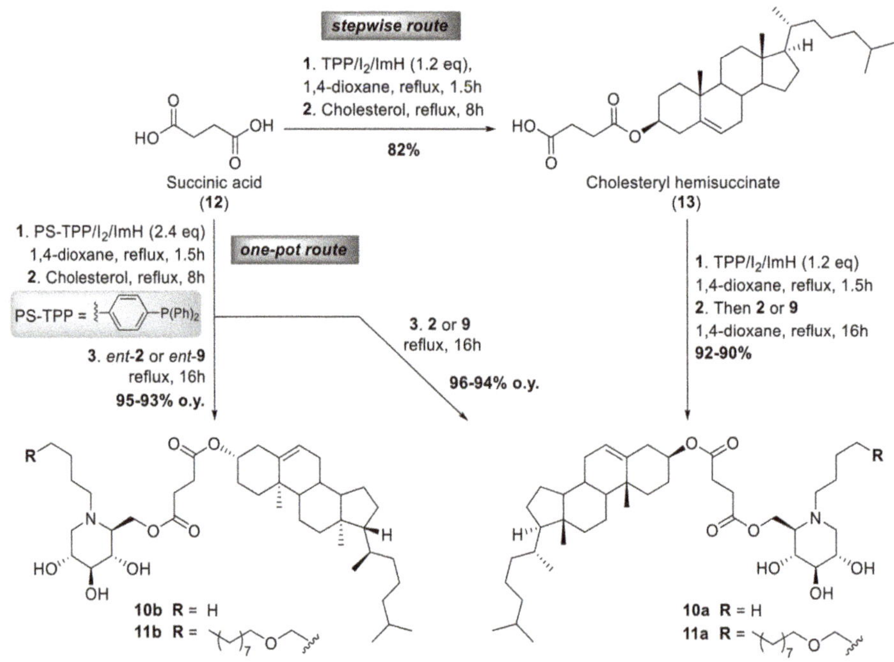

Scheme 2. Stepwise and one-pot routes to iminosugar conjugates **10a–b** and **11a–b**.

The subsequent coupling reaction of hemisuccinate **13** with unprotected iminosugars **2,9** and *ent*-**2,9** was then studied (Scheme 2). Successful activation of **13** with premixed TPP/I$_2$/ImH (1.2 eq) under previously described conditions (refluxing dioxane, 1.5 h) was suggested by the formation of a low-polarity, UV–visible [41] spot in the TLC analysis of the reaction mixture, which was a hint of covalent iodine incorporation by the starting material. The subsequent addition of D-NBDNJ (as model substrate) provided the desired cholesterol–iminosugar conjugate **10a** (stepwise route, Scheme 2). The complete conversion of the reagent (up to 92% yield) was achieved when heating the reaction mixture at reflux for 16 h. NMR analysis confirmed the occurred conjugation at the C6-OH group of D-NBDNJ, as indicated by the shift of diastereotopic methylene protons of the iminosugar moiety in DMSO-d_6 (**2**: H-6a, 3.54 ppm, H-6b, 3.72 ppm; **10a**: H-6a, 4.02 ppm, H-6b, 4.37 ppm). It is worth noting that more established coupling conditions were found as not efficient in this case. As an example, the reaction of hemisuccinate **13** and D-NBDNJ (**2**) with 2-(1H-benzotriazole-1-yl)-1,1,3,3-tetramethylaminium tetrafluoroborate (TBTU) and DIPEA, which has already been shown to succeed in the esterification of similar substrates [42], furnished conjugate **10a** only in moderate yield (49%; see Scheme S1 Supplementary Materials).

The above reactions pointed out the efficacy of the TPP/I$_2$/ImH system as coupling reagent for esterification reactions. Moreover, they indicated full compatibility and selectivity between alcohols and carboxylic acids of the reagents. However, it should be noted that only a slight excess of the iodinating complex (1.2 eq/eq of COOH) was always added to avoid further iodination reactions. The success of the stepwise approach enabled us to repeat the reaction using a one-pot procedure (Scheme 2). In this case, we chose to replace TPP with PS-TPP and to increase the synthetic potential of the procedure on a higher scale [15]. Accordingly, succinic acid was added to a stirring suspension of a 2-fold amount of PS-TPP/I$_2$/ImH (2.4 eq) to ensure the activation of both COOH functions for conjugation to both cholesterol and the iminosugar. After the addition of cholesterol, D-NBDNJ was added to the newly formed cholesteryl hemisuccinate iodide. In this case, the use of PS-TPP did not alter the overall reactivity; therefore, the reaction conditions tuned up with soluble TPP could be simply repeated. Filtration of the crude removed the residual phosphine and its oxide, while extractive work-up was required to remove the excess of imidazole. Eventually, the desired conjugate **10a** was recovered in a satisfying 96% yield. The same conditions were also effective when L-NBDNJ [24] was used as the iminosugar, leading to conjugate **10b** in 95% yield. Similarly, the reaction with D- and L-NPDNJ provided the corresponding conjugates **11a** and **11b** in 90% and 93% yields, respectively. As expected, ^{1}H NMR analysis of diastereomeric couples **10a–10b** and **11a–11b** indicated that the corresponding signals have wholly superimposable chemical shifts and multiplicities, presumably owing to the lack of interactions between the chiral centers of cholesterol and those of iminosugar enantiomers.

Preliminary biological evaluation of **10a–b** and **11a–b** was performed by broth microdilution assays, with the aim to measure the minimum inhibitory concentration (MIC) of iminosugar conjugates against *S. aureus*, a human pathogen that is responsible for a wide range of hospital-associated infections and has the capacity to develop multi-resistance to antibiotics. However, the assays were strongly hampered by the very limited solubility of the conjugates in the bacterial culture broth, which always precipitated even after pre-solubilization in DMSO. Studies are currently ongoing to overcome these limitations by searching for alternative solutions that will exploit the amphiphilic character of the glycomimetic agents.

3. Materials and Methods

3.1. Chemistry

All commercially available reagents and solvents were purchased at the highest degree of purity from commercial sources and used without purification. TLC analysis was carried out on precoated silica gel plate F254 (Merck), and products were visualized under UV radiation or by exposure to iodine vapor and chromic mixture. Column chromatography was performed with silica gel (70–230 mesh, Merck Kiesegel 60). CHNS analysis was performed to assess the purity of compounds and was ≥95% in all cases. NMR spectra were recorded on a Bruker AVANCE 400 MHz. Coupling constant values (J) were reported in Hz. Chemical synthesis and structural characterization of D-and L-NBDNJ (**2** and *ent*-**2**) and D-and L-NPDNJ (**9** and *ent*-**9**) were achieved as previously reported [15,24].

3.1.1. Procedure for the Synthesis of **10–11** through a Stepwise Route

Step 1: preparation of cholesteryl hemisuccinate **13**

I_2 (1.2 eq) and imidazole (2.4 eq) were added to a stirred solution of TPP (1.2 eq) in anhydrous 1,4-dioxane ($[I_2]$ = 62.5 mM) at rt. After 10 min, succinic acid (1.0 eq) was added to the slight yellow suspension, and the pH of the solution was adjusted to neutrality with the addition of imidazole. The resulting colorless solution was warmed to reflux temperature and stirred for 1.5 h. Cholesterol (1.0 eq) was then added, and the mixture was stirred at the same temperature for 8 h. The mixture was then cooled at rt, washed with brine and extracted with DCM. Organic layers were dried (Na_2SO_4) and evaporated under reduced pressure. Column chromatography of the crude residue over silica gel (hexane/EtOAc = 6:4) afforded the pure **13** (82% yield). ^{1}H and ^{13}C NMR spectra were fully in agreement with those reported in the literature [40,41].

Step 2: preparation of iminosugar conjugate **10a**

I_2 (1.2 eq) and imidazole (2.4 eq) were added to a stirred solution of TPP (1.2 eq) in anhydrous 1,4-dioxane at rt. After 10 min, hemisuccinate **13** (1.0 eq) was added to the slight yellow suspension, and the pH of the solution was adjusted to neutrality with the addition of imidazole. The resulting colorless solution was warmed to reflux temperature and stirred for 1.5 h. D-NBDNJ (1.0 eq) was then added, and the mixture was stirred at the same temperature for 8 h. The mixture was cooled at rt and diluted with DCM, and the organic layers were washed with brine, dried (Na_2SO_4) and concentrated under reduced pressure. Chromatography of the crude residue over silica gel (DCM/MeOH = 96:4) afforded the pure **10a** (92% yield) as a white solid.

3.1.2. General Procedure for the Synthesis of **10a–b** and **11a–b** through a One-Pot Route

I_2 (2.4 eq) and imidazole (4.8 eq) were added to a stirring solution of polymer-supported triphenylphosphine (PS-TPP; 100–200 mesh, extent of labeling: ~3 mmol/g triphenylphosphine loading) (2.4 eq) in anhydrous 1,4-dioxane at rt. After 10 min, succinic acid (1.0 eq) was added to the slight yellow suspension, and the pH of the solution was adjusted to neutrality by the addition of imidazole. The resulting colorless solution was warmed to reflux temperature and stirred for 1.5 h. Cholesterol (1.0 eq) was then added, and the mixture was stirred at the same temperature for 8 h. The appropriate iminosugar **2,9** and *ent*-**2,9** (1.0 eq) was then added, and the mixture was stirred for 16 h at the reflux temperature. The mixture was then cooled to rt, filtered (DCM) to remove triphenylphosphine oxide, washed with brine and extracted with DCM. Organic layers were dried (Na_2SO_4) and evaporated under reduced pressure, affording desired NBDNJ derivatives **10a** and **10b** (**10a**: 96% o.y.; **10b**: 95% o.y.) and NPDNJ derivatives **11a** and **11b** (**11a**: 94% o.y.; **11b**: 93% o.y.).

10a:^{1}H NMR (400 MHz, DMSO-d_6): δ 0.66 (s, 3H), 0.83–0.94 (m, 13H), 0.98 (s, 3H), 0.98–1.44 (m, 18H), 1.45–1.61 (m, 5H), 1.73–1.87 (m, 3H), 1.88–2.01 (m, 3H), 2.16 (ddd, *J* = 1.6, 4.3, 9.1, 1H, H-5′), 2.26 (td, *J* = 7.7, 1H, H-4), 2.25–2.36 (m, 1H, H-1″a), 2.53 (s, 4H, CH$_2$CH$_2$), 2.54–2.65 (m, 1H, H-1″b), 2.81 (dd, *J* = 4.8, 11.1, 1H, H-1′b), 2.92 (t, *J* = 9.1, 1H, H-3′), 3.01 (d, *J* = 5.3, 9.1, 1H, H-4′), 3.17–3.26 (m, 1H, H-2′), 4.05 (dd, *J* = 4.3, 12.0, 1H H-6′a), 4.39 (dd, *J* = 1.6, 12.0, 1H, H-6′b), 4.40–4.49 (m, 1H, H-3), 4.72 (d, *J* = 4.6, 1H, OH), 4.81 (bs, 1H, OH), 4.89 (d, *J* = 5.3, 1H, OH), 5.35 (d, *J* = 3.7, 1H, H-6). ^{13}C NMR (100 MHz, DMSO-d_6): δ 12.1, 14.4, 19.0, 19.4, 20.5, 21.0, 22.9, 23.1, 23.5, 24.2, 26.8, 27.8, 28.2, 29.2, 31.8; 35.6, 36.1, 36.5, 36.9, 38.1, 42.3, 49.9, 52.2, 56.0, 56.6, 57.2, 62.8, 64.3, 69.8, 70.9, 73.9, 79.4, 122.6, 139.9, 171.7, 172.3. $[\alpha]^{25}_D$ +27.0 (*c* 1.1, DMSO). Anal. calcd for $C_{41}H_{69}NO_7$: C, 71.58; H, 10.11; N, 2.04; O, 16.28. Found: C, 71.67; H, 10.07; N, 2.04.

10b: ^{1}H NMR (400 MHz, DMSO-d_6): δ 0.65 (s, 3H), 0.83–0.95 (m, 13H), 0.98 (s, 3H), 0.98–1.45 (m, 18H), 1.45–1.62 (m, 5H), 1.71–1.87 (m, 3H), 1.88–2.02 (m, 3H), 2.15 (d, *J* = 8.8, 1H, H-5′), 2.21–2.35 (m, 2H, H-4, H-1″a), 2.46 (s, 4H, CH$_2$CH$_2$), 2.54–2.65 (m, 1H, H-1″b), 2.81 (dd, *J* = 4.7, 11.0, 1H, H-1′b), 2.92 (t, *J* = 8.8, 1H, H-3′), 3.00 (t, *J* = 8.8, 1H, H-4′), 3.16–3.27 (m, 1H, H-2′), 4.04 (dd, *J* = 3.5, 12.0, 1H, H-6′a), 4.39 (d, *J* = 12.0, 1H, H-6′b), 4.42–4.52 (m, 1H, H-3), 4.65–4.94 (m, 3H, OH), 5.34 (d, *J* = 3.5, 1H, H-6). ^{13}C NMR data for compound **10b** were superimposable with those reported above for the corresponding diastereoisomer **10a**. $[\alpha]^{25}_D$ +35.0 (c 1.2, DMSO). Anal. calcd for $C_{41}H_{69}NO_7$: C, 71.58; H, 10.11; N, 2.04; O, 16.28. Found: C, 71.52;.H, 10.14; N, 2.04.

11a:^{1}H NMR (400 MHz, CD$_3$OD): δ 0.74 (s, 3H), 0.87–0.99 (m, 12H), 1.01–1.10 (m, 5H), 1.11–1.25 (m, 6H), 1.26–1.46 (m, 19H), 1.48–1.70 (m, 12H), 1.82–2.11 (m, 5H), 2.18 (t, *J* = 11.0, 1H, H-1′a), 2.28–2.39 (m, 3H), 2.44–2.54 (m, 1H, H-1″a), 2.59–2.67 (m, 4H, CH$_2$CH$_2$), 2.70–2.80 (m, 1H, H-1″a), 3.03 (dd, *J* = 4.9, 11.0, 1H, H-1b), 3.15 (t, *J* = 9.0, 1H, H-3′), 3.25–3.33 (m, 1H, H-4′), 3.40–3.45 (m, 4H), 3.46–3.54 (m, 1H, H-2′), 4.29 (dd, *J* = 3.5, 12.3, 1H, H-6′a), 4.51 (dd, *J* = 1.8, 12.3, 1H, H-6′b), 5.41 (d, *J* = 4.5, 1H, H-6). ^{13}C NMR (100 MHz, CD$_3$OD); δ 12.3, 14.5, 19.2, 19.7, 22.2, 22.9, 23.2, 23.8, 24.9, 25.2, 25.3, 27.3, 28.8, 29.1, 29.3, 29.9, 30.3, 30.7, 30.8, 33.0; 33.2, 37.1, 37.4, 37.8, 38.3, 39.2, 40.7, 41.1, 43.5, 51.6, 53.8, 57.6, 57.7, 58.1, 62.6, 65.4, 70.8, 71.9, 72.0, 72.1, 75.7, 80.4, 123.7, 141.0, 173.4, 174.1. $[\alpha]^{25}_D$ +46 (c 1.0, CH$_3$OH). Anal. calcd for $C_{51}H_{89}NO_8$: C, 72.55; H, 10.63; N, 1.66; O, 15.16. Found: C, 72.65; H, 10.59; N, 1.66.

11b:^{1}H NMR (400 MHz, CD$_3$OD): δ 0.75 (s, 3H), 0.87–1.00 (m, 12H), 1.01–1.10 (m, 5H), 1.11–1.25 (m, 6H), 1.26–1.47 (m, 19H), 1.48–1.70 (m, 12H), 1.83–2.10 (m, 5H), 2.18 (t, *J* = 10.9, 1H, H-1′a), 2.28–2.39 (m, 3H), 2.44–2.54 (m, 1H, H-1″a), 2.57–2.70 (m, 4H, CH$_2$CH$_2$), 2.70–2.81 (m, 1H, H-1″a), 3.03 (dd, *J* = 4.8, 11.3, 1H, H-1b), 3.15 (t, *J* = 9.0, 1H, H-3′), 3.25–3.33 (m, 1H, H-4′), 3.40–3.54 (m, 5H), 4.29 (dd, *J* = 3.4, 12.4, 1H, H-6′a), 4.52 (dd, *J* = 1.8, 12.4, 1H, H-6′b), 5.41 (d, *J* = 5.0, 1H, H-6). ^{13}C NMR data for compound **11b** were superimposable with those reported above for the corresponding diastereoisomer **11a**. $[\alpha]^{25}_D$ +40 (c 1.3, CH$_3$OH). Anal. calcd for $C_{51}H_{89}NO_8$: C, 72.55; H, 10.63; N, 1.66; O, 15.16. Found: C, 72.64; H, 10.60; N, 1.66.

3.2. Evaluation of Antibacterial Activity In Vitro

Broth microdilution assay was performed to determine the MIC value of compounds **10a–b** and **11a–b** as previously described [38]. Briefly, fresh overnight culture of *S. aureus* strains was diluted in Mueller–Hinton broth to 1×10^6 colony forming units per mL (CFU/mL). One hundred microliters of bacteria suspension (1×10^5 cfu) was dispensed into a 96-well microtiter plate containing the same volume of 2-fold serial dilutions of compounds **10a–b** and **11a–b**. The antibiotics gentamicin and oxacillin were used as positive controls. Following 16–24 h incubation at 37 °C, the optical density of each well was measured at 595 nm. The MICs were the lowest concentrations of compound to inhibit bacterial growth after incubation.

4. Conclusions

The synthesis of lipophilic iminosugars **10a–b** and **11a–b**, obtained by conjugation of NBDNJ and NPDNJ in both enantiomeric forms with cholesterol through a succinic acid linker, has been herein reported, using a one-pot procedure involving the use of PS-DPP/I$_2$/ImH as an activating system. Iminosugar conjugates **10a–b** and **11a–b** have been conceived to improve internalization within the bacterial cell compared to the corresponding unconjugated *N*-alkyl iminosugars **2** and **9**; thereby, they were expected to display more favorable antimicrobial properties. However, the marked increase in the lipophilicity of the synthesized iminosugars hampered the ability to perform in vitro assays because of the very limited solubility in water. To overcome these limitations, the focus in the future will be on the development of alternative strategies for in vitro assays, eventually exploiting the amphiphilic character of the glycomimetic agents.

Supplementary Materials: The following are available online at http://www.mdpi.com/1660-3397/18/11/572/s1. Scheme S1: Chemical synthesis of compound **10a** by an established procedure. Figure S1: ^{1}H spectrum of compound **13**; Figure S2: ^{1}H spectrum of compound **2**; Figure S3: ^{1}H and ^{13}C spectra of compound **10a**; Figure S4: ^{1}H and ^{1}H-^{1}H COSY spectra of compound **10b**; Figure S5: ^{1}H and ^{13}C spectra of compound **11a**; Figure S6: ^{1}H and ^{1}H-^{1}H COSY spectra of compound **11b**.

Author Contributions: Conceptualization, A.G.; methodology, A.E., D.D., S.D. and E.D.G.; investigation, E.D.G., A.E. and S.D.; data curation, S.D., D.D., A.E. and E.D.G.; writing—original draft preparation, A.G., D.D. and A.E.; supervision, A.G. and D.D.; funding acquisition, A.G. All authors have read and agreed to the published version of the manuscript.

Funding: This research was funded by the Italian Cystic Fibrosis Research Foundation, grant number FFC #13/2020 to AG.

Conflicts of Interest: The authors declare no conflict of interest.

References

1. Compain, P.; Martin, O.R. *Iminosugars: From Synthesis to Therapeutic Applications*; Compain, P., Martin, O.R., Eds.; John Wiley & Sons, Ltd.: Chichester, UK, 2007; ISBN 9780470517437.
2. D'Alonzo, D.; Guaragna, A.; Palumbo, G. Glycomimetics at the mirror: Medicinal chemistry of L-iminosugars. *Curr. Med. Chem.* **2009**, *16*, 473–505. [CrossRef] [PubMed]
3. Nash, R.J.; Kato, A.; Yu, C.Y.; Fleet, G.W. Iminosugars as therapeutic agents: Recent advances and promising trends. *Future Med. Chem.* **2011**, *3*, 1513–1521. [CrossRef] [PubMed]
4. Chen, X.; Wong, C.-H.; Ma, C. Targeting the bacterial transglycosylase: Antibiotic development from a structural perspective. *ACS Infect. Dis.* **2019**, *5*, 1493–1504. [CrossRef] [PubMed]
5. Hayes, J.M.; Kantsadi, A.L.; Leonidas, D.D. Natural products and their derivatives as inhibitors of glycogen phosphorylase: Potential treatment for type 2 diabetes. *Phytochem. Rev.* **2014**, *13*, 471–498. [CrossRef]
6. Skyler, J.S. Diabetes mellitus: Pathogenesis and treatment strategies. *J. Med. Chem.* **2004**, *47*, 4113–4117. [CrossRef] [PubMed]
7. Platt, F.M.; Jeyakumar, M. Substrate reduction therapy. *Acta Paediatr.* **2008**, *97*, 88–93. [CrossRef]

8. Cox, T.; Lachmann, R.; Hollak, C.; Aerts, J.; van Weely, S.; Hrebícek, M.; Platt, F.; Butters, T.; Dwek, R.; Moyses, C.; et al. Novel oral treatment of Gaucher's disease with N-butyldeoxynojirimycin (OGT 918) to decrease substrate biosynthesis. *Lancet* **2000**, *355*, 1481–1485. [CrossRef]

9. Coutinho, M.F.; Santos, J.I.; Alves, S. Less is more: Substrate reduction therapy for lysosomal storage disorders. *Int. J. Mol. Sci.* **2016**, *17*, 1065. [CrossRef]

10. Benjamin, E.R.; Della-Valle, M.C.; Wu, X.; Katz, E.; Pruthi, F.; Bond, S.; Bronfin, B.; Williams, H.; Yu, J.; Bichet, D.G.; et al. The validation of pharmacogenetics for the identification of Fabry patients to be treated with migalastat. *Genet. Med.* **2017**, *19*, 430–438. [CrossRef]

11. Markham, A. Migalastat: First global approval. *Drugs* **2016**, *76*, 1147–1152. [CrossRef]

12. Cox, T.M.; Platt, F.M.; Aerts, J.M.F.G. Medicinal use of iminosugars. In *Iminosugars*; John Wiley & Sons, Ltd.: Chichester, UK, 2008; pp. 295–326, ISBN 9780470033913.

13. Chang, J.; Block, T.M.; Guo, J.-T. Antiviral therapies targeting host ER alpha-glucosidases: Current status and future directions. *Antivir. Res.* **2013**, *99*, 251–260. [CrossRef] [PubMed]

14. Alonzi, D.S.; Scott, K.A.; Dwek, R.A.; Zitzmann, N. Iminosugar antivirals: The therapeutic sweet spot. *Biochem. Soc. Trans.* **2017**, *45*, 571–582. [CrossRef] [PubMed]

15. de Fenza, M.; D'Alonzo, D.; Esposito, A.; Munari, S.; Loberto, N.; Santangelo, A.; Lampronti, I.; Tamanini, A.; Rossi, A.; Ranucci, S.; et al. Exploring the effect of chirality on the therapeutic potential of N-alkyl-deoxyiminosugars: Anti-inflammatory response to Pseudomonas aeruginosa infections for application in CF lung disease. *Eur. J. Med. Chem.* **2019**, *175*, 63–71. [CrossRef] [PubMed]

16. Esposito, A.; D'Alonzo, D.; de Fenza, M.; de Gregorio, E.; Tamanini, A.; Lippi, G.; Dechecchi, M.C.; Guaragna, A. Synthesis and therapeutic applications of iminosugars in cystic fibrosis. *Int. J. Mol. Sci.* **2020**, *21*, 3353. [CrossRef]

17. Greimel, P.; Spreitz, J.; Stutz, A.; Wrodnigg, T. Iminosugars and relatives as antiviral and potential anti-infective agents. *Curr. Top. Med. Chem.* **2005**, *3*, 513–523. [CrossRef]

18. Islam, B.; Khan, S.N.; Haque, I.; Alam, M.; Mushfiq, M.; Khan, A.U. Novel anti-adherence activity of mulberry leaves: Inhibition of Streptococcus mutans biofilm by 1-deoxynojirimycin isolated from Morus alba. *J. Antimicrob. Chemother.* **2008**, *62*, 751–757. [CrossRef]

19. Hasan, S.; Singh, K.; Danisuddin, M.; Verma, P.K.; Khan, A.U. Inhibition of major virulence pathways of *Streptococcus mutans* by Quercitrin and Deoxynojirimycin: A synergistic approach of infection control. *PLoS ONE* **2014**, *9*, e91736. [CrossRef]

20. Segraves, N.L.; Crews, P. A Madagascar sponge *Batzella* sp. as a source of alkylated iminosugars. *J. Nat. Prod.* **2005**, *68*, 118–121. [CrossRef]

21. Vasconcelos, A.; Pomin, V. Marine carbohydrate-based compounds with medicinal properties. *Mar. Drugs* **2018**, *16*, 233. [CrossRef]

22. Guaragna, A.; D'Errico, S.; D'Alonzo, D.; Pedatella, S.; Palumbo, G. A general approach to the synthesis of 1-deoxy-L-iminosugars. *Org. Lett.* **2007**, *9*, 3473–3476. [CrossRef]

23. Guaragna, A.; D'Alonzo, D.; Paolella, C.; Palumbo, G. Synthesis of 1-deoxy-L-gulonojirimycin and 1-deoxy-L-talonojirimycin. *Tetrahedron Lett.* **2009**, *50*, 2045–2047. [CrossRef]

24. D'Alonzo, D.; de Fenza, M.; Porto, C.; Iacono, R.; Huebecker, M.; Cobucci-Ponzano, B.; Priestman, D.A.; Platt, F.; Parenti, G.; Moracci, M.; et al. N-Butyl-l-deoxynojirimycin (L-NBDNJ): Synthesis of an allosteric enhancer of α-glucosidase activity for the treatment of pompe disease. *J. Med. Chem.* **2017**, *60*, 9462–9469. [CrossRef] [PubMed]

25. D'Alonzo, D.; Guaragna, A.; van Aerschot, A.; Herdewijn, P.; Palumbo, G. L-Homo-DNA: Stereoselective de novo synthesis of β-L-erythro -hexopyranosyl nucleosides. *J. Org. Chem.* **2010**, *75*, 6402–6410. [CrossRef] [PubMed]

26. D'Alonzo, D.; Amato, J.; Schepers, G.; Froeyen, M.; Van Aerschot, A.; Herdewijn, P.; Guaragna, A. Enantiomeric selection properties of β-homoDNA: Enhanced pairing for heterochiral complexes. *Angew. Chem. Int. Ed.* **2013**, *52*, 6662–6665. [CrossRef] [PubMed]

27. Caso, M.F.; D'Alonzo, D.; D'Errico, S.; Palumbo, G.; Guaragna, A. Highly stereoselective synthesis of lamivudine (3TC) and emtricitabine (FTC) by a novel N-glycosidation procedure. *Org. Lett.* **2015**, *17*, 2626–2629. [CrossRef]

28. Esposito, A.; Giovanni, C.; de Fenza, M.; Talarico, G.; Chino, M.; Palumbo, G.; Guaragna, A.; D'Alonzo, D. A stereoconvergent tsuji–trost reaction in the synthesis of cyclohexenyl nucleosides. *Chem. Eur. J.* **2020**, *26*, 2597–2601. [CrossRef] [PubMed]

29. de Gregorio, E.; Esposito, A.; Vollaro, A.; de Fenza, M.; D'Alonzo, D.; Migliaccio, A.; Iula, V.D.; Zarrilli, R.; Guaragna, A. N-nonyloxypentyl-ʟ-deoxynojirimycin inhibits growth, biofilm formation and virulence factors expression of *Staphylococcus aureus*. *Antibiotics* **2020**, *9*, 362. [CrossRef]

30. Rautio, J.; Meanwell, N.A.; Di, L.; Hageman, M.J. The expanding role of prodrugs in contemporary drug design and development. *Nat. Rev. Drug Discov.* **2018**, *17*, 559–587. [CrossRef]

31. Wichitnithad, W.; Nimmannit, U.; Wacharasindhu, S.; Rojsitthisak, P. Synthesis, characterization and biological evaluation of succinate prodrugs of curcuminoids for colon cancer treatment. *Molecules* **2011**, *16*, 1888–1900. [CrossRef]

32. Capone, S.; Guaragna, A.; Palumbo, G.; Pedatella, S. Efficient synthesis of orthogonally protected anti-2,3-diamino acids. *Tetrahedron* **2005**, *61*, 6575–6579. [CrossRef]

33. Guaragna, A.; de Nisco, M.; Pedatella, S.; Pinto, V.; Palumbo, G. An expeditious procedure for the synthesis of isotopically labelled fatty acids: Preparation of 2,2-d2-nonadecanoic acid. *J. Label. Compd. Radiopharm.* **2006**, *49*, 675–682. [CrossRef]

34. Guaragna, A.; Amoresano, A.; Pinto, V.; Monti, G.; Mastrobuoni, G.; Marino, G.; Palumbo, G. Synthesis and proteomic activity evaluation of a new isotope-coded affinity tagging (ICAT) reagent. *Bioconjug. Chem.* **2008**, *19*, 1095–1104. [CrossRef] [PubMed]

35. Caputo, R.; Kunz, H.; Mastroianni, D.; Palumbo, G.; Pedatella, S.; Solla, F. Mild synthesis of protected α-D-glycosyl iodides. *Eur. J. Org. Chem.* **1999**, *1999*, 3147–3150. [CrossRef]

36. Caputo, R.; Guaragna, A.; Pedatella, S.; Palumbo, G. Mild and regiospecific phosphorylation of nucleosides. *Synlett* **1997**, *1997*, 917–918. [CrossRef]

37. Pedatella, S.; Guaragna, A.; D'Alonzo, D.; de Nisco, M.; Palumbo, G. Triphenylphosphine polymer-bound/iodine complex: A suitable reagent for the preparation of O-isopropylidene sugar derivatives. *Synthesis* **2006**, *18*, 305–308. [CrossRef]

38. Esposito, A.; de Gregorio, E.; de Fenza, M.; D'Alonzo, D.; Satawani, A.; Guaragna, A. Expeditious synthesis and preliminary antimicrobial activity of deflazacort and its precursors. *RSC Adv.* **2019**, *9*, 21519–21524. [CrossRef]

39. Garegg, P.J.; Regberg, T.; Stawiński, J.; Strömberg, R. A phosphorus nuclear magnetic resonance spectroscopic study of the conversion of hydroxy groups into iodo groups in carbohydrates using the iodine–triphenylphosphine–imidazole reagent. *J. Chem. Soc. Perkin Trans.* **1987**, *2*, 271–274. [CrossRef]

40. Klumphu, P.; Lipshutz, B.H. "Nok": A phytosterol-based amphiphile enabling transition-metal-catalyzed couplings in water at room temperature. *J. Org. Chem.* **2014**, *79*, 888–900. [CrossRef]

41. Kimura, K.; Nagakura, S. n → σ* Absorption spectra of saturated organic compounds containing bromine and iodine. *Spectrochim. Acta* **1961**, *17*, 166–183. [CrossRef]

42. Twibanire, J.K.; Grindley, T.B. Efficient and controllably selective preparation of esters using uronium-based coupling agents. *Org. Lett.* **2011**, *13*, 2988–2991. [CrossRef]

Publisher's Note: MDPI stays neutral with regard to jurisdictional claims in published maps and institutional affiliations.

marine drugs

Article

Design, Synthesis and In Vitro Experimental Validation of Novel TRPV4 Antagonists Inspired by Labdane Diterpenes

Sarah Mazzotta [1,2,†], Gabriele Carullo [1,3,†], Aniello Schiano Moriello [4,5], Pietro Amodeo [6], Vincenzo Di Marzo [4,7], Margarita Vega-Holm [2], Rosa Maria Vitale [6,*], Francesca Aiello [1,*], Antonella Brizzi [3,‡] and Luciano De Petrocellis [4,‡]

[1] Department of Pharmacy, Health and Nutritional Sciences, DoE 2018–2022, University of Calabria, Edificio Polifunzionale, 87036 Rende (CS), Italy; sarmaz1@alum.us.es (S.M.); gabriele.carullo@unisi.it (G.C.)
[2] Department of Organic and Medicinal Chemistry, Faculty of Pharmacy, University of Seville, Profesor García González 2, 41071 Seville, Spain; mvegaholm@us.es
[3] Department of Biotechnology, Chemistry and Pharmacy, DoE 2018–2022, University of Siena, Via Aldo Moro 2, 53100 Siena, Italy; brizzi3@unisi.it
[4] Endocannabinoid Research Group (ERG), Institute of Biomolecular Chemistry, National Research Council (ICB-CNR), Via Campi Flegrei 34, 80078 Pozzuoli (NA), Italy; aniello.schianomoriello@icb.cnr.it (A.S.M.); vincenzo.dimarzo@criucpq.ulaval.ca (V.D.M.); luciano.depetrocellis@icb.cnr.it (L.D.P.)
[5] Epitech Group SpA, 35030 Saccolongo (PD), Italy
[6] Institute of Biomolecular Chemistry, National Research Council (ICB-CNR), Via Campi Flegrei 34, 80078 Pozzuoli (NA), Italy; pamodeo@icb.cnr.it
[7] Canada Excellence Research Chair on the Microbiome-Endocannabinoidome Axis in Metabolic Health (CERC-MEND)-Université Laval, Québec City, QC G1V 0A6, Canada
*   Correspondence: rmvitale@icb.cnr.it (R.M.V.); francesca.aiello@unical.it (F.A.)
†   These authors contributed equally.
‡   Antonella Brizzi and Luciano De Petrocellis are joint senior authors.

Received: 3 September 2020; Accepted: 14 October 2020; Published: 18 October 2020

Abstract: Labdane diterpenes are widespread classes of natural compounds present in variety of marine and terrestrial organisms and plants. Many of them represents "natural libraries" of compounds with interesting biological activities due to differently functionalized drimane nucleus exploitable for potential pharmacological applications. The transient receptor potential channel subfamily V member 4 (TRPV4) channel has recently emerged as a pharmacological target for several respiratory diseases, including the severe acute respiratory syndrome coronavirus 2 (SARS-CoV-2) infection. Inspired by the labdane-like bicyclic core, a series of homodrimane-derived esters and amides was designed and synthesized by modifying the flexible tail in position 1 of (+)-sclareolide, an oxidized derivative of the bioactive labdane-type diterpene sclareol. The potency and selectivity towards rTRPV4 and hTRPV1 receptors were assessed by calcium influx cellular assays. Molecular determinants critical for eliciting TRPV4 antagonism were identified by structure-activity relationships. Among the selective TRPV4 antagonists identified, compound **6** was the most active with an IC$_{50}$ of 5.3 µM. This study represents the first report of semisynthetic homodrimane TRPV4 antagonists, selective over TRPV1, and potentially useful as pharmacological tools for the development of novel TRPV4 channel modulators.

Keywords: labdane scaffold; bioactive diterpenes; sclareolide; structure-activity relationships; TRPV4 channel; amides/esters; COVID-19; SARS-CoV-2

1. Introduction

Labdanes are bicyclic diterpenes found as secondary metabolites in marine organisms and in plants, characterized by a high chemical diversity. They exhibit a broad panel of pharmacological activities, ranging from antimicrobial and anti-inflammatory to cytotoxic and antitumoral ones [1].

The diversity of organisms, nature, extent and distribution of chemical modifications and functionalization and the corresponding spread of biological functions make the labdane diterpenes and their derivatives a set of naturally occurring libraries of bioactive compounds [2].

In the field of drug discovery, natural compounds represent useful tools for the development of transient receptor potential (TRP) channel modulators [3,4].

In particular, TRPV4 is a polymodal, non-selective cation channel, belonging to the vanilloid subfamily (V) member 4 of the TRP ion channels [5–7]. It is activated by a series of physical and chemical stimuli, including temperature, pH, hypotonicity, and stretch, as well as arachidonic acid and its metabolites. TRPV4 is a homo-tetramer sharing an overall architecture similar to that of other TRPV family members, featuring a transmembrane domain (TMD), consisting of helices S1 to S6, and a cytosolic region formed by the N- and C-terminal domains. The transmembrane helices S5–S6 and the pore loop form the pore channel, flanked by an S1–S4 voltage-sensor like domain (VSLD) showing a peculiar arrangement as emerged from the recently solved structures of TRPV4 from *Xenopus tropicalis* [8].

TRPV4 is implicated in various physiological processes due to it high expression in various tissues of the human body [8]. In particular, it is expressed in alveolo-capillary and immune cells of the immune system, such as alveolar macrophages and neutrophil granulocytes, which contribute to alveolo-capillary barrier function through proteases and cytokine release, as well as reactive oxygen species production [9].

TRPV4 has recently emerged as a pharmacological target for the treatment of pulmonary oedema caused by COVID-19 (coronavirus disease of 2019). TRPV4-evoked calcium uptake in lung endothelium has been associated with elevated pulmonary vascular pressure, lung congestion, and resulting dyspnea. Selective TRPV4 agonists have been shown to increase lung permeability in a dose-dependent manner in wild-type mice but not in TRPV4 knockout mice, suggesting the advantage of TRPV4 inhibition in lung oedema treatment [10].

To date, only a limited number of TRPV4 modulators have been identified; thus, the discovery and the development of new selective TRPV4 ligands represent an attractive challenge [11,12]. The first identified TRPV4 agonist was bisandrographolide A (BAA, EC_{50} 790–950 nM, Figure 1), a plant dimeric diterpenoid [13]. Among the antagonists, the quinoline-carboxamide GSK2193874, as well as 1-(4-piperidinyl)-benzimidazole amides [14], were developed for the treatment of pulmonary oedema associated with congestive heart failure [15].

The pyridine polyketide onydecalin A (Figure 1) was also validated as a TRPV4 antagonist (IC_{50} 45.9 μM), with a partial activity towards another TRPV channel, i.e., member 1 (TRPV1) [16].

The occurrence of a *trans*-decalin lipophilic moiety in two plant-derived TRPV4 modulators, i.e., bisandrographolide A, an agonist, and onydecalin A, an antagonist, as well as in labdane terpenes, such as (+)-yahazunol [17,18] and sclareol [19] (Figure 1), prompted us to exploit the potential of the homodrimane bicyclic nucleus for the development of a new class of semi-synthetic TRPV4 modulators.

Interestingly, the flexible tail of labdanescaffold is differently functionalized in natural compounds or even replaced by different functional group families, pointing to a suitable region to be varied for the creation of targeted libraries toward specific classes of pharmacologically interesting proteins.

In the present study, we focused on (+)-sclareolide (Figure 1), a fragrant compound found in *Salvia sclarea*, used as flavor additive in food. It represents the oxidized derivative of the bioactive labdane-type diterpene sclareol. (+)-Sclareolide gained attention due to its many biomedical properties, including anticancer and antiviral activities [20–22]. In addition, it also displays chemical versatility, since its lactone ring condensed with a *trans*-decalin-related homodrimane core can be easily opened and functionalized [23].

We performed the synthesis of a small, but diversified library of new derivatives, characterized by the homodrimane backbone bearing flexible tails of different nature and chemical properties at position 1 (Figure 1). We used the commercially available (+)-sclareolide as starting molecule. In particular, the substituent groups, bound to the bicyclic nucleus by either an amide or ester or ether functionality, differ in size, flexibility, and electronic properties.

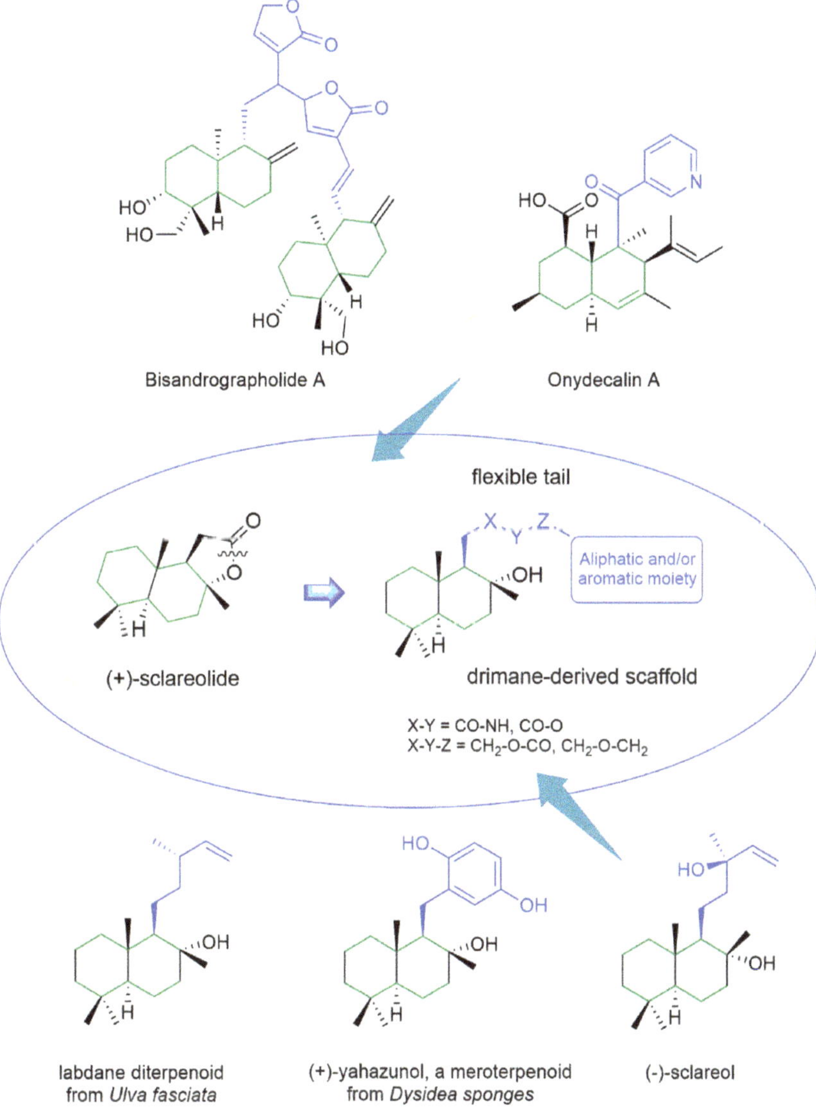

Figure 1. Design of new drimane-derived antagonists inspired by labdane diterpenoids starting from known natural TRPV4 ligands.

Therefore, recurring chemical motifs, such as variously substituted benzyl and phenylethyl residues, were inserted into the new molecules by proper choices of amines and alcohols.

Functional properties and selectivity on TRPV4 and TRPV1 channels of all final compounds were then evaluated by an in vitro calcium influx assay in intact living cells overexpressing either channel.

Among these new derivatives, some relative potent and selective TRPV4 antagonists were identified and their structure-activity relationship (SAR) study highlighted the crucial role of specific functional groups in eliciting TRPV4 antagonism that could be used for the development of a new class of selective TRPV4 modulators.

2. Results

2.1. Chemistry

With the aim of synthesizing a set of new compounds, we introduced several modifications in the (+)-sclareolide scaffold by lactone ring-opening, thus resulting in the homodrimane sesquiterpene moiety. Amide derivatives **1–16** were generated through two different procedures both carried out in THF (tetrahydrofuran) as the solvent, as depicted in Scheme 1. Derivatives **1–3** were obtained by DIBAL-H (diisobutylaluminium hydride) assisted amidation using aromatic amines while the direct aminolysis of (+)-sclareolide using aliphatic amines provided instead compounds **4–16**. Homodrimanyl acid esters **18** and **19** were prepared in two steps (Scheme 1).

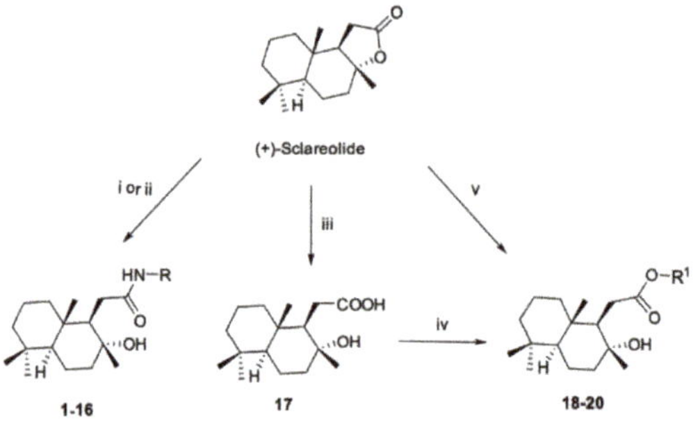

Scheme 1. Reagents and conditions. (**i**) Aromatic amine, DIBAL-H, dry THF, rt (room temperature), 3–5 h; (**ii**) Aliphatic amine, dry THF, 45 °C, 48 h; (**iii**) NaOH, MeOH, rt, 2 h; (**iv**) R1-Br, K_2CO_3, dry DMF, rt, 24 h; (**v**) MeOH, 45 °C, 72 h.

(+)-Sclareolide was first hydrolyzed with sodium hydroxide in methanol to give the free carboxylic acid **17** which was then reacted in anhydrous DMF (dimethylformamide) with the appropriate benzyl bromide in the presence of K_2CO_3, thus affording the desired compounds. Methyl ester **20** was obtained by alcoholysis, following a procedure previously described (see Supplementary Materials). Diol derivative **21** represented the key intermediate for the synthesis of reverted esters **22–24** and ether **25**, as shown in Scheme 2. Reduction of the carbonyl functionality of (+)-sclareolide with $LiAlH_4$ in anhydrous THF provided the desired diol **21** (in good yield), which was converted into the corresponding homodrimanyl alcohol esters **22–24** by reaction with the appropriate carboxylic acids under Steglich conditions. Differently, ether **25** was synthesized by reacting the diol derivative **21** with 3-chlorobenzyl bromide under basic condition in refluxing anhydrous THF.

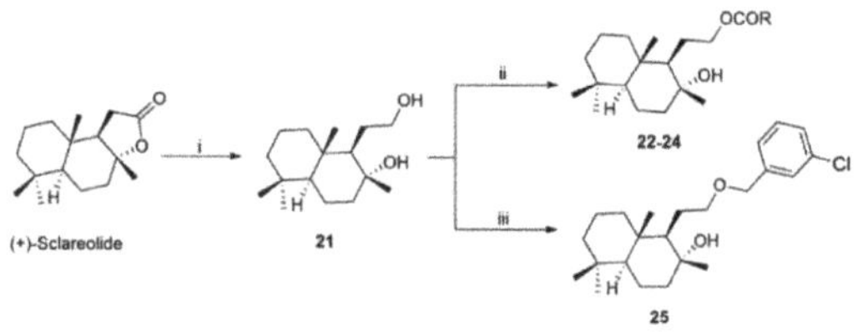

Scheme 2. Reagents and conditions. (**i**) LiAlH$_4$, dry THF, rt, 6 h; (**ii**) RCOOH, EDCI, DMAP, DCM, rt, 48–72 h; (**iii**) NaH, 3-chlorobenzyl bromide, dry THF, reflux, 48 h.

2.2. TRPV4 Assay

The interaction properties of all homodrimane-derived compounds **1–16**, **18–19**, and **22–25** with either TRPV type-4 or type-1 were studied in human embryonic kidney (HEK)-293 cells overexpressing either the rat recombinant type-4 (rTRPV4), which shares with the human orthologue ~95% of sequence identity, and the human recombinant type-1 (hTRPV1) channel, evaluating the changes in intracellular calcium levels. The values for both EC$_{50}$ (for activation) and IC$_{50}$ (for antagonism) were calculated and are summarized in Table 1.

All compounds share the homodrimane scaffold while they differ in the nature of the substituent at position 1. This latter includes an aliphatic and/or aromatic moiety, connected to C1 by a spacer of variable length containing one of the following functional groups: amide, ester, reverted ester, and ether. Accordingly, the final set of synthesized compounds featured homodrimanyl acid amides (compounds **1–16**), homodrimanyl acid esters (compounds **18** and **19**), homodrimanyl alcohol esters (compounds **22–24**), having a reverted ester group, and one ether derivative (compound **25**). Besides this library of trans-decalin-related homodrimane derivatives, also (+)-sclareolide, the methyl ester **20**, and homodrimanyl alcohol **21**, a key intermediate in the synthesis of the reverted ester and ether compounds (see Scheme 1), were tested for their ability to interact with rTRPV4 and hTRPV1. Homodrimanyl acid methyl ester **20**, reverted esters **22–24**, and ether compound **25** were found to be inactive toward both hTRPV1 and rTRPV4, regardless of the size, electronic properties or spacer length of the substituent at position 1 (see Table 1). Similarly, the natural starting compound, (+)-sclareolide, and compound **21**, characterized by the 2-hydroxyethyl group at position 1, were unable to modulate either channel. Instead, the homodrimanyl acid amide series **1–16** is by far more interesting, since most derivatives behaved as rTRPV4 ligands endowed with antagonistic activity. Moreover, all amides resulted completely selective for rTRPV4, showing no relevant activity versus hTRPV1 (see Supplementary Materials).

2.3. Cytotoxicity Assay

To exclude cytotoxic effects, the most active compounds within the series, namely **6** and **18**, were evaluated on both human cervical (HeLa) and human lung (A549) carcinoma cells at 1, 5, and 25 µM, using the 3-(4,5-dimethylthiazol-2-yl)-2,5-diphenyltetrazolium bromide (MTT) assay. As shown in Figure 2, neither compound caused statistically-significant cytotoxic effects at any tested concentration.

Table 1. Results of TRPV4 assay of compounds 1–16, 18–20, 22–25. [a]

Structure templates: **1-16** (HN–R amide), **18-20** (O–R ester), **22-24** (OCOR), **21, 25** (OR).

Cpd.	R	Efficacy[b] %	Potency EC50 (μM)	IC$_{50}$ (μM)[c] inh. TRPV4	Cpd.	R	Efficacy[b] %	Potency EC50 (μM)	IC$_{50}$ (μM)[c] inh. TRPV4
1	(structure)	<10	NA[d]	>100	14	(structure)	<10	NA	53.5 ± 1.8
2	(structure)	<10	NA	>100	15	(structure)	<10	NA	>100
3	(structure)	14.6 ± 1.5	1.1 ± 1.0	6.0 ± 0.1	16	(structure)	<10	NA	>100
4	(structure)	<10	NA	32.0 ± 0.8	18	(structure)	<10	NA	5.41 ± 0.07
5	(structure)	<10	NA	7.7 ± 0.3	19	(structure)	<10	NA	>100
6	(structure)	<10	NA	5.3 ± 0.3	20	Me	<10	NA	>100
7	(structure)	15.8 ± 0.8	13.4 ± 2.6	16.9 ± 0.8	21	H	<10	NA	>100
8	(structure)	<10	NA	29.7 ± 0.7	22	(structure)	<10	NA	NA
9	(structure)	<10	NA	18.1 ± 0.2	23	(structure)	<10	NA	>100
10	(structure)	<10	NA	15.6 ± 0.3	24	(structure)	<10	NA	>100
11	(structure)	<10	NA	7.0 ± 0.1	25	(structure)	<10	NA	>100
12	(structure)	<10	NA	11.4 ± 0.1	Scd[e]		11.3 ± 0.7	> 10	>100
13	(structure)	<10	NA	11.9 ± 0.4					

[a] Data are means ± SEM (standard error of the mean) of at least n = 3 determinations. [b] As percent of the effect of ionomycin (4 μM). Inh = inhibitory activity. [c] Determined against the effect of GSK1016790A (10 nM) after a 5-min pre-incubation with each compound. [d] NA = not active, if the efficacy is lower than 10%, the potency is not calculated, [e] Scd = (+)-Sclareolide. GSK1016790A efficacy 88.5 ± 1.1 EC$_{50}$ 3.5 ± 0.2 nM.

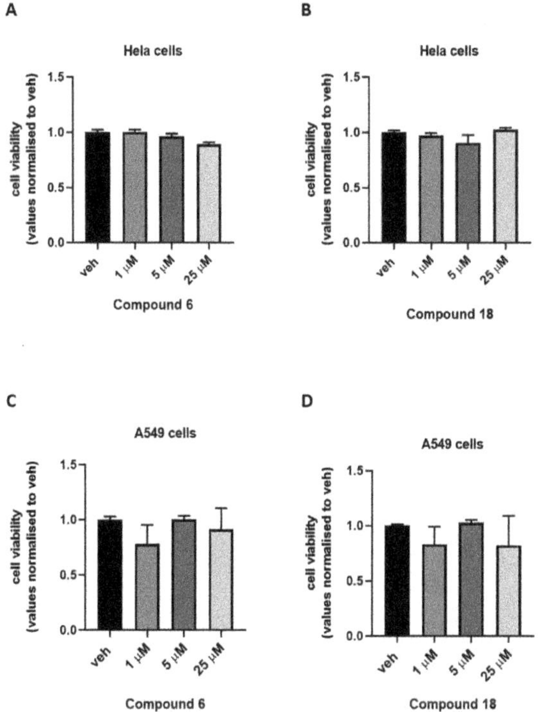

Figure 2. Cell viability measured in HeLa and A549 cells treated for 24 h with increasing concentration of compound **6** (panel **A**, **C**) and **18** (panel **B**, **D**). Bar graphs show cell viability measured using the 3-(4,5-dimethylthiazol-2-yl)-2,5-diphenyltetrazolium bromide (MTT) assay. Data are expressed as optical density (OD) at 595 nm, normalized to control (vehicle). The bar is the standard error of the mean (SEM) on 3 independent determinations. Statistically significant differences were accepted when the *p*-value was at least ≤0.05.

3. Discussion

Biological Activity and Structure-Activity Relationships (SARs)

Among the aromatic amides, characterized by a three-atom spacer and a phenyl head, only compound **3** behaved as a good rTRPV4 agonist, even though not very effective. This result suggests that a relatively bulky and electron-rich group (i.e., 2-(1H-pyrrol-1-yl), compound **3**), rather than a small and strongly electronegative substituent (i.e., 2-fluoro, compound **2**), is necessary at position 2 of the phenyl ring to produce rTRPV4 modulation. Amides characterized by a benzyl- or phenylethyl- moiety provided the most notable compounds, all behaving as rTRPV4 modulators with the only exception of compound **15**. Indeed, in the benzyl amide subset of compounds, a substituent in meta or para position was generally well tolerated. However, electron-attracting halogens (compounds **5**, **6** and **7**) performed better than an electron-donor methoxy group (compounds **8** and **9**); moreover, the 4-chloro derivative (compound **5**) was more active than the corresponding 4-fluoro one (compound **7**). This effect can be probably ascribed to composite factors including relative halogen-bonding propensity, lipophilicity and electron withdrawing properties. The introduction of a second chlorine at position 3 (compound **6**), further increased the activity, producing the most potent antagonist in this library of trans-decalin-related homodrimane derivatives. Conversely, the introduction of a bulky substituent, such as a second phenyl ring in para position (compound **15**) or the replacement of the benzyl-with a furan-2-yl-methyl- (compound **14**) moiety, was detrimental for rTRPV4 interaction, leading to derivatives either completely devoid of activity, i.e., compound **15**, or with very weak antagonist properties (compound **14**). Concerning the phenylethyl amide subset, compounds **11**, **12**, and **13** were more potent than compound **10**. Thus, a halogen atom in para- or meta-position of the phenyl nucleus caused a slight increase of the activity, consistently with previous observations. Anyway, these derivatives appeared to be less affected by the nature of the halogen atom, as indicated by their IC_{50} value trend (compounds **11–13**). Considering the influence of the length of the spacer connecting the homodrimane portion to the amide head, compounds **10**, **11**, and **12** performed better than compounds **4**, **5**, and **7**. In this view, experimental results suggested that a five-atom spacer, typical of the phenylethyl derivatives, provided a higher potency than the four-atom spacer present in the benzylic ones. However, a further extension to a six-atom spacer, as in compound **16**, characterized by a 3-(1H-imidazol-1-yl)propyl moiety, determined the complete loss of activity on rTRPV4. Since the amide series gave rise to the most interesting compounds within the tested panel, the role of the amide group was explored by synthesizing two benzyl analogs bearing the ester group in place of the amide, i.e., homodrimanyl acid esters **18** and **19**. Intriguingly, the biological activity results were quite surprising and not conclusive. Indeed, compound **18**, the benzyl ester analog of amide **4**, resulted in a considerably more potent antagonist (IC_{50} 5.41 μM and 32.0 μM, respectively), representing the most active synthesized derivative together with amide **6** (IC_{50} 5.30 μM). On the contrary, compound **19**, the 3,4-dichlorobenzyl ester analog of amide **6**, was completely devoid of activity on rTRPV4. A possible explanation for these results could be that the homodrimanyl amide and ester derivatives might occupy different receptor (sub)sites. The higher flexibility of the ester group compared to the amide group and the lack of the H-bond donor could facilitate its accommodation in a narrow pocket, filled by the phenyl group. The presence of two chlorine substituents, by increasing the bulkiness of the aromatic ring, may give rise to steric bumps within the pocket, thus preventing a correct ligand orientation. Finally, the presence of a carbonyl functionality (and its position in the spacer), appeared as stringent requirements, too, since its displacement, as in reverted esters **22–24**, or its removal, as in the ether derivative **25**, all resulted in inactive derivatives. Unfortunately, the combination of low resolution/missing regions of the available TRPV4 3D structures with the lack of structural insights in the binding site of known antagonists, do not allow at the present any reliable prediction and unequivocal identification of the putative binding sites of these compounds.

4. Materials and Methods

4.1. Chemicals, Materials, and Methods

All the reagents were purchased from Merck (Darmstadt, Germany) or Alfa Aesar (Tewksbury, MA, USA) and were used as received. Melting points were obtained using a Gallenkamp (G) (Fiorano Modenese, Italy) melting point apparatus. The structures of final compounds were unambiguously assessed by ^{1}H NMR (nuclear magnetic resonance) and ^{13}C NMR. Spectra were recorded in the indicated solvent (Chloroform- $CDCl_3$, Dimethyl Sulphoxide-DMSO-d_6) at 25 °C on a Bruker 300 MHz spectrometer (Bruker, Milano, Italy) or a Bruker Advance DPX400 employing TMS (tetramethyl silane)as internal standard. Chemical shifts are expressed in δ values (ppm) and coupling constants (*J*) in hertz (Hz). IR spectra were recorded on a PerkinElmer machine 10.4.00 (PerkinElmer, Milan, Italy). Reactions were monitored by TLC (thin layer chromatography) on silica gel plates Merck 60 F254 (Merck, Burlington, MA, USA). Final products were purified by a flash chromatography system with column chromatography, using Merck 60 silica gel, 230–400 mesh. Elemental analyses were performed on Leco Trunspec CHNS Micro elemental system (St. Joseph, MI, USA). The purity of final compounds was evaluated by C, H, N analysis, and it was confirmed to be ≥95%.

4.2. Chemistry

4.2.1. General Procedure for the Synthesis of Homodrimanyl Aryl Amides (**1–3**): DIBAL-H-Mediated Amidation From Anilines

According with Li D. et al. (2017) [24], a solution of DIBAL-H (1 M in toluene, 3.0 eq.) was added dropwise to a solution of appropriate aniline (1.0 mmoL, 2.5 eq.) in anhydrous THF (1.5 mL) at 0 °C, under argon flux and stirring. The reaction mixture was warmed to rt and stirred for the next 2 h. The prepared complex was used directly for the aminolysis. (+)-Sclareolide (100 mg, 0.4 mmoL, 1.0 eq.) was dissolved in anhydrous THF (1.0 mL), and the DIBAL-H-aniline complex solution was added. The mixture was stirred at rt until sclareolide spot disappeared on the TLC plate (about 3–5 h). Then, it was cooled to 0 °C, quenched with a 1 M $KHSO_4$ aqueous solution (2.0 mL), and extracted with DCM (3 × 10 mL). The combined organic layers were finally washed with brine and dried over anhydrous Na_2SO_4. After filtration, the evaporation of the solvent to dryness furnished the corresponding homodrimanyl amide. Column chromatography using a mixture of petroleum eter/ethyl acetate as eluent gave the pure compound in good yield.

2-((1*R*,2*R*,4a*S*,8a*S*)-2-Hydroxy-2,5,5,8a-tetramethyldecahydronaphthalen-1-yl)-*N*-phenylacetamide (**1**)

Ref. [21] The product was purified by column chromatography using PE/EtOAc (5/1) as eluent to give the compound as a white solid, mp 166–167 °C (G). Yield 84%. ^{1}H NMR (300 MHz, $CDCl_3$) δ (ppm): 8.77 (s, 1H, NH), 7.45 (d, *J* = 7.5 Hz, 2H, Ar), 7.21 (t, *J* = 7.4 Hz, 2H, Ar), 7.01 (t, *J* = 7.4 Hz, 1H, Ar), 3.10 (s, 1H, OH), 2.55 (dd, J_1 = 15.3 Hz, J_2 = 4.3 Hz, 1H, CH_2CO), 2.22 (dd, J_1 = 15.5 Hz, J_2 = 4.7 Hz, 1H, CH_2CO), 1.89 (dt, J_1 = 12.3 Hz, J_2 = 2.8 Hz, 1H), 1.77 (t, *J* = 4.1 Hz, 1H), 1.70–1.18 (m, 9H), 1.12 (s, 3H, CH_3), 1.10–0.85 (m, 1H, CH), 0.80 (s, 3H, CH_3), 0.74 (s, 6H, CH_3). ^{13}C NMR (75 MHz, $CDCl_3$) δ (ppm): 174.0, 138.4, 129.8, 127.7, 124.8, 120.7, 118.8, 73.6, 57.8, 55.1, 43.4, 41.7, 38.8, 35.9, 34.9, 33.2, 24.8, 23.3, 22.2, 20.5, 18.2, 143.6 Anal. Calcd. for $C_{22}H_{33}NO_2$: C, 76.92; H, 9.68; N, 4.08. Found: C, 77.05; H, 9.71; N, 4.07.

N-(2,5-difluorophenyl)-2-((1*R*,2*R*,4a*S*,8a*S*)-2-hydroxy-2,5,5,8a-tetramethyldecahydronaphthalen-1-yl)acetamide (**2**)

The product was purified by column chromatography using PE/EtOAc (8/1) as eluent to give the compound as a white solid, mp 165–166 °C. IR ν (cm^{-1}): 3266, 2925, 1680, 1630, 1542, 1441, 1189, 754. Yield 73%. ^{1}H NMR (400 MHz, $CDCl_3$) δ (ppm): 8.78 (brs, 2H), 8.24-8.14 (m, 1H, Ar), 6.99-6.91 (m, 1H, Ar), 6.67-6.61 (m, 1H, Ar), 2.59 (dd, J_1 = 15.0 Hz, J_2 = 4.6 Hz, 1H, CH_2CO), 2.26 (dd, J_1 = 15.0 Hz, J_2 = 4.0 Hz, 1H, CH_2CO), 1.92 (dt, J_1 = 12.3 Hz, J_2 = 3.1 Hz, 1H, $-CH_2-COH(CH_3)$), 1.78 (t, *J* = 4.2 Hz, 1H, $-CH-COH(CH_3)$), 1.71-1.66 (m, 2H), 1.60-1.50 (m, 2H), 1.48-1.24 (m, 4H), 1.20 (s, 3H, $COH(CH_3)$), 1.00-0.89 (m, 2H), 0.85 (s, 3H, CH_3), 0.79 (s, 3H, CH_3), 0.78 (s, 3H, CH_3). ^{13}C NMR (100 MHz, $CDCl_3$) δ (ppm): 173.6 (C = O), 158.7 (d, *J* = 240 Hz, Cq-F), 148.3 (d, *J* = 238 Hz, Cq-F), 128.1 (CqAr), 114.9 (dd, J_1 = 9.7 Hz, J_2 = 21.9 Hz, CHAr), 109.2 (d, *J* = 24.6 Hz, CHAr), 108.5 (d, *J* = 36.3 Hz, CHAr), 74.1 (Cq-OH(CH_3)), 58.2 (CH-CH_2CO), 56.0 (CH), 44.2, 41.8, 39.4, 38.9 (Cq-$(CH_3)2$), 34.7, 33.3, 29.7, 24.3, 21.4, 20.5, 18.2, 15.3. Anal. Calcd. for $C_{22}H_{31}F_2NO_2$: C, 69.63; H, 8.23; N, 3.69. Found: C, 69.50; H, 8.27; N, 3.68.

N-(2-(1*H*-pyrrol-1-yl)phenyl)-2-((1*R*,2*R*,4a*S*,8a*S*)-2-hydroxy-2,5,5,8a-tetramethyldecahydronaphthalen-1-yl)acetamide (**3**)

The product was purified by column chromatography using PE/EtOAc (4.5/1) as eluent to give the compound as a light yellow oil. Yield 80%. IR ν (cm^{-1}): 3669, 2970, 1681, 1525, 1451, 1215, 1069, 748, 666. ^{1}H NMR (400 MHz, $CDCl_3$) δ (ppm): 8.32 (d, *J* = 8.2 Hz, 1H, Ar), 8.10 (brs, 1H), 7.36 (t, *J* = 7.8 Hz, 1H, Ar), 7.21 (d, *J* = 7.4 Hz, 1H, Ar), 7.10 (d, *J* = 7.6 Hz, 1H, Ar), 6.79 (t, *J* = 1.8 Hz, 2H, Pyrrol), 6.36 (t, *J* = 1.8 Hz, 2H, Pyrrol), 2.41 (dd, J_1 = 14.9 Hz, J_2 = 4.5 Hz, 1H, CH_2CO), 2.11 (dd, J_1 = 14.9 Hz, J_2 = 4.1 Hz, 1H, CH_2CO), 1.84 (dt, J_1 = 11.8 Hz, J_2 = 2.9 Hz, 1H, $-CH_2-COH(CH_3)$),

1.69-1.63 (m, 2H), 1.61-1.48 (m, 2H), 1.45-1.33 (m, 3H), 1.28-1.10 (m, 2H), 1.05 (s, 3H, COH(CH$_3$)), 0.90-0.89 (m, 2H), 0.86 (s, 3H, CH$_3$) 0.76 (s, 3H, CH$_3$), 0.73 (s, 3H, CH$_3$). ^{13}C NMR (100 MHz, CDCl$_3$) δ (ppm): 173.8 (C=O), 134.8 (CqAr), 130.7 (CqAr), 129.0 (CHAr), 127.3 (CHAr), 123.7 (CHAr), 122.8 (×2, CHPyrrol), 121.8 (CHAr), 109.7 (×2, CHPyrrol), 73.4 (Cq-OH(CH$_3$)), 59.3 (CH-CH$_2$CO), 56.2 (CH), 43.7, 41.8, 39.4, 38.8 (Cq-(CH$_3$)$_2$), 34.7, 33.3, 33.2, 23.7, 21.4, 20.4, 18.2, 15.2. Anal. Calcd. for C$_{26}$H$_{36}$N$_2$O$_2$: C, 76.43; H, 8.88; N, 6.86. Found: C, 76.52; H, 8.91; N, 6.84.

4.2.2. General Procedure for the Synthesis of Homodrimanyl Aliphatic Amides (**4–16**): Aminolysis Reaction from Amines

According to a published procedure [21] with little modifications, a solution of (+)-sclareolide (100 mg, 0.4 mmoL, 1.0 eq.) and the opportune amine in dry THF (1.5 mL) was stirred at 45 °C for 48–72 h. The reaction mixture was then concentrated under reduced pressure and dispersed in water (15 mL). The inorganic phase was extracted twice with EtOAc (15 mL), and the collected organic layers were washed with brine (15 mL), dried over anhydrous Na$_2$SO$_4$ and filtered. The solvent was removed under vacuum, and the crude product was purified by flash chromatography using a mixture of PE/EtOAc as eluent to give the homodrimanyl aliphatic amide in good yield.

N-benzyl-2-((1*R*,2*R*,4a*S*,8a*S*)-2-hydroxy-2,5,5,8a-tetramethyldecahydronaphthalen-1-yl)acetamide (**4**)

The product was purified by column chromatography using PE/EtOAc (1.5/1) as eluent to give the compound as a colorless oil. Yield 98%. IR ν (cm^{-1}): 3014, 2926, 1650, 1214, 748, 666. ^{1}H NMR (300 MHz, CDCl$_3$) δ (ppm): 7.28–7.14 (m, 4H, Ar), 6.92 (t, *J* = 12.1 Hz, 1H, Ar), 4.38-4.21 (m, 2H, CH$_2$NH), 3.53 (s, 1H, OH), 2.36 (dd, *J*$_1$ = 15.2 Hz, *J*$_2$ = 4.5 Hz, 1H, CH$_2$CO), 2.09 (dd, *J*$_1$ = 15.5 Hz, *J*$_2$ = 4.7 Hz, 1H, CH$_2$CO), 1.83 (dt, *J*$_1$ = 12.1 Hz, *J*$_2$ = 2.7 Hz, 1H), 1.68 (t, *J* = 4.7 Hz, 1H), 1.62–1.08 (m, 9H, CH$_2$), 1.02 (s, 3H, CH$_3$), 0.94–0.85 (m, 1H, CH), 0.80 (s, 3H, CH$_3$), 0.71 (s, 6H, CH$_3$). ^{13}C NMR (75 MHz, CDCl$_3$) δ (ppm): 175.6, 138.4, 129.1, 128.6, 128.3, 128.1, 72.9, 57.3, 56.6, 55.1, 43.5, 41.7, 38.8, 34.1, 33.2, 32.5, 24.4, 22.9, 22.2, 20.5, 18.3, 16.2, 14.6. Anal. Calcd. for C$_{23}$H$_{35}$NO$_2$: C, 77.27; H, 9.87; N, 3.92. Found: C, 77.38; H, 9.90; N, 3.93.

N-(4-chlorobenzyl)-2-((1*R*,2*R*,4a*S*,8a*S*)-2-hydroxy-2,5,5,8a-tetramethyldecahydronaphthalen -1-yl)acetamide (**5**)

The product was purified by column chromatography using PE/EtOAc (2/1) as eluent to give the compound as a white solid, mp 154–155 °C (G). Yield 70%. IR ν (cm^{-1}): 3279, 2924, 1642, 1492, 1387, 1091, 1015, 938, 800. ^{1}H NMR (300 MHz, CDCl$_3$) δ: 7.25-7.23(m, 2H, Ar), 7.15-7.13 (m, 2H, Ar), 4.35-4.21 (m, 2H, CH$_2$NH), 3.69 (brs, 1H, OH), 2.42 (dd, *J*$_1$ = 15.4 Hz, *J*$_2$ = 4.5 Hz, 1H, CH$_2$CO), 2.16 (dd, *J*$_1$ = 15.4 Hz, *J*$_2$ = 4.6 Hz, 1H, CH$_2$CO), 1.90-1.85 (m, 1H, CH), 1.71-1.20 (m, 10H), 1.07 (s, 3H, CH$_3$), 0.94-0.90 (m, 1H, CH), 0.86 (s, 3H, CH$_3$), 0.77 (s, 3H, CH$_3$), 0.74 (s, 3H, CH$_3$). ^{13}C NMR (75 MHz, CDCl$_3$) δ: 175.8, 137.1, 132.9, 128.8 (×2), 128.6 (×2), 72.9, 57.9, 55.9, 44.1, 42.8, 41.7, 39.2, 38.7, 33.3, 33.2, 32.5, 23.7, 21.4, 20.4, 18.3, 15.4. Anal. Calcd. for C$_{23}$H$_{34}$ClNO$_2$: C, 70.48; H, 8.74; N, 3.57. Found: C, 70.22; H, 8.77; N, 3.56.

N-(3,4-dichlorobenzyl)-2-((1*R*,2*R*,4a*S*,8a*S*)-2-hydroxy-2,5,5,8a-tetramethyldecahydronaphthalen -1-yl)acetamide (**6**)

The product was purified by column chromatography using PE/EtOAc (1/1.5) as eluent to give the compound as a white solid, mp 149–150 °C (G). Yield 83%. IR ν (cm^{-1}): 3298, 2926, 1642, 1548, 1470, 1388, 1082, 1032, 754. ^{1}H NMR (400 MHz, DMSO) δ (ppm): 8.23 (t, 1H, *J* = 6.1 Hz, NH), 7.51 (d, *J* = 8.3 Hz, 1H, Ar), 7.44 (d, *J* = 1.8 Hz, 1H, Ar), 7.20 (dd, *J*$_1$ = 8.3 Hz, *J*$_2$ = 1.8 Hz, 1H, Ar), 4.26 (dd, *J*$_1$ = 15.5 Hz, *J*$_2$ = 6.2 Hz, 1H, CH$_2$NH), 4.15 (dd, *J*$_1$ = 15.5 Hz, *J*$_2$ = 5.8 Hz, 1H, CH$_2$NH), 2.34 (dd, *J*$_1$ = 15.4 Hz, *J*$_2$ = 2.8 Hz, 1H, CH$_2$CO), 2.05 (dd, *J*$_1$ = 15.4 Hz, *J*$_2$ = 7.1 Hz, 1H, CH$_2$CO), 1.76-1.66 (m, 2H), 1.54-1.02 (mm, 8H), 0.93 (s, 3H, COH(CH$_3$)), 0.87-0.82 (m, 2H), 0.80 (s, 3H, CH$_3$), 0.72 (s, 6H, CH$_3$). ^{13}C NMR (100 MHz, DMSO) δ (ppm): 174.6 (C=O), 141.7 (CqAr), 131.3 (CqAr), 130.8 (CHAr),

129.6 (CHAr + CqAr), 128.1 (CHAr), 71.6 (Cq-OH(CH$_3$)), 56.8 (CH), 56.0 (CH), 44.2, 42.0, 41.6, 39.2, 38.7, 33.7, 33.3, 31.6, 24.6, 21.8, 20.5, 18.3, 15.5. Anal. Calcd. for C$_{23}$H$_{32}$Cl$_2$NO$_2$: C, 64.78; H, 7.80; N, 3.28. Found: C, 65.02; H, 7.77; N, 3.29.

N-(4-fluorobenzyl)-2-((1*R*,2*R*,4a*S*,8a*S*)-2-hydroxy-2,5,5,8a-tetramethyldecahydronaphthalen -1-yl)acetamide (**7**)

The product was purified by column chromatography using PE/EtOAc (1.5/1) as eluent to give the compound as a white solid, mp 135–136 °C (G). Yield 97%. IR ν (cm^{-1}): 3675, 2987, 2907, 1510, 1214, 1057, 742, 666. ^{1}H NMR (400 MHz, CDCl$_3$) δ (ppm): 7.22-7.19 (m, 2H, Ar), 6.97 (t, *J* = 8.6 Hz, 2H, Ar), 6.38 (brs, 1H, NH), 4.37 (dd, *J*$_1$ = 15.0 Hz, *J*$_2$ = 6.0 Hz, 1H, CH$_2$-NH), 4.31 (dd, *J*$_1$ = 15.0 Hz, *J*$_2$ = 5.9 Hz, 1H, CH$_2$-NH), 2.46 (brs, 1H, OH), 2.39 (dd, *J*$_1$ = 15.4 Hz, *J*$_2$ = 5.2 Hz, 1H, CH$_2$CO), 2.14 (dd, *J*$_1$ = 15.4 Hz, *J*$_2$ = 4.1 Hz, 1H, CH$_2$CO), 1.90 (dt, *J*$_1$ = 12.5 Hz, *J*$_2$ = 3.0 Hz, 1H, -CH$_2$-COH(CH$_3$)), 1.76 (t, *J* = 4.6 Hz, 1H, -CH-COH(CH$_3$)), 1.68-1.52 (m, 2H), 1.50-1.32 (m, 4H), 1.29-1.12 (m, 2H), 1.10 (s, 3H, COH(CH$_3$)), 0.97-0.88 (m, 2H), 0.85 (s, 3H, CH$_3$), 0.76 (s, 3H, CH$_3$), 0.75 (s, 3H, CH$_3$). ^{13}C NMR (100 MHz, CDCl$_3$) δ (ppm): 175.2 (C=O), 162.1 (d, *J* = 267 Hz, Cq-F), 134.3 (CqAr), 129.4 (×2, *J* = 7.9 Hz, CHAr), 115.5 (×2, *J* = 21.3 Hz, CHAr), 73.2 (Cq-OH(CH$_3$)), 57.9 (CH), 56.0 (CH), 44.3, 43.0, 41.8, 39.4, 38.8 (Cq-(CH$_3$)$_2$), 33.3, 32.6, 29.7, 23.8, 21.4, 20.5, 18.4, 15.5. Anal. Calcd. for C$_{23}$H$_{24}$FNO$_2$: C, 73.56; H, 9.13; N, 3.73. Found: C, 73.76; H, 9.09; N, 3.72.

2-((1*R*,2*R*,4a*S*,8a*S*)-2-hydroxy-2,5,5,8a-tetramethyldecahydronaphthalen-1-yl)-*N* -(4-methoxybenzyl)acetamide (**8**)

The product was purified by column chromatography using PE/EtOAc (1/1) as eluent to give the compound as a yellow oil. Yield 78%. IR ν (cm^{-1}): 3675, 3289, 2920, 1512, 1214, 1038, 748, 666. ^{1}H NMR (400 MHz, CDCl$_3$) δ (ppm): 7.17 (d, 2H, *J* = 8.4 Hz, Ar), 6.83 (d, 2H, *J* = 8.6 Hz, Ar), 6.23 (brs, 1H, NH), 4.34 (dd, *J*$_1$ = 14.8 Hz, *J*$_2$ = 5.8 Hz, 1H, *CH$_2$*NH), 4.29 (dd, *J*$_1$ = 14.8 Hz, *J*$_2$ = 5.6 Hz, 1H, *CH$_2$*NH), 3.77 (s, 3H, OCH$_3$), 2.47 (brs, 1H, OH), 2.39 (dd, *J*$_1$ = 15.4 Hz, *J*$_2$ = 5.2 Hz, 1H, *CH$_2$*CO), 2.12 (dd, *J*$_1$ = 15.4 Hz, *J*$_2$ = 4.0 Hz, 1H, CH$_2$CO), 1.90 (dt, *J*$_1$ = 9.6 Hz, *J*$_2$ = 2.9 Hz, 1H, -CH$_2$-COH(CH$_3$)), 1.78 (t, *J* = 4.5 Hz, 1H, -*CH*-COH(CH$_3$)), 1.68-1.45 (mm, 4H), 1.42-1.12 (mm, 4H), 1.10 (s, 3H, COH(*CH$_3$*)), 0.97-0.93 (m, 2H), 0.85 (s, 3H, CH$_3$), 0.76 (s, 3H, CH$_3$), 0.75 (s, 3H, CH$_3$). ^{13}C NMR (100 MHz, CDCl$_3$) δ (ppm): 175.1 (C=O), 159.0 (*CqAr*), 130.5 (*CqAr*), 129.1 (×2, *CHAr*), 114.1 (×2, *CHAr*), 73.1 (*Cq*-OH(CH$_3$)), 57.8 (CH), 56.0 (CH), 55.3 (OCH$_3$), 44.3, 43.3, 41.8, 39.4, 38.8 (*Cq*-(CH$_3$)$_2$), 33.3, 33.2, 32.5, 23.7, 21.4, 20.5, 18.4, 15.5. Anal. Calcd. for C$_{24}$H$_{37}$NO$_3$: C, 74.38; H, 9.62; N, 3.61. Found: C, 74.42; H, 9.66; N, 3.62.

2-((1*R*,2*R*,4a*S*,8a*S*)-2-hydroxy-2,5,5,8a-tetramethyldecahydronaphthalen-1-yl)-*N* -(3 methoxybenzyl)acetamide (**9**)

The product was purified by column chromatography using PE/EtOAc (1/1) as eluent to give the compound as a yellow oil. Yield 85%. IR ν (cm^{-1}): 3291, 2945, 1642, 1264, 1214, 1051, 746, 666. ^{1}H NMR (300 MHz, CDCl$_3$) δ (ppm): 7.15-7.07 (m, 1H, Ar), 6.74-6.68 (m, 3H, Ar), 4.30-4.17 (m, 2H, *CH$_2$*NH), 3.68 (s, 3H, CH$_3$O), 2.36 (dd, *J*$_1$ = 15.4 Hz, *J*$_2$ = 4.4 Hz, 1H, CH$_2$CO), 2.09 (dd, *J*$_1$ = 15.3 Hz, mboxemph]$_2$ = 4.7 Hz, 1H, CH$_2$CO), 1.95-1.79 (m, 1H, CH), 1.67 (t, *J* = 4.5 Hz, 1H), 1.57-1.03 (m, 9H, CH$_2$), 1.00 (s, 3H, CH$_3$), 0.87-0.81 (m, 1H, CH), 0.78 (s, 3H, CH$_3$), 0.69 (s, 6H, CH$_3$). ^{13}C NMR (75 MHz, CDCl$_3$) δ (ppm): 178.9, 159.6, 140.1, 129.5, 119.7, 112.9, 112.7, 72.9, 57.8, 55.9, 44.1, 43.4, 41.7, 39.2, 38.7, 33.3, 33.2 (×2), 32.4, 23.6, 21.4, 20.4, 18.3, 15.3. Anal. Calcd. for C$_{24}$H$_{37}$NO$_3$: C, 74.38; H, 9.62; N, 3.61. Found: C, 74.48; H, 9.65; N, 3.60.

2-((1*R*,2*R*,4a*S*,8a*S*)-2-hydroxy-2,5,5,8a-tetramethyldecahydronaphthalen-1-yl) -*N*-phenethylacetamide (**10**)

The product was purified by column chromatography using PE/EtOAc (1/1) as eluent to give the compound as a light yellow oil. Yield 95%. ^{1}H NMR (300 MHz, CDCl$_3$) δ (ppm): 7.29-7.19 (m, 2H, Ar), 7.18-7.10 (m, 2H, Ar), 6.60-6.51 (m, 1H, Ar), 3.54 (s, 1H, OH), 3.48-3.32 (m, 2H, *CH$_2$*NH),

2.74 (t, J = 6.6 Hz, 2H, CH_2Ar), 2.02 (dd, J_1 = 14.5 Hz, J_2 = 3.4 Hz, 1H, CH_2CO), 1.85 (dd, J_1 = 12.3 Hz, J_2 = 5.0 Hz, 1H, CH_2CO), 1.91-1.80 (m, 1H), 1.65-1.10 (m, 10H), 1.04 (s, 3H, CH_3), 0.92-0.85 (m, 1H, CH), 0.80 (s, 3H, CH_3), 0.70 (s, 6H, CH_3). ^{13}C NMR (75 MHz, CDCl$_3$) δ (ppm): 175.6, 139.0, 129.8, 129.6, 127.8, 127.5, 125.3, 72.9, 57.2, 55.1, 40.7, 38.7, 35.5, 33.2, 32.6, 24.5, 23.1, 22.2, 20.5, 19.8, 19.1, 18.3, 16.2, 14.6. IR ν (cm^{-1}): 3021, 2930, 1655, 1214, 748, 666. Anal. Calcd. for $C_{24}H_{37}NO_2$: C, 77.58; H, 10.04; N, 3.77. Found: C, 77.60; H, 10.07; N, 3.77.

N-(4-chlorophenethyl)-2-((1*R*,2*R*,4a*S*,8a*S*)-2-hydroxy-2,5,5,8a-tetramethyldecahydronaphthalen -1-yl)acetamide (**11**)

The product was purified by column chromatography using PE/EtOAc (1/1) as eluent to give the compound as a white solid, mp 168 °C (G). Yield 75%. IR ν (cm^{-1}): 3298, 2977, 2914, 1634, 1214, 1056, 749, 666. ^{1}H NMR (400 MHz, DMSO) δ (ppm): 7.68 (m, 1H, NH), 7.31-7.28 (m, 2H, Ar), 7.22-7.19 (m, 2H, Ar), 4.23 (s, 1H, OH), 3.27-3.20 (m, 2H, CH_2NH), 2.69-2.64 (m, 2H, CH_2-Ar), 2.21 (d, J = 15.4 Hz, 1H, CH_2CO), 1.95 (dd, J_1 = 15.3 Hz, J_2 = 6.1 Hz, 1H, CH_2CO), 1.71-1.60 (m, 4H), 1.53-1.40 (m, 2H), 1.37-1.12 (m, 4H), 1.08-1.01 (m, 2H), 0.93 (s, 3H, COH(CH_3)), 0.82 (s, 3H, CH_3), 0.73 (s, 3H, CH_3), 0.69 (s, 3H, CH_3). ^{13}C NMR (100 MHz, DMSO) δ (ppm): 174.4 (C=O), 139.1 (*CqAr*), 131.0 (×2, *CHAr* + *CqAr*), 128.6 (×2, *CHAr*), 71.6 (*Cq*-OH(CH_3)), 56.8 (CH), 56.0 (CH), 44.2, 42.0, 39.4 (under DMSO), 38.9, 38.7 (*Cq*-(CH_3)$_2$), 34.7, 33.8, 33.3, 31.8, 24.6, 21.8, 20.5, 18.4, 15.5. Anal. Calcd. for $C_{24}H_{36}ClNO_2$: C, 71.00; H, 8.94; N, 3.45. Found: C, 71.12; H, 8.97; N, 3.46.

N-(4-fluorophenethyl)-2-((1*R*,2*R*,4a*S*,8a*S*)-2-hydroxy-2,5,5,8a-tetramethyldecahydronaphthalen -1-yl)acetamide (**12**)

The product was purified by column chromatography using PE/EtOAc (1/1) as eluent to give the compound as a white solid, mp 114–115 °C (G). Yield 83%. IR ν (cm^{-1}): 3298, 2970, 2933, 1642, 1509, 1215, 1057, 748, 666. ^{1}H NMR (400 MHz, CDCl$_3$) δ (ppm): 7.15-7.12 (m, 2H, Ar), 6.96 (t, 2H, J = 8.2 Hz, Ar), 6.03 (brs, 1H, NH), 3.52-3.37 (m, 2H, CH_2NH), 2.76 (t, J = 6.1 Hz, 2H, CH_2-Ar), 2.47 (brs, 1H, OH), 2.29 (dd, J_1 = 15.3 Hz, J_2 = 5.2 Hz, 1H, CH_2CO), 2.04 (dd, J_1 = 15.3 Hz, J_2 = 3.9 Hz, 1H, CH_2CO), 1.89 (dt, J_1 = 12.5 Hz, J_2 = 3.1 Hz, 1H, -CH_2-COH(CH_3)), 1.68-1.62 (m, 1H), 1.56-1.51 (m, 2H), 1.42-1.32 (m, 4H), 1.27-1.14 (m, 2H), 1.09 (s, 3H, COH(CH_3)), 0.93-0.90 (m, 2H), 0.85 (s, 3H, CH_3), 0.75 (s, 3H, CH_3), 0.73 (s, 3H, CH_3). ^{13}C NMR (100 MHz, CDCl$_3$) δ (ppm): 175.4 (C = O), 161.6 (d, J = 244 ppm, *Cq*-F), 134.7 (*CqAr*), 130.3 (×2, J = 7.5 Hz, *CHAr*), 115.4 (×2, J = 21.2 Hz, *CHAr*), 73.1 (*Cq*-OH(CH_3)), 57.9 (CH), 56.0 (CH), 44.3, 41.8, 40.7, 39.3, 38.7 (*Cq*-(CH_3)$_2$), 34.8, 33.3, 33.2, 32.6, 23.8, 21.4, 20.5, 18.3, 15.4. Anal. Calcd. for $C_{24}H_{36}FNO_2$: C, 74.00; H, 9.32; N, 3.60. Found: C, 73.88; H, 9.36; N, 3.59.

N-(3-fluorophenethyl)-2-((1*R*,2*R*,4a*S*,8a*S*)-2-hydroxy-2,5,5,8a-tetramethyldecahydronaphthalen -1-yl)acetamide (**13**)

The product was purified by column chromatography using PE/EtOAc (1/1) as eluent to give the compound as an amorphous solid. Yield 72%. IR ν (cm^{-1}): 3685, 3310, 2977, 1642, 1215, 1057, 747, 666. ^{1}H NMR (300 MHz, CDCl$_3$) δ (ppm): 7.27-7.18 (m, 1H, Ar), 7.00-6.84 (m, 2H, Ar), 6.66 (t, J = 5.4 Hz, 1H, Ar), 3.53-3.32 (m, 2H, CH_2NH), 2.77 (t, J = 6.9 Hz, 2H, CH_2), 2.34 (dd, J_1 = 15.3 Hz, J_2 = 4.6 Hz, 1H, CH_2), 2.06 (dd, J_1 = 15.3 Hz, J_2 = 5.3 Hz, 1H, CH_2), 1.91-1.86 (m, 1H, CH), 1.65-1.10 (m, 10H), 1.07 (s, 3H, CH_3), 0.95-0.87 (m, 1H, CH), 0.85 (s, 3H, CH_3), 0.73 (s, 6H, CH_3). ^{13}C NMR (75 MHz, CDCl$_3$) δ (ppm): 175.8, 164.5, 161.2, 141.6, 128.9, 125.6, 123.5, 114.8, 112.3, 72.9, 57.2, 56.6, 55.1, 40.5, 38.7, 35.2, 32.6, 31.7, 23.3, 20.9, 18.2, 16.2, 14.9, 14.5. Anal. Calcd. for $C_{24}H_{36}FNO_2$: C, 74.00; H, 9.32; N, 3.60. Found: C, 73.90; H, 9.34; N, 3.61.

N-(furan-2-ylmethyl)-2-((1*R*,2*R*,4a*S*,8a*S*)-2-hydroxy-2,5,5,8a-tetramethyldecahydronaphthalen -1-yl)acetamide (**14**)

The product was purified by column chromatography using PE/EtOAc (1/1) as eluent to give the compound as a yellow oil. Yield 95%. IR ν (cm^{-1}): 3306, 2926, 1648, 1214, 746, 666. ^{1}H NMR (400 MHz,

CDCl$_3$) δ (ppm): 7.31 (s, 1H, Fur), 6.32 (brs, 1H, NH), 6.28 (t, *J* = 1.3 Hz, 1H, Fur), 6.18 (d, *J* = 2.7 Hz, 1H, Fur), 4.40 (dd, J_1 = 15.5 Hz, J_2 = 5.5 Hz, 1H, *CH$_2$*NH), 4.36 (dd, J_1 = 15.5 Hz, J_2 = 5.4 Hz, 1H, *CH$_2$*NH), 2.92 (brs, 1H, OH), 2.38 (dd, J_1 = 15.4 Hz, J_2 = 5.1 Hz, 1H, *CH$_2$*CO), 2.12 (dd, J_1 = 15.4 Hz, J_2 = 4.2 Hz, 1H, *CH$_2$*CO), 1.90 (d, *J* = 12.5 Hz, 1H, -CH$_2$-COH(CH$_3$)), 1.76 (t, *J* = 4.5 Hz, 1H, -CH-COH(CH$_3$)), 1.67-1.51 (m, 2H), 1.49-1.31 (m, 4H), 1.29-1.19 (m, 2H), 1.10 (s, 3H, COH(*CH$_3$*)), 0.98-0.89 (m, 2H), 0.84 (s, 3H, CH$_3$), 0.76 (s, 6H, CH$_3$). ^{13}C NMR (100 MHz, CDCl$_3$) δ (ppm): 175.1 (C=O), 151.5 (*CqAr*), 142.1 (*CHFur*), 110.4 (*CHFur*), 107.2 (*CHFur*), 73.2 (*Cq*-OH(CH$_3$)), 57.8 (CH), 55.9 (CH), 44.3, 41.8, 39.3, 38.7 (*Cq*-(CH$_3$)$_2$), 36.8, 33.3, 33.2, 32.5, 23.7, 21.4, 20.5, 18.4, 15.5. Anal. Calcd. for C$_{21}$H$_{33}$NO$_3$: C, 72.58; H, 9.57; N, 4.03. Found: C, 72.70; H, 9.61; N, 4.03.

N-((1,1′-biphenyl)-4-ylmethyl)-2-((1*R*,2*R*,4a*S*,8a*S*)-2-hydroxy-2,5,5,8a-tetramethyldecahydronaphthalen-1-yl) acetamide (**15**)

The product was purified by column chromatography using PE/EtOAc (2/1) as eluent to give compound as an amorphous solid. Yield 40%. IR ν (cm^{-1}): 3288, 2924, 1637, 1548, 1386, 1123, 938, 759, 696. ^{1}H NMR (300 MHz, CDCl$_3$) δ (ppm): 7.58-7.52 (m, 4H, Ar), 7.46-7.41 (m, 2H, Ar), 7.37-7.31 (m, 2H, Ar), 6.60 (t, *J* = 5.2 Hz, 1H, Ar), 4.5-4.36 (m, 2H, *CH$_2$*NH), 3.15 (brs, 1H, OH), 2.46 (dd, J_1 = 15.4 Hz, J_2 = 4.9 Hz, 1H, CH$_2$CO), 2.17 (dd, J_1 = 15.4 Hz, J_2 = 4.3 Hz, 1H, CH$_2$CO), 1.94-1.89 (m, 1H), 1.81 (t, *J* = 4.6 Hz, 1H), 1.73-1.21 (m, 9H, CH$_2$), 1.12 (s, 3H, CH$_3$), 1.00-0.96 (m, 1H), 0.86 (s, 3H, CH$_3$), 0.77 (s, 6H, CH$_3$). ^{13}C NMR (75 MHz, CDCl$_3$) δ (ppm): 175.4, 140.7, 140.2, 137.5, 128.8 (×2), 128.1, 127.3 (×2), 127.0, 73.1, 57.8, 55.9, 44.2, 43.4, 41.7, 39.3, 38.7, 33.3, 33.2, 32.6, 31.6, 23.7, 22.7, 21.4, 20.5, 18.4, 15.5, 14.2. Anal. Calcd. for C$_{29}$H$_{39}$NO$_2$: C, 80.33; H, 9.07; N, 3.23. Found: C, 80.55; H, 9.11; N, 3.24.

N-(3-(1*H*-imidazol-1-yl)propyl)-2-((1*R*,2*R*,4a*S*,8a*S*)-2-hydroxy-2,5,5,8a-tetramethyldecahydronaphthalen-1-yl)acetamide (**16**)

Without any further purification, the product was obtained as a light yellow solid, mp 83–84 °C (G). Yield 93%. IR ν (cm^{-1}): 3291, 2926, 1644, 1390, 1214, 1082, 747, 666. ^{1}H NMR (400 MHz, CDCl$_3$) δ (ppm): 7.47 (s, 1H, imid), 7.00 (s, 1H, imid), 6.90 (s, 1H, imid), 6.79 (brt, *J* = 5.3 Hz, 1H, NH), 3.95 (t, *J* = 6.9 Hz, 2H, *CH$_2$*-Nimid), 3.23-3.11 (m, 2H, *CH$_2$*-NH), 2.72 (brs, 1H, OH), 2.34 (dd, J_1 = 15.2 Hz, J_2 = 5.3 Hz, 1H, *CH$_2$*CO), 2.12 (dd, J_1 = 15.2 Hz, J_2 = 4.0 Hz, 1H, *CH$_2$*CO), 1.97-1.88 (m, 3H), 1.69 (t, *J* = 4.6 Hz, 1H, -CH-COH(CH$_3$)), 1.66-1.32 (mm, 6H), 1.28-1.18 (m, 2H), 1.12 (s, 3H, COH(*CH$_3$*)), 0.95-0.89 (m, 2H), 0.84 (s, 3H, CH$_3$), 0.75 (s, 3H, CH$_3$), 0.74 (s, 3H, CH$_3$). ^{13}C NMR (100 MHz, CDCl$_3$) δ (ppm): 175.9 (C=O), 137.2 (*CHimid*), 129.2 (*CHimid*), 119.0 (*CHimid*), 73.3 (*Cq*-OH(CH$_3$)), 58.2 (CH), 56.1 (CH), 44.6, 44.3, 41.8, 39.5, 38.8 (*Cq*-(CH$_3$)$_2$), 36.6, 33.3, 33.2, 32.7, 31.0, 23.8, 21.4, 20.5, 18.4, 15.4. Anal. Calcd. for C$_{22}$H$_{37}$N$_3$O$_2$: C, 70.36; H, 9.93; N, 11.19. Found: C, 70.24; H, 9.97; N, 11.24.

4.2.3. General Procedure for the Synthesis of Homodrimanyl Acid Ester (**18**) and (**19**)

(+)-Sclareolide (100 mg, 0.4 mmol, 1.0 eq.) was dissolved in hot (60 °C) MeOH (1 mL), and NaOH (64 mg, 1.6 mmol, 4 eq.) was added under stirring. The resulting mixture was stirred for 2 h at 60 °C and then cooled to rt. Diluted HCl was then added until pH 5–6 and the formed precipitate was filtered under vacuum [25]. For the next esterification reaction, the obtained intermediate was dissolved in anhydrous DMF (2 mL); then, the appropriate benzyl bromide (0.4 mmol, 1.0 eq.) and solid K$_2$CO$_3$ (0.4 mmol, 1 eq.) were added. The reaction mixture was stirred at rt for 24 h and then quenched by addition of water (5 mL). The inorganic phase was extracted with EtOAc (3 × 10 mL) and the combined organic layers were washed with water (20 mL) and brine (20 mL). The whole organic phase was dried over anhydrous Na$_2$SO$_4$, filtered, and evaporated to dryness. The crude product was purified by flash chromatography on silica gel using the mixture PE/EtOAc as eluent to give the corresponding homodrimanyl acid ester in good yield.

Benzyl 2-((1*R*,2*R*,4a*S*,8a*S*)-2-hydroxy-2,5,5,8a-tetramethyldecahydronaphthalen-1-yl)acetate (**18**)

The product was purified by column chromatography using PE/EtOAc (6/1) as eluent to give the compound as a colorless oil. Yield 55%. IR ν (cm^{-1}): 3014, 2939, 1718, 1214, 907, 748, 730, 666. ^{1}H NMR (300 MHz, CDCl$_3$) δ (ppm): 7.45-7.30 (m, 5H, Ar), 5.20-5.10 (m, 2H, CH$_2$O), 4.72 (brs, 1H, OH), 2.60 (dd, J_1 = 15.2 Hz, J_2 = 4.4 Hz, 1H, CH$_2$CO), 2.38 (dd, J_1 = 15.4 Hz, J_2 = 4.6 Hz, 1H, CH$_2$CO), 1.99 (m, 2H), 1.62-1.08 (m, 8H, CH$_2$), 1.18 (s, 3H, CH$_3$), 1.09-0.98 (m, 1H, CH), 0.90 (s, 3H, CH$_3$), 0.81 (s, 6H, CH$_3$). ^{13}C NMR (75 MHz, CDCl$_3$) δ (ppm): 175.5, 136.0, 129.3, 129.2 (×2), 127.4 (×2), 73.1, 66.5, 57.7, 54.9, 43.1, 41.6, 38.5, 33.2, 32.5, 30.6, 29.7, 23.7, 22.3, 20.6, 18.3, 14.5. Anal. Calcd. for C$_{23}$H$_{34}$O$_3$: C, 77.05; H, 9.56. Found: C, 77.16; H, 9.60.

3,4-dichlorobenzyl 2-((1*R*,2*R*,4a*S*,8a*S*)-2-hydroxy-2,5,5,8a-tetramethyldecahydronaphthalen -1-yl)acetate (**19**)

The product was purified by column chromatography using PE/EtOAc (6/1) as eluent to give the compound as a colorless oil. Yield 50%. ^{1}H NMR (400 MHz, CDCl$_3$) δ (ppm): 7.44 (s, 1H, Ar), 7.40 (d, *J* = 8.2 Hz, 1H, Ar), 7.17 (d, *J* = 8.2 Hz, 1H, Ar), 5.02 (ABq, *J* = 12.8 Hz, 2H, CH$_2$O), 2.52 (dd, *J* = 16.2 Hz, *J* = 5.8 Hz, 1H, *CH$_2$*CO), 2.32 (dd, *J* = 16.2 Hz, *J* = 5.1 Hz, 1H, *CH$_2$*CO), 1.92 (dt, *J* = 12.4 Hz, *J* = 2.9 Hz, 1H, -*CH$_2$*-COH(CH$_3$)), 1.84 (t, *J* = 5.4 Hz, 1H, -*CH*-COH(CH$_3$)), 1.69-1.52 (m, 2H), 1.48-1.23 (mm, 6H), 1.12 (s, 3H, COH(*CH$_3$*)), 0.99-0.91 (m, 2H), 0.86 (s, 3H, CH$_3$), 0.77 (s, 6H, CH$_3$). ^{13}C NMR (100 MHz, CDCl$_3$) δ (ppm): 177.0 (C=O), 141.1 (*CqAr*), 132.6 (*CqAr*), 131.4 (*CqAr*), 130.5 (*CHAr*), 128.8 (*CHAr*), 126.0 (*CHAr*), 86.4 (*Cq*-OH(CH$_3$)), 63.9 (*OCH$_2$*-Ar), 59.1 (CH), 56.7 (CH), 42.2, 39.5, 38.7 (*Cq*-(CH$_3$)$_2$), 33.2 (×2), 28.7, 21.6 (×2), 20.9, 20.6, 18.1, 15.1.

4.2.4. General Procedure for the Synthesis of Homodrimanyl Diol Esters (**22–24**)

According to a published procedure with some modifications, homodrimanyl diol 21 (80 mg, 0.31 mmol, 1.0 eq.) was dissolved in anhydrous DCM (2 mL) under an inert atmosphere. The appropriate carboxylic acid (0.34 mmol, 1.1 eq.), 1-etil-3-(3-dimetilaminopropil) carbodiimide (EDCI) (0.37 mmol, 1.2 eq.), and 1-ethyl-3-(3-dimethylaminopropyl)carbodiimide (DMAP) (0.031 mmol, 0.1 eq.) was added under stirring to the solution. The mixture was stirred at rt, checking the reaction by TLC, until the starting material disappeared (48–72 h). The reaction was quenched by the addition of water (5 mL) and the inorganic layer was extracted with DCM (2 × 10 mL). The combined organic layers were washed with water (20 mL) and brine (20 mL), dried over anhydrous Na$_2$SO$_4$, and filtered. The solvent was removed under reduced pressure, and the resulting residue was purified by flash chromatography on silica gel using the mixture PE/EtOAc as eluent to give final ester in good yield [26].

2-((1*R*,2*R*,4a*S*,8a*S*)-2-Hydroxy-2,5,5,8a-tetramethyldecahydronaphthalen-1-yl)ethyl Benzo[*d*][1,3]dioxole-5-carboxylate (**22**)

The product was purified by column chromatography using PE/EtOAc (4/1) as eluent to give the compound as a white solid, mp 155–156 °C (G). Yield 59%. IR ν (cm^{-1}): 3675, 2977, 2901, 1705, 1441, 1258, 1214, 1076, 1041, 750, 666. ^{1}H NMR (400 MHz, CDCl$_3$) δ (ppm): 7.63 (d, *J* = 8.0 Hz, 1H, Ar), 7.45 (s, 1H, Ar), 6.81 (d, *J* = 8.1 Hz, 1H, Ar), 6.01 (s, 2H, OCH$_2$O), 4.36-4.29 (m, 2H, CH$_2$-OCO), 1.91-1.81 (m, 2H), 1.77-1.59 (m, 4H), 1.45-1.34 (m, 3H), 1.31-1.10 (mm, 7H), 0.97-0.90 (m, 2H), 0.85 (s, 3H, CH3), 0.79 (s, 3H, CH$_3$), 0.78 (s, 3H, CH$_3$). ^{13}C NMR (100 MHz, CDCl$_3$) δ (ppm): 166.1 (C=O), 151.5 (*CqAr*), 147.7 (*CqAr*), 125.3 (*CHAr*), 124.6 (*CqAr*), 109.6 (*CHAr*), 108.0 (*CHAr*), 101.8 (OCH$_2$O), 73.7 (*Cq*-OH(CH$_3$)), 67.1 (CH$_2$-O), 58.1 (CH), 56.1 (CH), 44.5, 41.9, 39.8, 38.8 (*Cq*-(CH$_3$)$_2$), 33.4, 33.3, 24.7, 24.0, 21.5, 20.5, 18.4, 15.3. Anal. Calcd for C$_{24}$H$_{34}$O$_5$: C, 71.61; H, 8.51. Found: C, 71.87; H, 8.55.

2-((1*R*,2*R*,4a*S*,8a*S*)-2-hydroxy-2,5,5,8a-tetramethyldecahydronaphthalen-1-yl)ethyl (*E*)-3-(benzo[*d*][1,3]dioxol-5-yl)acrylate (**23**)

The product was purified by column chromatography using PE/EtOAc (4/1) as eluent to give the compound as a white solid, mp 115–116 °C (G). Yield 65%. IR ν (cm^{-1}): 2983, 2901, 1214, 1057,

744, 668. ^{1}H NMR (400 MHz, CDCl$_3$) δ (ppm): 7.57 (d, J = 15.9 Hz, 1H, CH=CH-Ar), 7.00 (s, 1H, Ar), 6.98 (d, J = 8.0 Hz, 1H, Ar), 6.78 (d, J = 8.0 Hz, 1H, Ar), 6.22 (d, J = 15.9 Hz, 1H, CH=CH-Ar), 5.98 (s, 2H, OCH$_2$O), 4.28-4.17 (m, 2H, CH$_2$-OCO), 1.88 (dt, J = 12.3 Hz, J = 2.9 Hz, 1H, -CH_2-COH(CH$_3$)), 1.83-1.53 (m, 4H), 1.46-1.34 (m, 4H), 1.31-1.20 (m, 2H), 1.16 (s, 3H, COH(CH_3)), 1.14-1.09 (m, 2H), 0.95-0.90 (m, 2H), 0.85 (s, 3H, CH$_3$), 0.79 (s, 3H, CH$_3$), 0.77 (s, 3H, CH$_3$). ^{13}C NMR (100 MHz, CDCl$_3$) δ (ppm): 167.3 (C=O), 149.6 ($CqAr$), 148.4 ($CqAr$), 144.5 (=CH-Ar), 128.9 ($CqAr$), 124.4 ($CHAr$), 116.2 (CH=CHAr), 108.6 ($CHAr$), 106.6 ($CHAr$), 101.6 (OCH$_2$O), 73.6 (Cq-OH(CH$_3$)), 66.6 (CH$_2$-O), 58.1 (CH), 56.1 (CH), 44.4, 41.9, 39.7, 38.8 (Cq-(CH$_3$)$_2$), 33.4, 33.3, 24.7, 24.0, 21.5, 20.5, 18.4, 15.3. Anal. Calcd for C$_{26}$H$_{36}$O$_5$: C, 72.87; H, 8.47. Found: C, 73.01; H, 8.49.

2-((1*R*,2*R*,4a*S*,8a*S*)-2-hydroxy-2,5,5,8a-tetramethyldecahydronaphthalen-1-yl)ethyl 4-(thiophen-2-yl)butanoate (**24**)

The product was purified by column chromatography using PE/EtOAc (3/1) as eluent to give the compound as a colorless oil. Yield 70%. IR ν (cm^{-1}): 3675, 2958, 1724, 1390, 1214, 1082, 748, 692, 667. ^{1}H NMR (400 MHz, CDCl$_3$) δ (ppm): 7.10 (d, J = 5.0 Hz, 1H, *thienyl*), 6.90 (t, J = 4.2 Hz, 1H, *thienyl*), 6.78 (d, J = 2.4 Hz, 1H, *thienyl*), 4.17-4.06 (m, 2H, CH$_2$-OCO), 2.86 (t, J = 7.5 Hz, 2H, COCH$_2$-CH$_2$-CH$_2$), 2.34 (t, J = 7.5 Hz, 2H, COCH_2-CH$_2$-CH$_2$), 2.03-1.95 (m, 2H, COCH$_2$-CH_2-CH$_2$), 1.87 (d, J = 12.3 Hz, 1H, -CH_2-COH(CH$_3$)), 1.77-1.69 (m, 1H), 1.67-1.52 (mm, 6H), 1.43-1.35 (m, 3H), 1.30-1.19 (m, 1H), 1.14 (s, 3H, COH(CH_3)), 1.12-1.07 (m, 1H), 0.92-0.88 (m, 2H), 0.86 (s, 3H, CH$_3$), 0.77 (s, 6H, CH$_3$). ^{13}C NMR (100 MHz, CDCl$_3$) δ (ppm): 173.4 (C=O), 144.1 ($Cqthienyl$), 126.8 ($CHthienyl$), 124.5 ($CHthienyl$), 123.2 ($CHthienyl$), 73.6 (Cq-OH(CH$_3$)), 66.6 (CH$_2$-O), 58.0 (CH), 56.1 (CH), 44.4, 41.9, 39.7, 38.8 (Cq-(CH$_3$)$_2$), 33.5, 33.4, 33.3, 29.2, 26.8, 24.5, 24.0, 21.5, 20.5, 18.4, 15.3. Anal. Calcd for C$_{24}$H$_{38}$O$_3$S: C, 70.89; H, 9.42. Found: C, 70.75; H, 9.44.

Synthesis of (1*R*,2*R*,4a*S*,8a*S*)-1-(2-((3-Chlorobenzyl)oxy)ethyl)-2,5,5,8a-tetramethyldecahydronaphthalen -2-ol (homodrimanyl diol ether) (**25**)

Homodrimanyl diol **20** (90 mg, 0.35 mmol, 1.0 eq.) was dissolved in anhydrous THF (10 mL) under argon atmosphere and NaH, previously purified, (67.0 mg, 0.4 mmol, 1.2eq) was added. The reaction mixture was refluxed for 30 min and then, after cooling at rt, 3-chlorobenzyl bromide (67.6 mg, 0.42 mmol, 1.2 eq.) was added. The mixture was still heated to reflux for 48 h, cooled to rt and quenched with water and saturated NH$_4$Cl solution (pH 7). Afterward, the aqueous layer was extracted with EtOAc (3 × 15 mL) and the combined organic phases were washed with water (30 mL) and brine (30 mL), dried over anhydrous Na$_2$SO$_4$, filtered, and evaporated to dryness to obtain the final homodrimanyl diol ether. Column chromatography with PE/EtOAc (8/1) as eluent gave the pure compound as a light yellow oil. Yield 27%. IR ν (cm^{-1}): 2926, 1214, 1077, 750, 667. ^{1}H NMR (300 MHz, CDCl$_3$) δ (ppm): 7.39-7.18 (m, 4H), 4.49 (s, 2H, CH2-Ar), 3.68-3.57 (m, 1H), 3.41-3.30 (m, 1H), 3.19 (brs, 1H, OH), 1.96-1.85 (m, 1H), 1.83-1.70 (m, 1H), 1.69-1.48 (m, 3H), 1.44-1.17 (m, 8H), 1.14 (s, 3H, CH$_3$), 0.96-0.90 (m, 1H), 0.88 (s, 3H, CH$_3$), 0.78 (s, 6H, CH3). ^{13}C NMR (75 MHz, CDCl$_3$) δ (ppm): 139.9, 129.8, 127.8, 127.8, 126.5, 125.7, 72.5, 72.3, 72.2, 59.0, 56.0, 43.9, 41.8, 39.5, 38.9, 33.4, 33.2, 25.2, 24.3, 21.5, 20.4, 18.4, 15.3. Anal. Calcd for C$_{23}$H$_{35}$ClO$_2$: C, 72.89; H, 8.31. Found: C, 72.56; H, 8.34.

4.3. TRPV1 and TRPV4 Channel Assays

Compound effects on intracellular Ca^{2+} concentration ([Ca^{2+}]$_i$) were determined using the selective intracellular fluorescent probe for Ca^{2+} Fluo-4 and assays were performed as described [27]. Briefly, human embryonic kidney (HEK-293) cells, stably transfected with recombinant rat TRPV4 or human TRPV1 (selected by Geneticin 600 μg/mL) or not transfected, were cultured in EMEM (Eagle's Minimum Essential Medium) +2 mM Glutamine +1% Non-Essential Amino Acids +10% FBS (fetal bovine serum) and maintained at 37 °C with 5% CO$_2$. On the day of the experiment the cells were loaded in the dark at rt for 1 h with Fluo-4 AM (4 μM in DMSO containing 0.02% Pluronic F-127). After that the cells were rinsed and resuspended in Tyrode's solution (145 mM NaCl, 2.5 mM KCl,

1.5 mM $CaCl_2$, 1.2 mM $MgCl_2$, 10 mM D-glucose, and 10 mM HEPES (4-(2-hydroxyethyl)-1-piperazine ethane sulfonic acid), pH 7.4) then transferred to a quartz cuvette of a spectrofluorimeter (Perkin-Elmer LS50B Beaconsfield UK; λEX = 488 nm, λEM = 516 nm) under continuous stirring. Cell fluorescence before and after the addition of various concentrations of test compounds was measured normalizing the effects against the response to ionomycin (4 μM). The values of the effect on $[Ca^{2+}]_i$ in not-transfected HEK-293 cells are used as a baseline and subtracted from the values obtained from transfected cells. The potency of the compounds (EC_{50} values) is determined as the concentration required to produce half-maximal increases in $[Ca^{2+}]_i$. Antagonist behavior is evaluated against the GSK1016790A agonist of TRPV4 (10 nM) [28] and analyzed by adding the compounds directly in the quartz cuvette 5 min before stimulation of cells with the agonist. IC_{50} is expressed as the concentration exerting a half-maximal inhibition of agonist effect, taking as 100% the effect on $[Ca^{2+}]_i$ exerted by GSK1016790A (10 nM) alone. Similarly, for TRPV1 using agonist capsaicin 0.1 μM. Dose-response curve fitting (sigmoidal dose-response variable slope) and parameter estimation were performed with Graph-Pad Prism8® (GraphPad Software Inc., San Diego, CA, USA). All determinations were performed at least in triplicate.

4.4. MTT Assay

HeLa and A549 cells were seeded at 2×10^3 cells/cm^2 density in 24-well plastic plates. One day after plating, compounds **6** and **18** were added to the culture medium for 24 h. Cell viability was evaluated with the 3-(4,5-dimethylthiazol-2-yl)-2,5-diphenyltetrazolium bromide (MTT; 0.5 mg/mL; Sigma-Aldrich, Milan, Italy) reduction assay, and formazan salt formation upon MTT reduction by the mitochondria of living cells was detected spectrophotometrically at 595 nm according to published procedures [29].

5. Conclusions

In this study, 22 novel trans-decalin-related homodrimane molecules were designed and synthesized, inspired by the natural diversity of labdane diterpenes, bioactive compounds largely occurring in terrestrial and marine sources. The validity of this idea was attested by the identification of a new class of TRPV4 antagonists, active at low micromolar concentration, and endowed with selectivity over TRPV1. The most interesting compounds belong to the homodrimanyl acid amide series, in particular, benzyl and phenylethyl amides. From the SAR study, several critical determinants for activity emerged: (1) the length of the spacer connecting the homodrimane skeleton to the aromatic moiety; (2) the nature of the functional group inside the spacer and, in particular, the presence and position of the carbonyl functionality; and (3) the size and electronic properties of substituents decorating the phenyl ring. The optimal spacer length ranged from four to five atoms, containing a carbonyl group. Thus, this feature represented a quite stringent structural requirement. The presence of an aromatic core is mandatory for activity since both natural (+)-sclareolide and compounds **20** and **21** were inactive. These results are consistent with the observed antagonism of onydecalin A, also featuring an aromatic-substituted *trans*-decalin scaffold and a carbonyl linker. Some of the newly synthesized derivatives exhibited a 9-fold potency increase on onydecalin A activity, an effect probably ascribable to the optimization of the spacer connecting the homodrimane skeleton to the aromatic moiety. In particular, compound **6**, the 3,4-dichlorobenzyl homodrimanyl acid amide, and compound **18**, the benzyl homodrimanyl acid ester, behaved as potent and selective TRPV4 antagonists. The biological results presented herein confirm that the bicyclic *trans*-decalin nucleus is a valuable scaffold for the design of new TRPV4 modulators and prompts us to further investigate this field with the aim to obtain new molecules potentially useful for the treatment of pain and pulmonary oedema linked to COVID-19 syndrome, and to search for other scaffolds in the available and continuously increasing libraries of natural labdane derivatives and related compounds.

Supplementary Materials: The following are available online at http://www.mdpi.com/1660-3397/18/10/519/s1. Procedure for the synthesis and, physical and chemical data of compounds **20** and **21**. Representations of the ^{1}H-NMR and ^{13}C-NMR spectra of compounds **1–16**, **18–19**, **22–25**. Biological data on TRPV1 assay.

Author Contributions: Conceptualization, S.M., G.C., F.A., A.B., R.M.V. and L.D.P.; methodology, S.M., G.C., F.A., A.B., L.D.P. and A.S.M.; formal analysis, G.C., S.M., F.A., A.B., L.D.P. and A.S.M.; investigation, S.M., G.C., F.A., A.B., R.M.V., L.D.P., A.S.M. and P.A.; resources, F.A., A.B., L.D.P. and V.D.M.; writing—original draft preparation, S.M., G.C., F.A., A.B., R.M.V., and L.D.P.; writing—review and editing, S.M., G.C., F.A., A.B., R.M.V., L.D.P., P.A., V.D.M. and M.V.-H.; visualization, S.M., G.C., and P.A.; funding acquisition, F.A., A.B. and V.D.M. All authors have read and agreed to the published version of the manuscript.

Funding: This research was funded by "Dipartimento di Eccellenza grant 2018–2022", Ministero dell'Istruzione, dell'Università e della Ricerca (Rome), received by Department of Pharmacy, Health and Nutritional Sciences, University of Calabria and Department of Biotechnology, Chemistry and Pharmacy, University of Siena.

Acknowledgments: We are very grateful to Marco Allarà and Fabio Arturo Iannotti for the help with the cell culture and MTT assay.

Conflicts of Interest: The authors declare no conflict of interests.

References

1. Singh, M.; Pal, M.; Sharma, R.P. Biological activity of the labdane diterpenes. *Planta Med.* **1999**, *65*, 2–8. [CrossRef]

2. Pal, M.; Mishra, T.; Kumar, A.; Tewari, S.K. Medicinal plants: Biological evaluation of terrestrial and marine plant originated labdane diterpenes (a review). *Pharm. Chem. J.* **2016**, *50*, 558–567. [CrossRef]

3. Aiello, F.; Carullo, G.; Badolato, M.; Brizzi, A. TRPV1–FAAH–COX: The Couples Game in Pain Treatment. *ChemMedChem* **2016**, *11*, 1686–1694. [CrossRef]

4. Premkumar, L.S. Transient receptor potential channels as targets for phytochemicals. *ACS Chem. Neurosci.* **2014**, *5*, 1117–1130. [CrossRef] [PubMed]

5. Vitale, R.M.; Schiano Moriello, A.; De Petrocellis, L. Natural compounds and synthetic drugs targeting the ionotropic cannabinoid members of transient receptor potential (TRP) channels. In *New Tools to Interrogate Endocannabinoid Signalling—From Natural Compounds to Synthetic Drugs*; Maccarrone, M., Ed.; RSC: London, UK, 2020; in press.

6. Liedtke, W.; Choe, Y.; Marti-Renom, M.A.; Bell, A.M.; Denis, C.S.; Šali, A.; Hudspeth, A.J.; Friedman, J.M.; Heller, S. Vanilloid receptor-related osmotically activated channel (VR-OAC), a candidate vertebrate osmoreceptor. *Cell* **2000**, *103*, 525–535. [CrossRef]

7. Strotmann, R.; Harteneck, C.; Nunnenmacher, K.; Schultz, G.; Plant, T.D. OTRPC4, a nonselective cation channel that confers sensivity to extracellular osmolarity. *Nat. Cell Biol.* **2000**, *2*, 695–702. [CrossRef] [PubMed]

8. Deng, Z.; Paknejad, N.; Maksaev, G.; Sala-Rabanal, M.; Nichols, C.G.; Hite, R.K.; Yuan, P. Cryo-EM and X-ray structures of TRPV4 reveal insight into ion permeation and gating mechanisms. *Nat. Struct. Mol. Biol.* **2018**, *25*, 252–260. [CrossRef]

9. Kuebler, W.M.; Jordt, S.E.; Liedtke, W.B. Urgent reconsideration of lung edema as a preventable outcome in COVID-19: Inhibition of TRPV4 represents a promising and feasible approach. *Am. J. Physiol. Lung Cell. Mol. Physiol.* **2020**, *318*, L1239–L1243. [CrossRef]

10. Jian, M.Y.; King, J.A.; Al-Mehdi, A.B.; Liedtke, W.; Townsley, M.I. High vascular pressure-induced lung injury requires P450 epoxygenase-dependent activation of TRPV4. *Am. J. Respir. Cell Mol. Biol.* **2008**, *38*, 386–392. [CrossRef]

11. Vincent, F.; Duncton, M.A.J. TRPV4 Agonists and Antagonists. *Curr. Top. Med. Chem.* **2011**, *11*, 2216–2226. [CrossRef]

12. Lawhorn, B.G.; Brnardic, E.J.; Behm, D.J. Recent advances in TRPV4 agonists and antagonists. *Bioorg. Med. Chem. Lett.* **2020**, *30*, 127022. [CrossRef] [PubMed]

13. Smith, P.L.; Maloney, K.N.; Pothen, R.G.; Clardy, J.; Clapham, D.E. Bisandrographolide from Andrographis paniculata activates TRPV4 channels. *J. Biol. Chem.* **2006**, *281*, 29897–29904. [CrossRef]

14. Hilfiker, M.A.; Hoang, T.H.; Cornil, J.; Eidam, H.S.; Matasic, D.S.; Roethke, T.J.; Klein, M.; Thorneloe, K.S.; Cheung, M. Optimization of a novel series of TRPV4 antagonists with in vivo activity in a model of pulmonary edema. *ACS Med. Chem. Lett.* **2013**, *4*, 293–296. [CrossRef] [PubMed]

15. Cheung, M.; Bao, W.; Behm, D.J.; Brooks, C.A.; Bury, M.J.; Dowdell, S.E.; Eidam, H.S.; Fox, R.M.; Goodman, K.B.; Holt, D.A.; et al. Discovery of GSK2193874: An Orally Active, Potent, and Selective Blocker of Transient Receptor Potential Vanilloid 4. *ACS Med. Chem. Lett.* **2017**, *8*, 549–554. [CrossRef]

16. Lin, Z.; Phadke, S.; Lu, Z.; Beyhan, S.; Abdel Aziz, M.H.; Reilly, C.; Schmidt, E.W. Onydecalins, Fungal Polyketides with Anti- Histoplasma and Anti-TRP Activity. *J. Nat. Prod.* **2018**, *81*, 2605–2611. [CrossRef]

17. Chakraborty, K.; Lipton, A.P.; Paul Raj, R.; Vijayan, K.K. Antibacterial labdane diterpenoids of Ulva fasciata Delile from southwestern coast of the Indian Peninsula. *Food Chem.* **2010**, *119*, 1399–1408. [CrossRef]

18. Pérez-García, E.; Zubía, E.; Ortega, M.J.; Carballo, J.L. Merosesquiterpenes from two sponges of the genus Dysidea. *J. Nat. Prod.* **2005**, *68*, 653–658. [CrossRef] [PubMed]

19. Caniard, A.; Zerbe, P.; Legrand, S.; Cohade, A.; Valot, N.; Magnard, J.L.; Bohlmann, J.; Legendre, L. Discovery and functional characterization of two diterpene synthases for sclareol biosynthesis in *Salvia sclarea* (L.) and their relevance for perfume manufacture. *BMC Plant. Biol.* **2012**, *12*, 1–13. [CrossRef]

20. Coricello, A.; El-Magboub, A.; Luna, M.; Ferrario, A.; Haworth, I.S.; Gomer, C.J.; Aiello, F.; Adams, J.D. Rational drug design and synthesis of new α-Santonin derivatives as potential COX-2 inhibitors. *Bioorg. Med. Chem. Lett.* **2018**, *28*, 993–996. [CrossRef]

21. Coricello, A.; Adams, J.D.; Lien, E.J.; Nguyen, C.; Perri, F.; Williams, T.J.; Aiello, F. A Walk in Nature: Sesquiterpene Lactones as Multi-Target Agents Involved in Inflammatory Pathways. *Curr. Med. Chem.* **2018**, *27*, 1501–1514. [CrossRef]

22. Chen, Q.; Tang, K.; Guo, Y. Discovery of sclareol and sclareolide as filovirus entry inhibitors. *J. Asian Nat. Prod. Res.* **2020**, *22*, 464–473. [CrossRef] [PubMed]

23. Dixon, D.D.; Lockner, J.W.; Zhou, Q.; Baran, P.S. Scalable, divergent synthesis of meroterpenoids via "borono-sclareolide". *J. Am. Chem. Soc.* **2012**, *134*, 8432–8435. [CrossRef] [PubMed]

24. Li, D.; Zhang, S.; Song, Z.; Wang, G.; Li, S. Bioactivity-guided mixed synthesis accelerate the serendipity in lead optimization: Discovery of fungicidal homodrimanyl amides. *Eur. J. Med. Chem.* **2017**, *136*, 114–121. [CrossRef] [PubMed]

25. Koga, T.; Aoki, Y.; Hirose, T.; Nohira, H. Resolution of sclareolide as a key intermediate for the synthesis of Ambrox®. *Tetrahedron Asymmetry* **1998**, *9*, 3819–3823. [CrossRef]

26. Aiello, F.; Badolato, M.; Pessina, F.; Sticozzi, C.; Maestrini, V.; Aldinucci, C.; Luongo, L.; Guida, F.; Ligresti, A.; Artese, A.; et al. Design and Synthesis of New Transient Receptor Potential Vanilloid Type-1 (TRPV1) Channel Modulators: Identification, Molecular Modeling Analysis, and Pharmacological Characterization of the N-(4-Hydroxy-3-methoxybenzyl)-4-(thiophen-2-yl)butanamide. *ACS Chem. Neurosci.* **2016**, *7*, 737–748. [CrossRef]

27. Schiano Moriello, A.; De Petrocellis, L. *Endocannabinoid Signaling: Methods and Protocols*; Springer: Berlin/Heidelberg, Germany, 2016; Volume 1412, ISBN 9781493935376.

28. Thorneloe, K.S.; Sulpizio, A.C.; Lin, Z.; Figueroa, D.J.; Clouse, A.K.; McCafferty, G.P.; Chendrimada, T.P.; Lashinger, E.S.R.; Gordon, E.; Evans, L.; et al. N-((1S)-1-{[4-((2S)-2-{[(2,4-dichlorophenyl)sulfonyl]amino} -3-hydroxypropanoyl)-1-piperazinyl]carbonyl}-3-methylbutyl)-1-benzothiophene-2-carboxamide (GSK1016790A), a novel and potent transient receptor potential vanilloid 4 channel agonist induces urin. *J. Pharmacol. Exp. Ther.* **2008**, *326*, 432–442. [CrossRef] [PubMed]

29. Iannotti, F.A.; De Maio, F.; Panza, E.; Appendino, G.; Taglialatela-Scafati, O.; De Petrocellis, L.; Amodeo, P.; Vitale, R.M. Identification and Characterization of Cannabimovone, a Cannabinoid from Cannabis sativa, as a Novel PPAR Agonist via a Combined Computational and Functional Study. *Molecules* **2020**, *25*, 1119. [CrossRef]

Publisher's Note: MDPI stays neutral with regard to jurisdictional claims in published maps and institutional affiliations.

 *marine drugs* 

Article

A Concise Route for the Synthesis of Tetracyclic Meroterpenoids: (±)-Aureol Preparation and Mechanistic Interpretation

Antonio Rosales Martínez [1],*, Lourdes Enríquez [2], Martín Jaraíz [2] , Laura Pozo Morales [1], Ignacio Rodríguez-García [3] and Emilio Díaz Ojeda [1]

[1]   Department of Chemical Engineering, Escuela Politécnica Superior, University of Sevilla, 41011 Sevilla, Spain; lauratar@us.es (L.P.M.); emidi@us.es (E.D.O.)
[2]   Department of Electricity and Electronics, University of Valladolid, 47011 Valladolid, Spain; louenr@tel.uva.es (L.E.); mjaraiz@ele.uva.es (M.J.)
[3]   Organic Chemistry, CeiA3, University of Almería, 04120 Almería, Spain; irodrigu@ual.es
*   Correspondence: arosales@us.es

Received: 30 July 2020; Accepted: 25 August 2020; Published: 26 August 2020

Abstract: A new concise general methodology for the synthesis of different tetracyclic meroterpenoids is reported: (±)-aureol (**1**), the key intermediate of this general route. The synthesis of (±)-aureol (**1**) was achieved in seven steps (28% overall yield) from (±)-albicanol. The key steps of this route include a C–C bond-forming reaction between (±)-albicanal and a lithiated arene unit and a rearrangement involving 1,2-hydride and 1,2-methyl shifts promoted by $BF_3 \bullet Et_2O$ as activator and water as initiator.

Keywords: aureol; tetracyclic meroterpenoids; natural products synthesis

1. Introduction

Marine sponges appear to have become an almost inexhaustible source of new natural compounds, showing a broad spectrum of biological activities and different structural patterns. Among these compounds there is a structurally unique class of natural products, the meroterpenoids, which are constituted by a sesquiterpene unit linked to a phenolic or quinone moiety [1]. Important examples of tetracyclic meroterpenoids (Figure 1) include (+)-aureol (**1**) [2,3], (+)-strongylin A (**2**) [4], (−)-cyclosmenospongine (**3**) [5] and (+)-smenoqualone (**4**) [6].

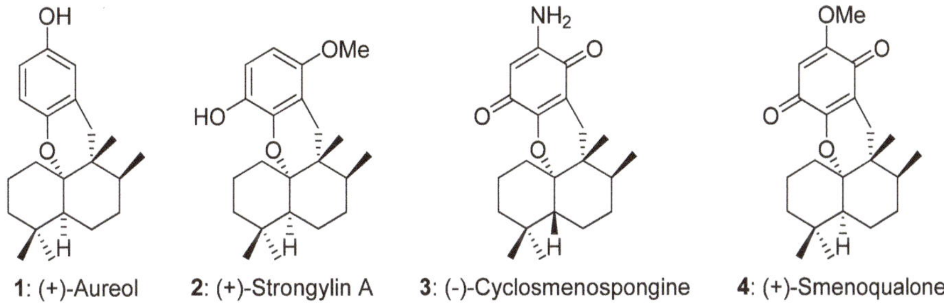

Figure 1. Selected members of tetracyclic meroterpenoids.

(+)-Aureol (**1**) was initially isolated and characterized by Faulker et al. [2] from the Caribbean sponge *Smeonspongia aurea*. It was later also found in some other species of Caribbean sponges,

Verongula gigantea and *Smenospongia* sp. [7]. (+)-Aureol (**1**) is a tetracyclic meroterpenoid with a unique structure that combines a *cis*-decalin system with a substituted benzopyran moiety. It shows anti-influenza-A virus activity [8] and selective cytotoxicity against human tumor cells, including colon adenocarcinoma HT-29 cells [9] and nonsmall cell lung cancer A549 [9].

Although the tetracyclic meroterpenoids have exclusive structural features and a wide assortment of biological activities, only one highly modular and robust platform for the synthesis of this class of natural products has been reported to date [10]. The rest of the reported routes are synthetic operations (10–27 linear steps) that have not enabled straightforward access to the whole family of these interesting natural products [11–20].

2. Results and Discussion

As a continuation of our research on the synthesis of marine natural bioactive compounds [18,21–23], we have developed a new concise route for the synthesis of tetracyclic meroterpenoids. In this new synthetic route, aureol (**1**) is the key intermediate from which other tetracyclic meroterpenoids, such as **2, 3** and **4**, can be easily synthesized by simple functional modification of its aromatic ring.

We thought the synthesis of **1** could be achieved through a coupling of albicanal (**6**) with 2-lithiohydroquinone dimethyl ether and a biogenetic-type rearrangement (previously explored by us) as pivotal steps (Scheme 1).

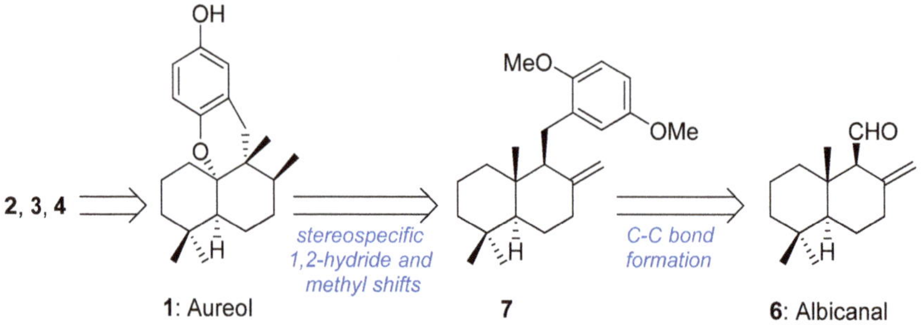

Scheme 1. Retrosynthesis of tetracyclic meroterpenoids.

The synthesis of (±)-aureol ((±)-**1**) (Scheme 2) used as starting material (±)-albicanol (**5**), which was prepared through Cp$_2$TiCl-catalyzed radical cascade cyclization of epoxy-farnesyl acetate, as previously reported by us and others [24,25]. Dess–Martin oxidation of **5** almost quantitatively afforded (±)-albicanal (**6**). The first key step was the coupling of (±)-albicanal (**6**) with the lithiated arene unit. For this purpose, an efficient and economical methodology previously reported by Seifert et al. [26] was used. In our hands, the addition of 2-lithiohydroquinone dimethyl ether to (±)-albicanal (**6**) gave a mixture of diastereomeric benzylic alcohols. In order to remove the free hydroxy group, the reaction crude was treated with lithium in liquid NH$_3$/THF followed by NH$_4$Cl. In this way, *trans*-decaline **7** was obtained in 90% yield (two steps).

The second key step in our synthesis of (±)-aureol ((±)-**1**) was based on a biogenetic-type rearrangement of **7** to give **8** that was previously reported by us [18]. In this way, a BF$_3$•Et$_2$O-mediated rearrangement of **7** leads to the formation of the desired product **8** as a single stereoisomer in a 62% yield, together with a minor tetracyclic compound **9** in a 28% yield. Demethylation of **8** following the conditions reported by Wright et al. [27] in the synthesis of natural compound (+)-frondosin gave **10** in an 82% yield over the two steps. Finally, cyclization of phenolic compound **10** was carried out with BF$_3$•Et$_2$O. This reaction afforded (±)-aureol ((±)-**1**) in a 62% yield. Physical and spectroscopic properties of synthetic (±)-aureol ((±)-**1**) matched those previously reported for the natural compound [2]. Thus, the synthesis of (±)-aureol ((±)-**1**) from (±)-albicanol (**5**) was completed in only seven steps and a global 28% yield, substantially improving the synthetic procedures previously published [11–20]. Moreover, a simple epimerization of aureol (**1**) to 5-*epi*-aureol (**11**) has already been reported [10]. From these two compounds, aureol (**1**) and 5-*epi*-aureol (**11**), adequate functionalization sequences can lead to (−)-cyclomenospongine (**3**), (+)-strongylin A (**2**) and (+)-smenoquealone (**4**), sequences that can be considered alternative formal syntheses of these tetracyclic compounds [9,28]. In this way, the methodology here described can be considered a general method for the synthesis of tetracyclic meroterponoids.

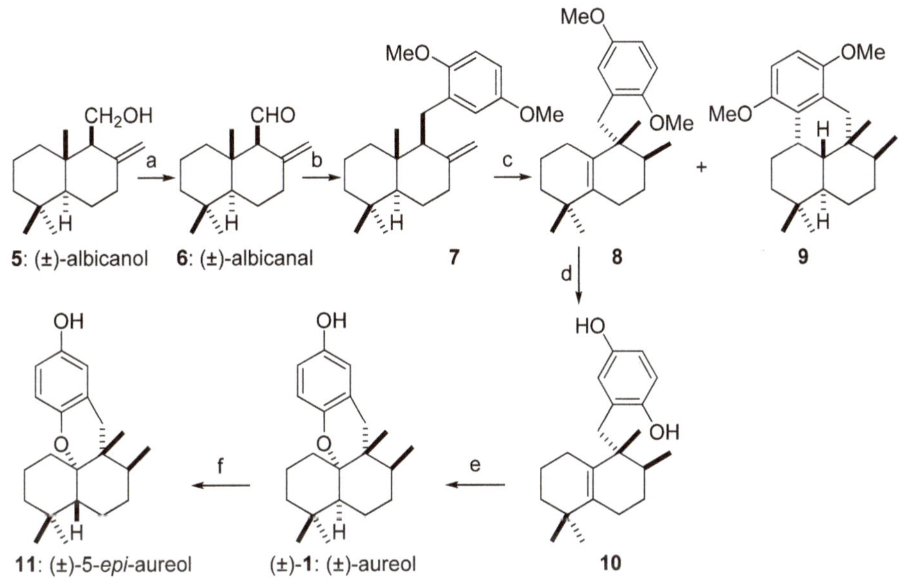

Scheme 2. *Reagents and conditions:* (**a**) Dess–Martin, 99.7%; (**b**) (i) Hydroquinone dimethyl ether (3 equiv), Et$_2$O, *sec*-BuLi (2 equiv), 5 min at 0 °C, 3 h at room temperature (rt). Then, **6** (1 equiv), Et$_2$O, 5 min, rt, quantitative; (ii) Liquid NH$_3$, THF, Li (5.3 equiv), 15 min, −78 °C. Then, mixture of benzylalcohols (1 equiv), THF, 15 min, −78 °C. Finally, NH$_4$Cl (13.6 equiv), 30 min, −78 °C, 90% (two steps); (**c**) **7** (1 equiv), BF$_3$•Et$_2$O (5.0 equiv), CH$_2$Cl$_2$, 5 h, −50 to −5 °C, 62% (**8**), 28% (**9**); (**d**) (i) **8** (1 equiv), AgO (2.0 equiv), 6N HNO$_3$ (3.0 equiv), 1,4-dioxane, rt, 15 min; (ii) 10% Pd/C (0.05 equiv), H$_2$ (1 atm), CHCl$_3$, 25 min, rt, 82%; (**e**) **10** (1 equiv), BF$_3$•Et$_2$O (4.5 equiv), CH$_2$Cl$_2$, −60 to −20 °C, 3 h, 62%; (**f**) HI, benzene, 90 °C, ref. 10, 87%.

The transformation of the exocyclic alkene **7** into the rearranged products **8** and **9** can be rationalized as depicted in Scheme 3. It is known that pure Lewis acids, such as boron trifluoride, are not effective initiators in alkene cationic polymerization [29], which makes more likely a pathway involving a proton transfer. On the other hand, it is well known that BF$_3$•Et$_2$O is very moisture-sensitive, and inevitably over time the HF that forms from the hydrolysis of BF$_3$ will react with excess BF$_3$ to form HBF$_4$, which is a strong acid and possibly triggers the cationic rearrangement. Thus, when the exocyclic alkene group in the bicyclic compound **7** is activated by a proton, the tertiary carbocation intermediate **I** is formed. Since the cleavage of a C–H bond is usually easier than a C–C bond, the hydrogen on C9 has a higher migratory aptitude than the alkyl group. In addition, migration of any of the hydrogens on C7 would lead to a secondary carbocation, less stable. In this way, the carbocationic intermediate **II** would be formed. From the stereochemical point of view, the configuration of C9 facilitates a 1,2-hydrogen shift on the α-face of the carbocation intermediate **I** to form carbocation intermediate **II**. Subsequently, the configuration of C10 facilitates a 1,2-methyl shift on the β-face of the carbocation intermediate **II** to form the carbocation intermediate **III**, which leads (pathway a, Scheme 3), after losing a H$^+$, to the major compound **8**. On the other hand, the intermediate **III** could suffer a 1,2-hydride shift from the C1 position to the carbocation on C10 to form the carbocationic intermediate **IV** (pathway b, Scheme 3), which can react with the aromatic ring by electrophilic substitution to generate the minor tetracyclic by-product **9**. In both pathways, a H^+ is liberated, which can react with more alkene **7** to continue the catalytic cycle. On the other hand, the simultaneous formation of **8** and **9** suggests that the all of the abovementioned rearrangements leading from **7** to **8** are not part of a concerted process, but proceed through a series of rapidly interconverting carbocations.

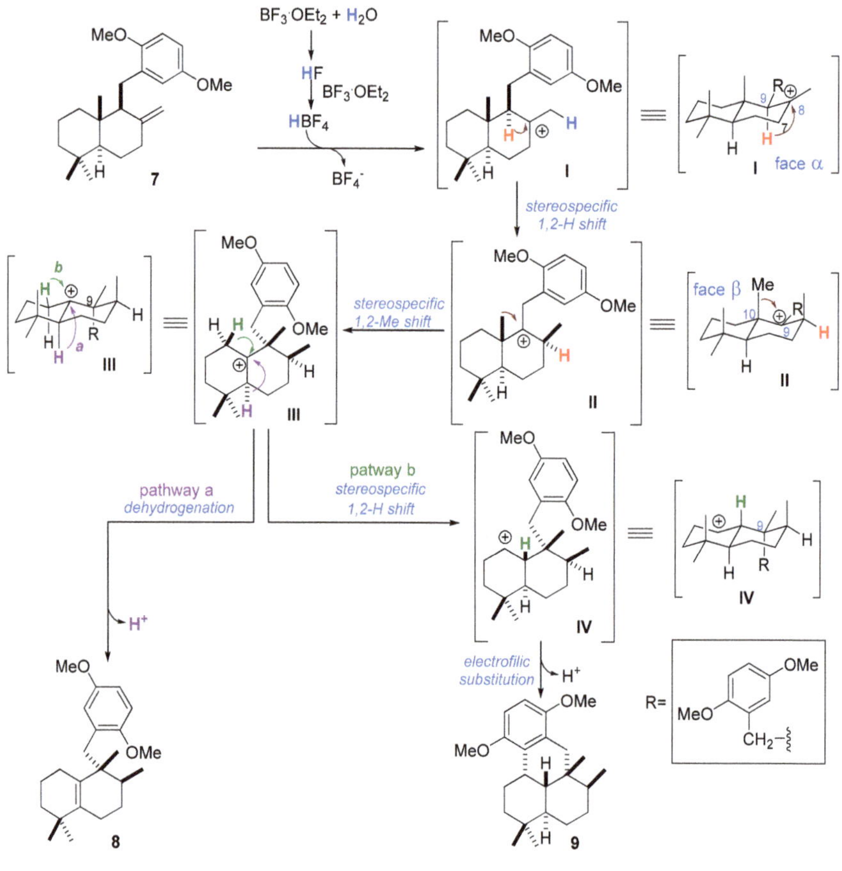

Scheme 3. Proposed reaction mechanisms for the formation of tetrasubstituted alkene **8** and by-product **9**.

3. Experimental Section

3.1. General Methods

All reagents were used as received from commercial sources. All solvents were distilled before use. THF was refluxed over Na and CH_2Cl_2 over calcium hydride before being distilled under an Ar atmosphere. Reaction products were purified by conventional column chromatography on Merck silica gel 50. Analytical thin-layer chromatography (TLC) was performed on 0.2 mm DC-Fertigfolien Alugram® Xtra Sil G/UV254 silica gel plates and visualized under a UV lamp or by immersion in an ethanol solution of phosphomolybdic acid (7%) followed by heating. ^{1}H and ^{13}C NMR spectra were recorded in Varian spectrometers operating at 300, 500 or 600 MHz. $CDCl_3$ was always used as NMR solvent. (±)-Albicanol was prepared from commercial farnesol according to a known procedure [22,24]. Copies of ^{1}H and ^{13}C NMR spectra of relevant known compounds are provided in Supplementary Materials.

3.2. Dess–Martin Oxidation of (±)-Albicanol 5

To a CH_2Cl_2 (35 mL) solution of compound **5** (1.85 g, 8.32 mmol), 5.3 g of Dess–Martin periodinane (12.5 mmol) was added and the mixture stirred for 1 h at room temperature until completion by TLC. The mixture was then washed with $NaHCO_3$ (sat. soln. 3 × 20 mL) and the organic phase dried over $MgSO_4$, filtered and the solvent removed in vacuo. Chromatographic purification of the crude residue (silica gel column, Hexane/AcOEt 9:1) yielded (±)-albicanal (**6**) (1.83 g, 8.30 mmol, 99.7%) as a colorless oil. ^{1}H and ^{13}C NMR were identical to those previously reported [30].

3.3. Synthesis of Cis-Decaline 7

Hydroquinone dimethyl ether (0.83 g, 6.0 mmol) was dissolved in Et_2O (13 mL) and *sec*-BuLi (3.1 mL, 1.3 M in cyclohexane) was added at 0 °C. After stirring the mixture for 3 h at room temperature, a solution of (±)-albicanal (**6**) (440 mg, 2.0 mmol) in Et_2O (3 mL) was dropwise added. The reaction was stirred for 5 min before dropwise addition of NH_4Cl (0.3 mL of saturated solution). To the mixture was then added 3 mL of saturated NaCl-solution, the organic phase dried over anhydrous Na_2SO_4 and the solvent removed in vacuo.

A mixture of liquid NH_3 (24 mL), THF (13 mL) and Li (70 mg, 10 mmol, granulate, Merck) at −78 °C was prepared and stirred for 15 min. To this mixture was added a solution of the former reaction crude in THF (7 mL). The reaction was then stirred for 15 min at the same temperature. After that, NH_4Cl (1.4 g) was added in portions (a change in color was observed from dark blue to colorless). Next, the mixture was allowed to reach room temperature to allow the evaporation of NH_3 (2 h) and finally the reaction mixture was extracted with EtOAc. The combined organic layers were washed with brine, dried (anhydrous Na_2SO_4) and the solvent removed in vacuo. Column chromatography (Hexane/AcOEt 9:1) of the residue yielded the coupling product **7** (618 mg, 1.8 mmol) (90%), isolated as a colorless solid, m.p. 74–75 °C. IR (ATR) v (cm^{-1}) 3000, 2940, 2860, 2830, 1640, 1605, 1495, 1460, 1440, 1210, 1050. ^{1}H NMR (500 MHz, $CDCl_3$) δ (ppm) 6.75–6.60 (m, 3H), 4.74 (s, 1H), 4.61 (s, 1H), 3.79 (s, 3H), 3.74 (s, 3H), 2.75 (d, *J* = 15 Hz, 2H), 2.36 (m, 1H), 2.22 (m, 1H), 2.01 (m, 1H), 1.88 (m, 1H), 1.80–1.20 (m, 9H), 0.90 (s, 3H), 0.84 (s, 3H), 0.82 (s, 3H). ^{13}C NMR (125 MHz, $CDCl_3$) δ (ppm) 153.2 (C), 151.7 (C), 148.3 (C),132.1 (C), 116,2 (CH), 110.8 (CH), 109.6 (CH), 107.6 (CH_2), 55.9 (CH), 55.8 (CH_3), 55.7 (CH), 55.5 (CH_3), 42.2 (CH_2), 39.9 (C), 39.1 (CH_2), 38.3 (CH_2), 33.6 (C), 33.6 (CH_3), 24.4 (CH_2), 23.2 (CH_2), 21.8 (CH_3), 19.5 (CH_2), 14.6 (CH_3). HRMS (ESI/Q-TOF) *m/z*: [M + H]$^+$ calcd for $C_{23}H_{35}O_2$ 343.2632; found 343.2629. ^{1}H and ^{13}C NMR data match with those previously reported [26].

3.4. Synthesis of Tetrasubstituted Olefin 8

BF_3•Et_2O (0.35 mL, 2.5 mmol) was added to a chilled solution (−50 °C) of **7** (171 mg, 0.5 mmol) in CH_2Cl_2 (50 mL). The mixture was slowly warmed up to −5 °C and stirred for 5 h. Then, the solvent was removed and the residue suspended in Et_2O. The solution was washed with brine, dried over Na_2SO_4 and the solvent was removed in vacuo. Column chromatography of the residue (cyclohexane) yielded **8** (106 mg, 0.31 mmol, 62%) together with the by-product **9** (48 mg, 0.14 mmol, 28%). Compound **8** as a white solid; m.p. 58–61 °C.

IR (ATR) v (cm^{-1}) 3020, 2930, 2850, 1620, 1592, 1495, 1240. ^{1}H NMR (500 MHz, $CDCl_3$) δ (ppm) 6.87 (d, *J* = 3 Hz, 1H), 6.75 (d, *J* = 9 Hz, 1H), 6.68 (dd, *J* = 9, 3.1 Hz, 1H), 3.76 (s, 3H), 3.73 (s, 3H), 2.93 (d, *J* = 15 Hz, 1H), 2.62 (d, *J* = 15 Hz, 1H), 2.09–2.01 (m, 4H), 1.96–1.90 (m, 1H), 1.69–1.58 (m, 4H), 1.39–1.32 (m, 2H), 1.01 (s, 3H), 1.00 (s, 3H), 0.92 (s, 3H), 0.79 (d, *J* = 7 Hz, 3H). ^{13}C NMR (125 MHz, $CDCl_3$) δ (ppm) 152.9 (C), 152.2 (C), 135.6 (C), 132.6 (C), 129.6 (C), 116.4 (CH), 110.8 (CH), 110.7 (CH), 55.7 (CH_3), 55.5 (CH_3), 41.4 (C), 39.7 (CH_2), 34.5 (CH_2), 34.2 (C), 33.3 (CH), 28.2 (CH_3), 28.0 (CH_3), 26.6 (CH_2), 26.2 (CH_2), 23.4 (CH_2), 21.9 (CH_3), 19.8 (CH_2), 15.9 (CH_3). HRMS (ESI/Q-TOF) *m/z*: [M + H]$^+$ calcd for $C_{23}H_{35}O_2$ 343.2632; found 343.2630. Compound **9** as a colorless solid, m.p. 111–113 °C. IR (ATR) v

(cm^{-1}) 3010, 2950, 2870, 1610, 1592, 1461, 1249. ^{1}H NMR (500 MHz, CDCl$_3$) δ (ppm) 6.67–6.63 (m, 2H), 3.79 (s, 3H), 3.74 (s, 3H), 3.08–3.00 (m, 3H), 2.12–2.06 (m, 2H), 1.60–1.10 (m, 9H), 1.01 (d, *J* = 13 Hz, 3H), 0.85 (s, 3H), 0.79 (s, 3H), 0.75 (s, 3H). ^{13}C NMR (125 MHz, CDCl$_3$) δ (ppm) 153.1 (C), 151.8 (C), 128.3 (C), 128.2 (C), 108.4 (CH), 106.5 (CH), 55.7 (CH$_3$), 55.4 (CH$_3$), 42.3 (CH), 39.0 (CH), 38.3 (CH), 38.1 (CH$_2$), 34.5 (C), 33.5 (CH), 32.7 (CH$_2$), 32.5 (C), 30.5 (CH$_3$), 28.9 (CH$_2$), 25.0 (CH$_3$), 24.0 (CH$_2$), 21.5 (CH$_2$), 20.3 (CH$_3$), 14.6 (CH$_3$). HRMS (ESI/Q-TOF) *m/z*: [M + H]$^+$ calcd for C$_{23}$H$_{35}$O$_2$ 343.2632; found 343.2629. ^{1}H and ^{13}C NMR data for compounds **8** [18] and **9** [31] were in agreement with those previously reported.

3.5. Preparation of **10** by Methyl Ether Deprotection of **8**

A solution of **8** (171 mg, 0.5 mmol) in dioxane (13 mL) was placed in a flame-dried flask under Ar. AgO (125 mg, 1.0 mmol) followed by 6N HNO$_3$ (0.24 mL, 1.5 mmol) were added and the mixture stirred for 15 min at room temperature. Then, NaHCO$_3$ (aq. sat. soln., 5 mL) was added and the mixture extracted with Et$_2$O (20 mL + 2 × 5 mL). The combined organic layers were washed with H$_2$O (3 × 10 mL) and brine (2 × 10 mL), dried over Na$_2$SO$_4$ and the solvent removed in vacuo. The crude quinone was used without purification in the next step. In this way, the residue was dissolved in CHCl$_3$ (15 mL), 55 mg added of 10% Pd/C (0.025 mmol) and the flask evacuated and backfilled with H$_2$ (3 cycles). After stirring the reaction mixture under an atmosphere of H$_2$ (balloon) for 15 min, it was filtered through a short pad of SiO$_2$ with the aid of Et$_2$O (3.0 mL). Finally, the solvent was removed in vacuo and the residue purified by column chromatography (Hexane/AcOEt, 95:5) to give 138 mg of the product **10** (82%) as a white foam. IR (ATR) *v* (cm^{-1}) 3375, 3082, 2925, 2873, 1541, 1490, 1192. ^{1}H NMR (500 MHz, CDCl$_3$) δ (ppm) 6.67 (d, *J* = 9 Hz, 1H), 6.65 (d, *J* = 3 Hz, 1H), 6.55 (dd, *J* = 9, 3 Hz, 1H), 4.91 (s, 1H), 2.93 (d, *J* = 15 Hz, 1H), 2.50 (d, *J* = 15 Hz, 1H), 2.13–2.08 (m, 1H), 2.00–1.95 (m, 1H), 1.91–1.86 (m, 2H), 1.76–1.73 (m, 1H), 1.66–1.63 (m, 1H), 1.59–1.39 (m, 5H), 1.05 (s, 3H), 1.00 (s, 3H), 0.98 (s, 3H), 0.84 (d, *J* = 7 Hz, 3H). ^{13}C NMR (125 MHz, CDCl$_3$) δ (ppm) 148.9 (C), 148.7 (C), 137.8 (C), 132.7 (C), 127.7 (C), 118.4 (CH), 116.5 (CH), 113.7 (CH), 41.7 (C), 40.5 (CH$_2$), 39.6 (CH$_2$), 39.5 (C), 35.7 (CH$_2$), 34.6 (CH), 28.5 (CH$_3$), 28.1 (CH$_2$), 27.1 (CH$_3$), 26.2 (CH$_2$), 22.3 (CH$_3$), 19.7 (CH$_2$), 15.8 (CH$_3$). HRMS (ESI/Q-TOF) *m/z*: [M + H]$^+$ calcd for C$_{21}$H$_{31}$O$_2$ 315.2319; found 315.2315. NMR data of compound **10** were consistent with those of the original isolation literature [2].

3.6. Synthesis of (±)-Aureol ((±)-**1**)

Hydroquinone **10** (157 mg, 1.0 mmol) was dissolved in anhydrous CH$_2$Cl$_2$ (50 mL) and the solution cooled to −60 °C. Then, BF$_3$•Et$_2$O (0.28 mL, 2.25 mmol) was added and the mixture stirred for 3 h at −60 °C. After that, it was warmed to −20 °C and the reaction stopped by addition of NH$_4$Cl (aqueous saturated solution). The mixture was extracted with CH$_2$Cl$_2$ (3 × 10 mL) and the combined organic layers dried (Na$_2$SO$_4$) and the solvent removed in vacuo. Column chromatography of the residue (Hexane/AcOEt 9:1) yielded (±)-aureol ((±)-**1**) as a white solid (195 mg, 62%), m.p. 143–144 °C. IR (ATR) v (cm^{-1}): 3312, 3005, 3296, 2938, 2869, 1492, 1458, 1208, 948. ^{1}H NMR (CDCl$_3$, 500 MHz): δ 6.60 (d, *J* = 9 Hz, 1H), 6.56 (dd, *J* = 9, 3 Hz, 1H), 6.49 (d, *J* = 3 Hz, 1H), 4.26 (br s, 1H), 3.37 (d, *J* = 17 Hz, 1H), 2.11–1.99 (m, 2H), 1.97 (d, *J* = 17 Hz, 1H), 1.85–1.75 (m, 2H), 1.70–1.65 (m, 2H), 1.60–1.50 (m, 1H), 1.49–1.30 (m, 5H), 1.11 (d, *J* = 7 Hz, 3H), 1.07 (s, 3H), 0.92 (s, 3H), 0.78 (s, 3H). ^{13}CNMR (CDCl$_3$, 125 MHz): δ 148.3 (C), 145.8 (C), 122.2 (C), 117.3 (CH), 115.1 (CH), 114.0 (CH), 82.4 (C), 44.0 (CH), 39.3 (CH), 38.1 (C), 37.4 (CH$_2$), 33.9 (CH$_2$), 33.8 (C), 31.9 (CH$_3$), 29.8 (CH$_3$), 29.3 (CH$_2$), 27.9 (CH$_2$), 22.2 (CH$_2$), 20.2 (CH$_3$), 18.4 (CH$_2$), 17.3 (CH$_3$). HRMS (ESI/Q-TOF) *m/z*: [M + H]$^+$ calcd for C$_{21}$H$_{31}$O$_2$ 315.2319; found 315.2312. Physical and spectroscopic data of (±)-aureol ((±)-**1**) matched those reported in the original isolation literature [2].

4. Conclusions

We devised a short and efficient synthetic route for the synthesis of (±)-aureol (**1**) and (±)-5-*epi*-aureol (**11**). Our strategy relies on a C–C bond-forming reaction between (±)-albicanal (**6**) and an aryllithium derivative and a sequence of 1,2-hydride and 1,2-methyl shifts mediated by

BF$_3$•Et$_2$O as activator and water as initiator. We are currently engaged in a computational study of the reaction mechanism, which will be published in due course. (±)-Aureol (**1**) and (±)-5-*epi*-aureol (**5**) obtained by this route are key intermediates for the synthesis of a large number of natural and synthetic derivative tetracyclic meroterpenoids, which will be used for further analysis as antitumor and antiviral agents.

Supplementary Materials: The following are available online at http://www.mdpi.com/1660-3397/18/9/441/s1: Figures S2–S13: ^{1}H NMR of compounds **1**, **5–10** and ^{13}C NMR of **1**, **7–10**.

Author Contributions: A.R.M.: design and coordination of the project, experimental work and manuscript preparation; L.E. and M.J.: preliminary studies and computational calculations; L.P.M. and E.D.O.: part of the experimental work; I.R.-G.: part of the experimental work and manuscript revision. All authors have read and agreed to the published version of the manuscript.

Funding: This work was supported by the Vicerrectorado de Investigación (Project 2020/00001014) of the University of Sevilla (Spain).

Acknowledgments: A. Rosales Martínez acknowledges the University of Sevilla for his position as professor and for financial support (Project 2020/00001014).

Conflicts of Interest: The authors declare no conflict of interest.

References

1. Capon, R.J. *Studies in Natural Products Chemistry*; Atta-ur-Rahman, Ed.; Elsevier Science: New York, NY, USA, 1995; Volume 15, p. 89.

2. Djura, P.; Stierle, D.B.; Sullivan, B.; Faulkner, D.J.; Arnold, E.; Clardy, J. Some metabolites of the marine sponges *Smenospongia aurea* and *Smenospongia (polyfibrospongia) echina*. *J. Org. Chem.* **1980**, *45*, 1435–1441. [CrossRef]

3. Ciminiello, P.; Dell'Aversano, C.; Fattorusso, E.; Magno, S.; Pansini, M. Chemistry of verongida sponges. Secondary metabolite composition of the Caribbean sponge *Verongula gigantea*. *J. Nat. Prod.* **2000**, *63*, 263–266. [CrossRef] [PubMed]

4. Wright, A.E.; Rueth, S.A.; Cross, S.S. An antiviral sesquiterpene hydroquinone from the marine sponge *Strongylophora hartmani*. *J. Nat. Prod.* **1991**, *54*, 1108–1111. [CrossRef] [PubMed]

5. Utkina, N.K.; Denisenko, V.A.; Scholokova, O.V.; Virovaya, M.V.; Prokof'eva, N.G. Cyclosmenospongine, a new sesquiterpenoid aminoquinone from an Australian marine sponge *Spongia* sp. *Tetrahedron Lett.* **2003**, *44*, 101–102. [CrossRef]

6. Bourguet-Kondracki, M.-L.; Martin, M.-T.; Guyot, M. Smenoqualone a novel sesquiterpenoid from the marine sponge *Smenospongia* sp. *Tetrahedron Lett.* **1992**, *33*, 8079–8080. [CrossRef]

7. Shen, Y.-C.; Liaw, C.-C.; Ho, J.-R.; Khalil, A.T.; Kuo, Y.-H. Isolation of aureol from *smenospongia* sp. And cytotoxic activity of some aureol derivatives. *Nat. Prod. Res.* **2006**, *20*, 578–585. [CrossRef]

8. Wright, A.E.; Cross, S.S.; Burres, N.S.; Koehn, F. Antiviral and Antitumor Terpene Hydroquinones from Marine Sponge and Methods of Use. Harbor Branch Oceanographic Institution, Inc., Fort Pierce, FL, USA. U.S. Patent PCT WO 9112250 A1, 22 August 1991.

9. Longley, R.E.; McConnell, O.J.; Essich, E.; Harmody, D. Evaluation of marine sponge metabolites for cytotoxicity and signal transduction activity. *J. Nat. Prod.* **1993**, *56*, 915–920. [CrossRef]

10. Wildermuth, R.; Speck, K.; Haut, F.-L.; Mayer, P.; Karge, B.; Brönstrup, M.; Magauer, T. A modular synthesis of tetracyclic meroterpenoid antibiotics. *Nat. Commun.* **2017**, *8*, 2083. [CrossRef]

11. Taishi, T.; Takechi, S.; Mori, S. First total synthesis of (±)-stachyflin. *Tetrahedron Lett.* **1998**, *39*, 4347–4350. [CrossRef]

12. Nakamura, M.; Suzuki, A.; Nakatani, M.; Fuchikami, T.; Inoue, M.; Katoh, T. An efficient synthesis of (+)-aureol via boron trifluoride etherate-promoted rearrangement of (+)-arenarol. *Tetrahedron Lett.* **2002**, *43*, 6929–6932. [CrossRef]

13. Watanabe, K.; Sakurai, J.; Abe, H.; Katoh, T. Total synthesis of (+)-stachyflin: A potential anti-influenza A virus agent. *Chem. Commun.* **2010**, *46*, 4055–4057. [CrossRef] [PubMed]

14. Marcos, I.S.; Conde, A.; Moro, R.F.; Basabe, P.; Diez, D.; Urones, J.G. Synthesis of quinone/hydroquinone sesquiterpenes. *Tetrahedron* **2010**, *66*, 8280–8290. [CrossRef]

15. Sakurai, J.; Kikudhi, T.; Takahasi, O.; Watanabe, K.; Katoh, T. Enantioselective total synthesis of (+)-stachyflin; a potential anti-influenza A virus agente isolated from a microorganism. *Eur. J. Org. Chem.* **2011**, *16*, 2948–2957. [CrossRef]

16. Kuan, K.K.W.; Pepper, H.P.; Bloch, W.M.; George, J.H. Total synthesis of (+)-aureol. *Org. Lett.* **2012**, *14*, 4710–4713. [CrossRef] [PubMed]

17. Kamishima, T.; Kikuchi, T.; Katoh, T. Total synthesis of (+)-strongylin A, a rearranged sesquiterpenoid hydroquinone from a marine sponge. *Eur. J. Org. Chem.* **2013**, *21*, 4558–4563. [CrossRef]

18. Rosales, A.; Muñoz-Bascón, J.; Roldan-Molina, E.; Rivas-Bascón, N.; Padial, N.M.; Rodríguez-Maecker, R.; Rodríguez-García, I.; Oltra, J.E. Synthesis of (±)-aureol by bionspired rearrangements. *J. Org. Chem.* **2015**, *80*, 1866–1870. [CrossRef]

19. Katoh, T.; Atsumi, S.; Saito, R.; Narita, K.; Katoh, T. Unified synthesis of the marine sesquiterpene quinones (+)-smenoqualone, (−)-ilimaquinone, (+)-smenospongine, and (+)-sospongiaquinone. *Eur. J. Org. Chem.* **2017**, *26*, 3837–3849. [CrossRef]

20. Wang, J.-L.; Li, H.-J.; Wang, M.; Wang, J.-H.; Wu, Y.-C. A six-step approach to marine natural product (+)-aureol. *Tetrahedron Lett.* **2018**, *59*, 945–948. [CrossRef]

21. Gansäuer, A.; Rosales, A.; Justicia, J. Catalytic epoxypolyene cyclization via radicals: Highly diastereoselective formal synthesis of puupehedione and 8-*epi*-puupehedione. *Synlett* **2006**, *6*, 927–929. [CrossRef]

22. Rosales, A.; López-Sánchez, C.; Álvarez-Corral, M.; Muñoz-Dorado, M.; Rodríguez-García, I. Total synthesis of (±)-euryfuran through Ti(III) catalyzed radical cyclization. *Lett. Org. Chem.* **2007**, *4*, 553–555. [CrossRef]

23. Rosales, A.; Muñoz-Bascón, J.; Moráles-Álcazar, V.M.; Castilla-Alcalá, J.A.; Oltra, J.E. Ti(III)-catalyzed, concise synthesis of marine furanospongian diterpenes. *RSC Adv.* **2012**, *2*, 12922–12925. [CrossRef]

24. Rosales Martinez, A.; Pozo Morales, L.; Diaz Ojeda, E. Cp$_2$TiCl-catalyzed, concise synthetic approach to marine natural product (±)-cyclozonarone. *Synth. Commun.* **2019**, *49*, 2554–2560. [CrossRef]

25. Goehl, M.; Seifert, K. Synthesis of the sesquiterpenes albicanol, drimanol, and drimanic acid, and the marine sesquiterpene hydroquinone deoxyspongiaquinol. *Eur. J. Org. Chem.* **2014**, *31*, 6975–6982. [CrossRef]

26. Laube, T.; Schröder, J.; Stehle, R.; Seifert, K. Total synthesis of yahazunol, zonarone and isozonarone. *Tetrahedron* **2002**, *58*, 4299–4309. [CrossRef]

27. Oblak, E.Z.; VanHyst, M.D.; Li, J.; Wiemer, A.J.; Wright, D.L. Cyclopropene cycloadditions with annulated furans: Total synthesis of (+)- and (−)-frondosin B and (+)-frondosin A. *J. Am. Chem. Soc.* **2014**, *136*, 4309–4315. [CrossRef] [PubMed]

28. Speck, K.; Wildermuth, R.; Magauer, T. Convergent assembly of the tetracyclic meroterpenoid (−)-cyclosmenospongine by a non-biomimetic polyene cyclization. *Angew. Chem. Int. Ed.* **2016**, *55*, 14131–14135. [CrossRef]

29. Carraher, C.E., Jr. Ionic Chain-Reaction and Complex Coordinative Polymerization (Addition Polymerization). In *Polymer Chemistry*, 6th ed.; Marcel Dekker: New York, NY, USA, 2003.

30. Poigny, S.; Huor, T.; Guyot, M.; Samadi, M. Synthesis of (−)-Hyatellaquinone and Revision of Absolute Configuration of Naturally Occurring (+)-Hyatellaquinone. *J. Org. Chem.* **1999**, *64*, 9318–9320. [CrossRef]

31. Urban, S.; Capon, R.J. Marine sesquiterpene quinones and hydroquinones: Acid-catalyzed rearrangements and stereochemical investigations. *Aust. J. Chem.* **1994**, *47*, 1023–1029. [CrossRef]

 *marine drugs*

Communication

First Total Synthesis of 5′-*O*-α-D-Glucopyranosyl Tubercidin

Wenliang Ouyang, Haiyang Huang * , Ruchun Yang, Haixin Ding and Qiang Xiao *

Jiangxi Key Laboratory of Organic Chemistry, Institute of Organic Chemistry, Jiangxi Science and Technology Normal University, Nanchang 330013, China; ouyangwl93@yeah.net (W.O.); ouyangruchun@163.com (R.Y.); dinghaixin_2010@163.com (H.D.)
* Correspondence: huanghaiyang1209@163.com (H.H.); xiaoqiang@tsinghua.org.cn (Q.X.)

Received: 11 July 2020; Accepted: 28 July 2020; Published: 29 July 2020

Abstract: The first total synthesis of 5′-*O*-α-D-glucopyranosyl tubercidin was successfully developed. It is a structurally unique disaccharide 7-deazapurine nucleoside exhibiting fungicidal activity, and was isolated from blue-green algae. The total synthesis was accomplished in eight steps with 27% overall yield from commercially available 1-*O*-acetyl-2,3, 5-tri-*O*-benzoyl-β-D-ribose. The key step involves stereoselective α-*O*-glycosylation of the corresponding 7-bromo-6-chloro-2′, 3′-*O*-isopropylidene-β-D-tubercidin with 2,3,4,6-tetra-*O*-benzyl-glucopyranosyl trichloroacetimidate. All spectra are in accordance with the reported data for natural 5′-*O*-α-D-glucopyranosyl tubercidin. Meanwhile, 5′-*O*-β-D-glucopyranosyl tubercidin was also prepared using the same strategy.

Keywords: total synthesis; natural product; 7-deazapurine nucleoside; disaccharide nucleoside; tubercidin

---

1. Introduction

As the first isolated naturally occurring 7-deazapurine nucleoside, tubercidin (Figure 1, **1**) is a biogenic analogue of adenosine, in which the *N*-7 atom is replaced by CH (purine numbering is used throughout this work) [1–3]. It showed significant cytostatic activity in various cancer cell lines. Although the application of tubercidin in clinical trials has been attempted, its use was eventually halted because of significant toxicity [4–7]. Since then, 7-deazapurine has been recognized as a privileged scaffold in developing new antitumor and antiviral nucleosides [8]. Consequently, many 7-deazapurine nucleosides have been synthesized and their biological activities have been screened in the past 50 years [9–11]. Among them, several promising lead compounds have been discovered [12–14].

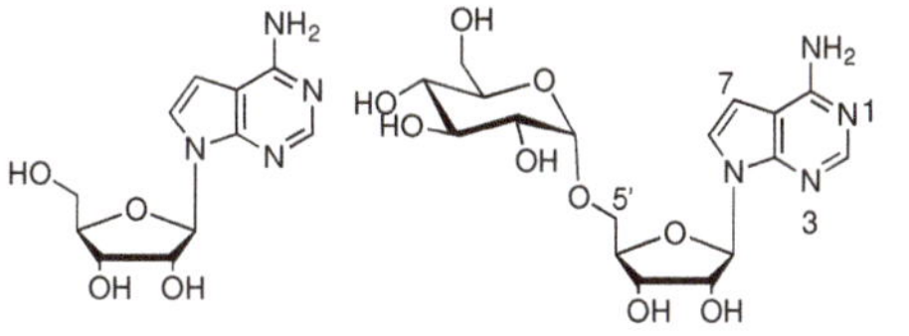

Figure 1. Chemical structures of tubercidin (**1**) and 5′-*O*-α-D-glucopyranosyl tubercidin (**2**).

5′-*O*-α-D-Glucopyranosyl tubercidin 1 is a disaccharide nucleoside isolated from blue-green algae in 1988 (Figure 1, **2**) [15]. It is characterized by a unique 5′-*O*-α-D-glucopyranosyl moiety attached to tubercidin. Preliminary biological evaluation indicated that it exhibited moderate cytotoxicity and fungicidal activity. Although many disaccharide nucleosides have been identified as naturally occurring products [16,17], disaccharide 7-deazapurine nucleosides are rarely reported.

Our group has a long-standing interest in total synthesis and biological activity evaluation of naturally occurring 7-deazapurine nucleosides and their biological activities [18–22]. Because of the unique disaccharide structure and a great need for structure-based biological study, structural confirmation of 5′-*O*-α-D-glucopyranosyl tubercidin **2** and synthesis of its analogues in a concise and modular fashion are preferred. Herein, we report the first total synthesis of 5′-*O*-α-D-glucopyranosyl tubercidin **1**.

Our retrosynthetic analysis is depicted in Scheme 1. From a synthetic point of view, there are two possible approaches. The first approach is Vorbrüggen glycosylation of disaccharide **3** with nucleobase **4** directly (Scheme 1, Path a). Because of the labile α-*O*-glycosylic bond of disaccharide **3**, the isomerization might occur during Vorbrüggen glycosylation. The second approach is postglycosylation of properly protected nucleoside **5** with glycosyl donor **6** (Scheme 1, Path b). The postglycosylation approach has been widely used for synthesizing disaccharide nucleosides [23–25]. Considering that a modular strategy is needed, we employed the postglycosylation approach herein.

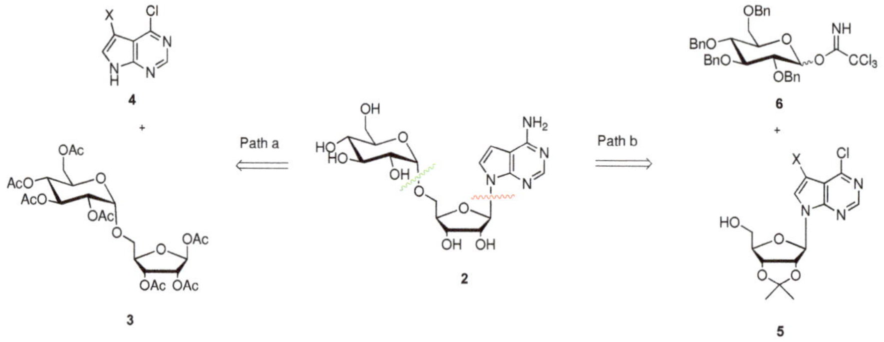

Scheme 1. Retrosynthetic analysis of 5′-*O*-α-D-glucopyranosyl tubercidin **2**.

2. Results

First, 7-deazpurine nucleoside **9** was synthesized based on a previous report by Seela and coworkers (Scheme 2) [26–29]. After silylation of 6-chloro-7-bromo-7-deazapurine **8** with *N*,*O*-bis(trimethylsilyl)acetamide (BSA) in freshly distilled acetonitrile (CH₃CN), 1-*O*-acetyl-2,3,5-tri-*O*-benzoyl-β-D-ribose **7** (2 eq.) was added followed by TMSOTf (3 eq.) in one pot. Then, the resulting reaction mixture was heated at 80 °C for 8 h to afford 7-deazpurine nucleoside **9** with a 76% yield. The 7-bromo substitute is critical for the above Vorbrüggen glycosylation, otherwise the reaction cannot proceed. The reason may be that the 7-bromo substituent can reduce the reactivity of *N*-9, which results in the deazapurine being more purine-like [20,26–29].

Then, all benzoyl groups of nucleoside **9** were removed in saturated ammonia in methanol at 0 °C to give nucleoside **10** in a 94% yield. It should be noted that the C-6 chloride was retained to avoid extra manipulation of protecting groups in the following steps. Next, *p*-toluenesulfonic acid-catalyzed reaction of nucleoside **10** with 2, 2-dimethoxy propane in acetone afforded isopropylidene-protected nucleoside **11** with a 98% yield, which left the 5′-OH free for further glycosylation.

With nucleoside **11** in hand, we focused on carrying out the key glycosylation reaction. In general, 1,2-*trans* β-*O*-glucoside can be prepared by the neighboring group participation of the 2-*O*-acyl glucose donor, which is very reliable and highly stereoselective [29]. In contrast, the construction of our desired 1,2-*cis* α-*O*-glucoside is much more challenging. It requires glycosyl donors having nonassisting functionality at C-2 position, and even so, the reaction normally produces a mixture of α and β isomers [30–32]. In this work, we chose 2,3,4,6-tetra-*O*-benzyl-glucopyranosyl trichloroacetimidate **6** as the glucosyl donor, which was synthesized from 2,3,4,6-tetra-*O*-benzyl glucose with trichloroacetonitrile in dichloromethane (DCM) using anhydrous potassium carbonate as catalysis (98% yield, α:β $\approx$ 1:4) [33,34]. After optimization (Supplementary Materials), TMSOTf catalyzed glycosylation of nucleoside **11** with glycosyl donor **6** at −30 °C gave a mixture of α isomer **12** and β isomer **13** (4:1) in 79% overall yield. Fortunately, they can be separated by careful silica gel chromatography to afford the desired nucleoside **12** with α glycosylic configuration. These two isomers can be clearly identified by the $J_{1''\text{-}2''}$ coupling constant (α isomer 3.5 Hz, β isomer 8.0 Hz) and ^{13}C chemical shift of C1$''$ (α isomer 96.4 ppm, β isomer 102.9 ppm).

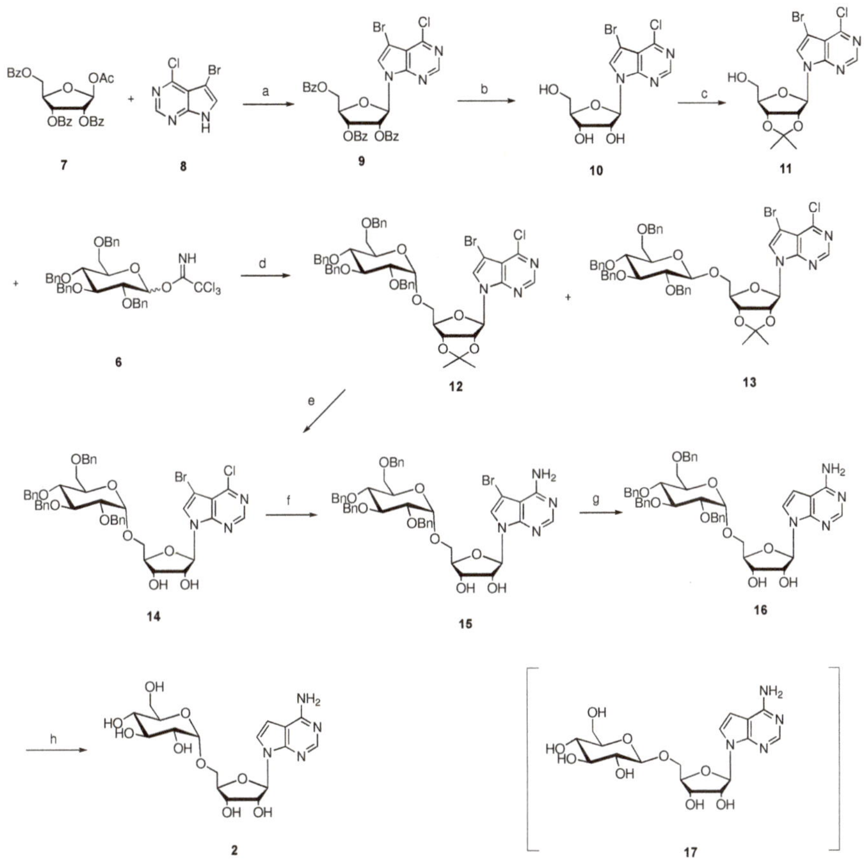

Scheme 2. Total synthesis of 5′-*O*-α-D-glucopyranosyl tubercidin **2**. Reagents and conditions: (**a**) bis(trimethylsilyl)acetamide (BSA), TMSOTf, CH$_3$CN, 80 °C, 1 h, 75%; (**b**) NH$_3$, MeOH, 0 °C, 12 h, 94%; (**c**) 2,2-dimethoxy propane, p-toluenesulfonic acid, acetone, rt., 4 h, 98%; (**d**) TMSOTf, CH$_2$Cl$_2$, −20 °C, 4 h, 79% (α:β, 4:1); (**e**) 80% aqueous acetic acid, 50 °C, 12 h, 85%; (**f**) NH$_3$, MeOH, 130 °C, 12 h, 95%; (**g**) H$_2$, 10% Pd/C, Et$_3$N, THF/MeOH, rt. 5 h, 83.5%; (**h**) H$_2$, 20% Pd(OH)$_2$/C, THF/MeOH, 40 °C, 12 h, 80%.

Next, the isopropylidene-protecting group was removed in 80% aqueous acetic acid to give nucleoside **14** with an 85% yield. Then, the substitution of C-6 chloride with freshly prepared saturated ammonium in methanol at 130 °C afforded nucleoside **15** with a 95% yield. Then, we tried to remove the C-7 bromide atom and all benzyl-protecting groups by hydrogenation with 10% Pd/C at the same time. However, the presence of triethylamine in the reaction due to the need to neutralize HBr, which is generated through bromide reduction, prevented the removal of the benzyl group. Only bromide was reduced, to afford nucleoside **16** in 84% yield. Finally, further hydrogenation with 20% Pd(OH)$_2$/C gave 5'-*O*-α-D-glucopyranosyl tubercidin **2** with 80% yield. All spectra were in accordance with the reported data of the authentic naturally occurring product.

Since the corresponding 5'-*O*-β-D-glucopyranosyl tubercidin has not been previously reported, its synthesis provides more evidence for structural elucidation and biological activity. 5'-*O*-β-D-Glucopyranosyl tubercidin **17** was also prepared according to a similar procedure for the synthesis of nucleoside **2**.

3. Materials and Methods

All reagents were purchased from commercial sources and used without purification unless specified. Acetonitrile, pyridine and CH$_2$Cl$_2$ were refluxed with CaH$_2$ and distilled prior to use. THF was dried with LiAlH$_4$ and distilled prior to use. Thin-layer chromatography was performed using silica gel GF-254 plates (Qingdao Chemical Company, Qingdao) detected by UV (254 nm) or charting with 10% sulfuric acid in ethanol. Column chromatography was performed on silica gel (200–300 mesh, Qingdao Chemical Company, Qingdao). NMR spectra were recorded on a Bruker AV400 spectrometer. Chemical shifts (δ) are reported in ppm and coupling constants (J) are reported in Hz. Then, ^{1}H and ^{13}C NMR spectra were calibrated with TMS as an internal standard. The specific rotation was measured on a Rudolph autopol IV polarimeter. ESI-MS was acquired with a Bruker Dalton microTOFQ II spectrometer.

3.1. H NMR, ^{13}C NMR and HRMS of 2,3,4,6-O-Tetrabenzyl-D-glucopyranose

R_f = 0.32 (PE/EA = 3:1); mp 153–155 °C; ^{1}H NMR (400 MHz, CDCl$_3$): δ 7.36–7.28 (m, 18H, Bn), 7.18–7.12 (m, 2H, Bn), 5.31–5.19 (m, 1H, H-1), 4.98–4.48 (m, 8H, 4CH$_2$), 4.09–3.94 (m, 2H, H-4,5), 3.74–3.58 (m, 4H, H-2,3,6), 3.06 (d, J = 2.2 Hz, 1H, 1-OH); ^{13}C NMR (101 MHz, CDCl$_3$): δ 138.7 (C), 138.2 (C), 137.8 (2C), 128.5 (CH), 128.4 (CH), 128.2 (CH), 128.1 (CH), 128.0 (CH), 127.9 (CH), 127.7 (CH), 127.6 (CH), 91.3 (CH-1), 81.7 (CH-5), 80.0 (CH-2), 77.7 (CH-3), 75.7 (CH$_2$), 75.0 (CH$_2$), 73.5 (CH$_2$), 73.3 (CH$_2$), 70.3 (CH-4), 68.6 (CH$_2$-6); HRMS Calcd. For C$_{34}$H$_{36}$O$_6$Na$^+$ [M + Na]$^+$: 563.2404, found: 563.2410.

3.2. Synthesis of 2,3,4,6-Tetra-O-benzyl-glucopyranosyl Trichloroacetimidate (6)

To the suspension of anhydrous K$_2$CO$_3$ (10 g, 101 mmol) in dry dichloromethane (100 mL) was added 2,3,4,6-*O*-tetrabenzyl-D-glucopyranose (10 g, 18.5 mmol) and trichloroacetonitrile (11.3 mL, 112 mmol) sequentially. The mixture was stirred at room temperature for 4 h until the reaction was completed (monitored by TLC, PE/EA = 6:1, R_f = 0.21). After evaporation of the solvent under reduced pressure, the crude product was directly purified by flash column chromatography on silica gel (Et$_3$N/PE/EA = 0.1:3:1) to provide 12.4 g compound **6** as a white solid (98% yield, α:β $\approx$ 1:4); mp: 77–79 °C; ^{1}H NMR (400 MHz, CDCl$_3$): δ 8.73 (s, 0.8H, β-NH), 8.60 (s, 0.2H, -NH), 7.32 (m, 18H), 7.20 (m, 2H), 6.55 (d, J = 3.2 Hz, 0.2H, α-H-1), 5.84 (d, J = 6.1 Hz, 0.8H, β-H-1), 5.01–4.46 (m, 9H), 3.81–3.76 (m, 4H), 3.67 (d, J = 4.6 Hz, 1H); ^{13}C-NMR (101 MHz, CDCl$_3$): δ 161.2 (C), 138.4 (C), 138.1 (C), 138.0 (C), 138.0 (C), 128.5 (CH), 128.4 (CH), 128.4 (CH), 128.4 (CH), 128.1 (CH), 128.0 (CH), 128.0 (CH), 127.9 (CH), 127.8 (CH), 127.7 (CH), 127.6 (CH), 127.6 (CH), 98.4 (CH), 84.6 (CH), 81.0 (CH), 76.8 (CH), 75.9 (CH$_2$), 75.6 (CH$_2$), 75.0 (CH$_2$), 74.9 (CH$_2$), 73.4 (CH), 68.2 (CH); HRMS Calcd. For C$_{36}$H$_{36}$Cl$_3$NO$_6$Na$^+$ [M + Na]$^+$: 706.1500, found: 706.1509.

3.3. Synthesis of 7-Bromo-6-chloro-pyrrole [2,3-d]pyrimidine (8)

To the stirred suspension of 6-chloro-pyrrole[2,3-d]pyrimidine (12.0 g, 78.2 mmol) in dry dichloromethane (500 mL) was added NBS (16.1 g, 91.1 mmol) in batches. After the addition was finished, the mixture was stirred at room temperature for another 10 h until completion (monitored by TLC, DCM/EA = 5:1, R_f = 0.41) and then poured into ice water (500 mL). The resulting brown precipitate was filtered under reduced pressure and the filter cake was washed with water (500 × 3 mL) and dried to afford compound **8** as a brown solid (17.2 g, 95% yield); mp: >300 °C; ^{1}H NMR (400 MHz, DMSO): δ 12.95 (s, 1H, NH), 8.59 (s, 1H, H-2), 7.91 (s, 1H, H-8); ^{13}C NMR (100 MHz, DMSO): δ 151.5 (C), 151.4 (CH), 150.7 (C), 129.1 (CH), 114.1 (C), 86.3 (C); HRMS Calcd. For $C_6H_4BrClN_3^+$ [M + H]$^+$: 231.9272, found: 231.9270.

3.4. Synthesis of 7-Bromo-6-chloro-9-(2′,3′,5′-O-tribenzoyl-β-D-ribofuranose-yl)-pyrrole[2,3-d]pyrimidine (9)

To the stirred suspension of 7-bromo-6-chloro-pyrrole[2,3-d]pyrimidine **8** (5.0 g, 21.5 mmol) in dry acetonitrile (40 mL) was added BSA (*N,O*-bis(trimethylsiyl)acetamide, 5.4 g, 26.0 mmol) at 0 °C under argon. The resulting mixture was stirred at room temperature for 15 min, and then 1-*O*-ethanoyl-2,3,5-*O*-tribenzoyl-β-D-ribofuranose (16.0 g, 32.5 mmol) and TMSOTf (8.0 g, 46 mmol) were added sequentially. After the addition was completed, the mixture was stirred at 80 °C for 1 h (monitored by TLC, CH_2Cl_2: PE, 5:1, R_f 0.27). After the mixture was cooled to room temperature, water (200 mL) was added to quench the reaction and the resulting mixture was extracted with ethyl acetate (200 × 3 mL). The organic layers were separated and washed with saturated $NaHCO_3$ (200 mL) and saturated brine (200 mL) and dried with anhydrous Na_2SO_4. After the solvent was removed under reduced pressure, the crude product was purified by flash column chromatography on silica gel to provide 10.1 g compound **9** (75.5% yield) as a white solid. mp: 142–143 °C; $[\alpha]_D^{25}$ = −110 (c = 0.30, CHCl$_3$); ^{1}H-NMR (400 MHz, CDCl$_3$): δ (ppm) 8.57 (s, 1H), 8.10 (d, J = 8.0 Hz, 2H), 7.99 (d, J = 8.0 Hz, 2H), 7.91 (d, J = 7.9 Hz, 2H), 7.61–7.32 (m, 10H), 6.67 (d, J = 5.2 Hz, 1H), 6.12 (m, 2H), 4.89 (d, J = 12.1 Hz, 1H), 4.80 (d, J = 3.1 Hz, 1H), 4.67 (dd, J = 12.2, 3.1 Hz, 1H); ^{13}C NMR (101MHz, CDCl$_3$): δ (ppm) 166.1 (C), 165.4 (C), 165.1 (C), 152.7 (C), 151.6 (CH), 150.7 (C), 133.9 (CH), 133.9 (CH), 133.7 (CH), 129.9 (4CH), 129.7 (2CH), 129.2 (C), 128.8 (2CH), 128.6 (2CH; C), 128.6 (2CH), 128.3 (C), 126.5 (CH), 115.9 (C), 90.2 (C), 86.7 (CH), 80.6 (CH), 74.1 (CH), 71.4 (CH), 63.5 (CH), HRMS Calcd. For $C_{32}H_{23}BrClN_3O_7Na^+$ [M + Na]$^+$: 698.0300, found: 698.0305.

3.5. Synthesis of 7-Bromo-6-chloro-9-(β-D-ribofuranose-yl)-pyrrole[2,3-d]pyrimidine (10)

7-Bromo-6-chloro-9-(2′,3′,5′-O-tribenzoyl-β-D-ribofuranose-yl)-pyrrole[2,3-d]pyrimidine **9** (10 g, 14.77 mmol) was dissolved in methanolic ammonia (methanol was saturated with NH$_3$ at 0 °C, 300 mL) and the mixture was stirred at 0 °C for 12 h until the reaction was completed (monitored by TLC, CH_2Cl_2:CH$_3$OH, 50:1, R_f 0.45). The mixture was concentrated and the crude product was purified by flash column chromatography on silica gel (CH_2Cl_2/CH$_3$OH, 70:1) to provide 5.1 g compound **10** (94.0% yield) as a white solid. mp: 179–180 °C; $[\alpha]_D^{25}$ = −64 (c = 0.1, CH$_3$OH); ^{1}H NMR (400 MHz, DMSO): δ 8.72 (s, 1H, H-2), 8.27 (s, 1H, H-8), 6.24 (d, J = 5.8 Hz, 1H, H-1′), 5.49 (d, J = 6.1 Hz, 1H, 2′-OH), 5.25 (d, J = 4.9 Hz, 1H, 3′-OH), 5.15 (t, J = 5.3 Hz, 1H, 5′-OH), 4.40 (dd, J = 11.1, 5.6 Hz, 1H, H-2′), 4.13 (dd, J = 8.3, 4.5 Hz, 1H, H-3′), 3.96 (d, J = 3.4 Hz, 1H, H-4′), 3.71–3.56 (m, 2H, H-5′); ^{13}C NMR (101 MHz, DMSO): δ 151.6 (CH-2), 151.0 (C-6), 151.0 (C-4), 128.75(CH-8), 114.9 (C-5), 87.7 (C-7; CH-1′), 86.0 (CH-4′), 74.9 (CH-2′), 70.8 (CH-3′), 61.7 (CH$_2$-5′); HRMS Calcd. For $C_{11}H_{11}BrClN_3O_4Na^+$ [M + Na]$^+$: 385.9514, found: 385.9511.

3.6. Synthesis of 7-Bromo-6-chloro-9-(2′,3′-O-isopropylidene-β-D-ribofuranose-yl)-pyrrole[2,3-d] pyrimidine (11)

To the stirred suspension of compound **10** (4.0 g, 11 mmol) in dry acetone (200 mL) was added *p*-toluenesulfonic acid (200 mg, 1.2 mmol) and 2,2-dimethoxy propane (6.0 g, 55 mmol). The mixture was stirred at room temperature for 4 h until the reaction was completed (monitored by TLC, CH_2Cl_2:CH$_3$OH, 50:1, R_f 0.57). After that, the mixture was neutralized by triethylamine (70.0 mg, 0.7 mmol) and concentrated under reduced pressure. The crude product was purified by flash column chromatography on silica gel (CH_2Cl_2/CH$_3$OH, 70:1) to provide 4.4 g compound **11** (98.0% yield) as a white solid. mp: 73–74 °C; $[\alpha]_D^{25}$ = −63 (c = 0.05, CH$_3$OH); ^{1}H NMR (400 MHz, DMSO): δ 8.73 (s,

1H, H-2), 8.24 (s, 1H, H-8), 6.36 (d, *J* = 2.1 Hz, 1H, H-1′), 5.32–5.04 (m, 2H, H-2′,5′-OH), 5.03–4.83 (m, 1H, H-3′), 4.21 (s, 1H, H-4′), 3.56 (s, 2H, H-5′), 1.54 (s, 3H, Me), 1.31 (s, 3H, Me); ^{13}C NMR (101 MHz, DMSO): δ 151.8 (CH-2), 151.2 (C-4), 150.5 (C-6), 129.2 (CH-8), 115.0 (C-5), 113.6 (C), 90.0 (CH-1′), 87.8 (C-7), 86.9 (CH-4′), 84.4 (CH-2′), 81.4 (CH-3′), 61.9 (CH$_2$-5′), 27.5 (Me), 25.6 (Me); HRMS Calcd. For $C_{14}H_{15}BrClN_3O_4Na^+$ [M + Na]$^+$: 425.9827, found: 425.9811.

3.7. Synthesis of 7-Bromo-6-chloro-(2′,3′-O-isopropylidene-5′-O-(2″,3″,4″,6″-O-tetrabenzyl-α-D-glucopyranosyl)tubercidin (12) and 7-bromo-6-chloro-(2′,3′-O-isopropylidene-5′-O-(2″,3″,4″,6″-O-tetrabenzyl-β-D-glucopyranosyl)tubercidin (13)

Compound **6** (6.4 g, 9.3 mmol) and compound **11** (2.5 g, 6.2 mmol) were dissolved in dry dichloromethane (300 mL) and the resulting mixture was stirred at room temperature for 5 min under argon. After that, TMSOTf (16.5 mg, 0.07 mmol) was added to the mixture at −20 °C. The mixture was slowly warmed to room temperature and stirred for another 4 h until the reaction was completed (monitored by TLC, PE: EA, 3:1, R$_f$ 0.5). The mixture was concentrated and the crude product was purified by flash column chromatography on silica gel (CH$_2$Cl$_2$/CH$_3$OH, 70:1) to provide 8.0 g of compounds **12** and **13** (79% yield) as a light yellow oil.

12: $[\alpha]_D^{25}$ = −7 (*c* = 0.1, CH$_3$OH); ^{1}H NMR (400 MHz, DMSO): δ 8.71 (s, 1H), 8.23 (s, 1H), 7.34–7.22 (m, 18H), 7.13–7.04 (m, 2H), 6.41 (d, *J* = 2.6 Hz, 1H), 5.20 (dd, *J* = 6.1, 2.7 Hz, 1H), 4.97 (dd, *J* = 6.1, 3.2 Hz, 1H), 4.90 (m, 2H), 4.72–4.61 (m, 4H), 4.40 (m, 4H), 3.78–3.67 (m, 2H), 3.61 (dd, *J* = 10.9, 4.1 Hz, 1H), 3.56–3.37 (m, 5H), 1.54 (s, 3H), 1.29 (s, 3H); ^{13}C NMR (101 MHz, DMSO): δ 151.8 (CH), 151.3 (C), 150.4 (C), 139.3 (C), 138.9 (C), 138.7 (C), 138.6 (CH), 129.2 (CH), 128.7 (CH), 128.6 (CH), 128.2 (CH), 128.0 (CH), 128.0 (CH), 127.9 (CH), 127.8 (CH), 115.1 (C), 114.0 (C), 96.4 (CH), 89.6 (CH), 88.3 (C), 84.5 (CH), 84.3 (CH), 81.5 (CH), 81.4 (CH), 79.8 (CH), 77.7 (CH), 75.0 (CH$_2$), 74.4 (CH$_2$), 72.8 (CH$_2$), 71.8 (CH$_2$), 70.4 (CH), 68.8 (CH$_2$), 68.0 (CH$_2$), 27.5 (Me), 25.8 (Me); HRMS Calcd. For $C_{48}H_{49}BrClN_3O_9Na^+$ [M + Na]$^+$: 948.2233, found: 948.2236.

13: $[\alpha]_D^{25}$ = −19 (*c* = 0.1, CH$_3$OH); ^{1}H NMR (400 MHz, DMSO): δ 8.67 (s, 1H), 8.17 (s, 1H), 7.24 (m, 14H), 7.14 (m, 6H), 6.36 (d, *J* = 2.5 Hz, 1H), 5.19 (dd, *J* = 6.1, 2.6 Hz, 1H), 4.92 (dd, *J* = 6.1, 2.8 Hz, 1H), 4.78–4.64 (m, 4H), 4.56–4.30 (m, 6H), 4.04–3.98 (m, 1H), 3.76–3.38 (m, 6H), 3.24 (m, 1H), 1.50 (s, 3H), 1.23 (s, 3H); ^{13}C NMR (101 MHz, DMSO): δ 151.7 (CH), 151.2 (C), 150.3 (C), 139.0 (C), 138.9 (C), 138.7 (C), 138.6 (C), 129.3 (CH), 128.7 (CH), 128.6 (CH), 128.4 (CH), 128.2 (CH), 128.0 (CH), 128.0 (CH), 127.9 (CH), 127.9 (CH), 127.8 (CH), 115.1 (C), 113.8 (C), 102.9 (CH), 90.4 (CH), 87.9 (C), 85.0 (CH), 84.4 (CH), 84.1 (CH), 82.1 (CH), 81.4 (CH), 78.1 (CH), 74.9 (CH$_2$), 74.4 (CH$_2$), 74.4 (CH$_2$), 74.0 (CH$_2$), 72.7 (CH), 69.7 (CH$_2$), 69.2 (CH$_2$), 27.4 (Me), 25.6 (Me); MS (ESI): HRMS Calcd. For $C_{48}H_{49}BrClN_3O_9Na^+$ [M + Na]$^+$: 948.2233, found: 948.2235.

3.8. Synthesis of 7-Bromo-6-chloro-5′-O-(2″,3″,4″,6″-O-tetrabenzyl-α-D-glucopyranosyl) Tubercidin (14)

Compound **12** (1.0 g, 1.08 mmol) was added to 80% acetic acid aqueous solution (50 mL) and the solution was stirred at room temperature for 30 min. After that, the mixture was transferred to preheated oil and stirred at 50 °C for another 12 h until the reaction was completed (monitored by TLC, CH$_2$Cl$_2$:CH$_3$OH,20:1, R$_f$ 0.35). The mixture was concentrated and the crude product was purified by flash column chromatography on silica gel (CH$_2$Cl$_2$/CH$_3$OH, 50:1) to provide 0.9 g compound **14** (85.0% yield) as a white solid. mp: 49–50 °C; $[\alpha]_D^{25}$ = −26° (*c* = 0.3, CH$_3$OH); ^{1}H NMR (400 MHz, DMSO): δ 8.71 (s, 1H, H-2), 8.37 (s, 1H, H-8), 7.42–7.21 (m, 18H, H-Ph), 7.16–7.09 (m, 2H, H-Ph), 6.35 (d, *J* = 6.4 Hz, 1H, H-1′), 5.60 (d, 1H, 2′-OH), 5.37 (s, 1H, 3′-OH), 5.01 (d, *J* = 11.1 Hz, 1H, H-1″), 4.84 (m, 2H, H-CH$_2$), 4.77–4.65 (m, 3H, H-CH$_2$; H-2′), 4.45 (m, 4H, H-CH$_2$), 4.19 (d, *J* = 2.3 Hz, 1H, H-3′), 4.12 (s, 1H, H-5′), 3.82–3.62 (m, 3H, H-5′; H-4′), 3.58–3.56 (m, 3H, H-3″; H-6″), 3.52–3.40 (m, 2H, H-2″; H-4″); ^{13}C NMR (101 MHz, DMSO): δ 151.7 (CH-2), 151.3 (C-6), 151.1 (C-4), 139.3 (C), 138.8 (C), 138.6 (2C), 129.6 (CH), 128.7 (CH), 128.6 (CH), 128.5 (CH), 128.4 (CH), 128.2 (CH), 128.1 (CH), 127.9 (CH), 127.89 (CH), 114.9 (C-5), 96.4 (CH-1″), 88.4 (C-7), 87.3 (CH-1′), 83.8 (CH-5″), 82.0 (CH-4″), 79.5 (CH-4′), 77.8 (CH-3″),

75.3 (CH$_2$), 75.1 (CH-2′), 74.6 (CH$_2$), 72.8 (CH$_2$), 71.9 (CH$_2$), 71.6 (CH-3′), 70.6 (CH-2″), 69.0 (CH$_2$-6″), 67.9 (CH$_2$-5′); HRMS Calcd. For C$_{45}$H$_{46}$BrClN$_3$O$_9$$^+$ [M + H]$^+$: 886.2100, found: 886.2098.

3.9. Synthesis of 7-Bromo-6-chloro-5′-O-(2″,3″,4″,6″-O-tetrabenzyl-β-D-glucopyranosyl) Tubercidin (14′)

Compound **13** was converted into **14′** (79.0% yield) as described for the synthesis of **14**: R$_f$ 0.34 (CH$_2$Cl$_2$/CH$_3$OH, 20:1); mp: 49–50 °C; $[\alpha]_D^{25}$ = −12 (c = 0.3, CH$_3$OH); ^{1}H NMR (400 MHz, DMSO): δ 8.67 (s, 1H), 8.22 (s, 1H), 7.24 (s, 14H), 7.17–7.05 (m, 6H), 6.23 (d, J = 5.4 Hz, 1H), 5.54 (d, J = 6.1 Hz, 1H), 5.34 (s, 1H), 4.78 (d, J = 11.1 Hz, 2H), 4.69 (d, J = 8.8 Hz, 2H), 4.60–4.37 (m, 7H), 4.12 (m, 3H), 3.86–3.72 (m, 1H), 3.72–3.50 (m, 4H), 3.45 (m, 1H), 3.29–3.23 (m, 1H); ^{13}C NMR (101 MHz, DMSO) δ 151.6 (CH-2), 151.0 (C-6), 150.9 (C-4), 139.1 (C), 138.8 (C), 138.7 (C), 138.6 (C), 128.8 (CH), 128.7 (CH), 128.6 (CH), 128.6 (CH), 128.5 (CH), 128.4 (CH), 128.2 (CH), 128.1 (CH), 128.0 (CH), 127.9 (CH), 127.9 (CH), 127.8 (CH), 127.8 (CH), 115.0 (C-5), 103.0 (CH-1″), 88.2 (C-7), 87.9 (CH-1′), 84.1 (CH-4′), 83.7 (CH-5″), 82.4 (CH-4″), 78.1 (CH-3″), 74.9 (CH-2′), 74.7 (CH-3′), 74.5 (CH$_2$), 74.4 (CH$_2$), 74.1 (CH$_2$), 72.7 (CH$_2$), 71.0 (CH-2″), 70.0 (CH$_2$-5′), 69.2 (CH$_2$-6″); HRMS Calcd. For C$_{45}$H$_{46}$BrClN$_3$O$_9$$^+$ [M + H]$^+$: 886.2100, found: 886.2099.

3.10. Synthesis of 7-Bromo-5′-O-(2″,3″,4″,6″-O-tetrabenzyl-α-D-glucopyranosyl)tubercidin (15)

Compound **14** (100 mg, 0.11 mmol) was dissolved in methanolic ammonia (methanol was saturated with NH$_3$ at 0 °C, 50 mL) and the mixture was stirred at 130 °C for 12 h (monitored by TLC, CH$_2$Cl$_2$:CH$_3$OH, 25:1, R$_f$ 0.36). After cooling to room temperature, the mixture was concentrated and the crude product was purified by flash column chromatography on silica gel (CH$_2$Cl$_2$/CH$_3$OH, 30:1) to provide 92.9 mg compound **15** (94.9% yield) as a white solid. mp: 68–69 °C, $[\alpha]_D^{25}$ = −33 (c = 0.1, CH$_3$OH); ^{1}H NMR (400 MHz, DMSO): δ 8.10 (s, 1H, H-2), 7.87 (s, 1H, H-8), 7.40–7.18 (m, 18H, H-Bn), 7.18–7.07 (m, 2H, H-Bn), 6.76 (br s, 2H, H-NH$_2$), 6.19 (d, J = 6.6 Hz, 1H, H-1′), 5.46 (d, J = 6.3 Hz, 1H, 2′-OH), 5.26 (d, J = 4.1 Hz, 1H, 3′-OH), 4.98 (d, J = 11.1 Hz, 1H, H-1″), 4.81–4.66 (m, 5H, H-2′; 2CH$_2$), 4.49–4.32 (m, 4H, 2CH$_2$), 4.10–4.09 (m, 1H, H-3′), 4.06–4.04 (m, 1H, H-5″), 3.78–3.50 (m, 6H, H-4′,5′,3″,6″), 3.48–3.39 (m, 2H, H-2″,4″); ^{13}C NMR (101 MHz, DMSO): δ 157.4 (C-4), 153.0 (C-6), 150.6 (CH-2), 139.3 (C), 138.8 (C), 138.7 (2C), 128.7 (CH), 128.7 (CH), 128.4 (CH), 128.2 (CH), 128.1 (CH), 128.0 (CH), 127.9 (CH), 121.7 (C-5), 96.3 (CH-1″), 87.9 (C-7), 86.5 (CH-1′), 83.3 (CH-5″), 82.0 (CH-4′), 79.5 (CH-4″), 77.8 (CH-3″), 75.3 (CH$_2$), 74.9 (CH-2′), 74.6 (CH$_2$), 72.8 (CH$_2$), 71.8 (CH$_2$), 71.7 (CH-3′), 70.6 (CH-2″), 69.0 (CH$_2$-6′), 68.1 (CH$_2$-5′); HRMS Calcd. For C$_{45}$H$_{48}$BrN$_4$O$_9$$^+$ [M + H]$^+$: 867.2599, found: 867.2598.

3.11. Synthesis of 7-Bromo-5′-O-(2″,3″,4″,6″-O-tetrabenzyl-β-D-glucopyranosyl) Tubercidin (15′)

Compound **14′** was converted into **15′** (79.0% yield) as described for the synthesis of **15**; R$_f$ 0.40 (CH$_2$Cl$_2$/CH$_3$OH, 20:1); mp: 145–147 °C; $[\alpha]_D^{25}$ = −15 (c = 0.1, CH$_3$OH); ^{1}H NMR (400 MHz, DMSO) δ 8.11 (s, 1H, H-2), 7.68 (s, 1H, H-8), 7.40–7.19 (m, 14H, H-Bn), 7.19–7.10 (m, 6H, H-Bn), 6.77 (br s, 2H, NH$_2$), 6.13 (d, J = 5.4 Hz, 1H, H-1′), 5.44 (d, J = 6.1 Hz, 1H, 2′-OH), 5.26 (d, J = 5.0 Hz, 1H, 3′-OH), 4.81 (m, 2H, H-2′,1″), 4.71 (d, J = 11.0 Hz, 2H, CH$_2$), 4.60–4.08 (m, 6H, 3CH$_2$), 4.15–4.04 (m, 3H, H-3′,5′,5″), 3.78–3.42 (m, 6H, H-4′,5′,2″,3″,6″), 3.29 (s, 1H, H-4″); ^{13}C NMR (101 MHz, DMSO) δ 157.4 (C-6), 153.0 (CH-2), 150.2 (C-4), 139.1 (CH), 138.8 (CH), 138.7 (CH), 138.6 (CH), 128.7 (CH), 128.7 (CH), 128.6 (CH), 128.5 (CH), 128.5 (CH), 128.2 (CH), 128.1 (CH), 128.0 (CH), 127.8 (CH), 122.0 (C-5), 103.0 (C-7), 101.5 (CH-1″), 87.5 (CH-1′), 87.4 (CH-4′), 84.1 (CH-5″), 83.0 (CH-4″), 82.3 (CH-3″), 78.1 (CH-2′), 74.9 (CH-3′), 74.4 (CH$_2$), 74.4 (CH$_2$), 74.2 (CH$_2$), 72.7 (CH$_2$), 71.0 (CH-2″), 70.3 (CH$_2$-5′), 69.2 (CH$_2$-6″); HRMS Calcd. For C$_{45}$H$_{48}$BrN$_4$O$_9$$^+$ [M + H]$^+$: 867.2599, found: 867.2599.

3.12. Synthesis of 5′-O-(2″,3″,4″,6″-O-Tetrabenzyl-α-D-glucopyranosyl) Tubercidin (16)

To the stirred suspension of compound **15** (200 mg, 0.23 mmol) in THF (15 mL) and methanol (15 mL) was added triethylamine (0.3 mL) and 20% Pd(OH)$_2$/C (50 mg, 0.04 mmol), and the mixture was stirred at room temperature for 5 h under a continuous hydrogen environment until the reaction was completed (monitored by TLC, $CH_2Cl_2:CH_3OH$, 20:1, R_f 0.24). The mixture was filtered, and the filtrate was concentrated to provide the crude product, which was purified by flash column chromatography on silica gel (CH_2Cl_2/CH_3OH, 20:1) to provide 151.8 mg compound **16** (83.5% yield) as a white solid. mp: 74–75 °C; $[\alpha]_D^{25}$ = 1.8 (*c* = 0.1, CH_3OH); ^{1}H NMR (400 MHz, DMSO) δ 8.06 (s, 1H, H-2), 7.63 (d, *J* = 3.6 Hz, 1H, H-8), 7.41–7.21 (m, 18H, H-Bn), 7.17–7.10 (m, 2H, H-Bn), 7.00 (s, 2H, NH_2), 6.53 (d, *J* = 3.5 Hz, 1H, H-7), 6.16 (d, *J* = 5.9 Hz, 1H, H-1′), 5.41 (d, *J* = 6.2 Hz, 1H, 2′-OH), 5.22 (s, 1H, 3′-OH), 4.95–4.41 (m, 9H, H-1″,Bn), 4.36–4.32 (m, 1H, H-2′), 4.12–4.08 (m, 2H, H-3′,5′), 3.83–3.77 (m, 2H, H-4′,5′), 3.70–3.41 (m, 6H, H-2″,3″,4″,5′,6″); ^{13}C NMR (101 MHz, DMSO): δ 157.5 (C-6), 151.7 (CH-2), 150.8 (C-4), 139.2 (CH), 138.9 (CH), 138.7 (CH), 128.7 (CH), 128.7 (CH), 128.2 (CH), 128.1 (CH), 128.1 (CH), 128.0 (CH), 127.9 (CH), 122.1 (CH-8), 103.1 (C-5), 100.4 (CH-7), 96.3 (CH-1″), 87.0 (CH-1′), 82.7 (CH-5″), 82.0 (CH-4′), 79.9 (CH-4″), 77.8 (CH-3″), 75.2 (CH_2), 74.9 (CH-2′), 74.6 (CH_2), 72.8 (CH_2), 72.0 (CH_2), 71.4 (CH-3′), 70.5 (CH-2″), 69.0 (CH_2-6″), 68.0 (CH_2-5′); HRMS Calcd. For $C_{45}H_{49}N_4O_9{}^+$ [M + H]$^+$: 789.3494, found: 789.3499.

3.13. Synthesis of 5′-O-(2″,3″,4″,6″-O-Tetrabenzyl-β-D-glucopyranosyl) Tubercidin (**16′**)

Compound **15′** was converted into **16′** (79.0% yield) as described for the synthesis of **16**; R_f 0.23 (CH_2Cl_2/CH_3OH, 20:1); mp: 166–167 °C; $[\alpha]_D^{25}$ = −21 (*c* = 0.1, CH_3OH); ^{1}H NMR (400 MHz, DMSO): δ 8.03 (s, 1H, H-2), 7.38–7.18 (m, 15H, H-8, Bn), 7.19–7.05 (m, 6H, H-Bn), 6.98 (s, 2H, NH_2), 6.59 (d, *J* = 3.6 Hz, 1H, H-7), 6.09 (d, *J* = 5.2 Hz, 1H, H-1′), 5.36 (d, *J* = 6.3 Hz, 1H, 2′-OH), 5.22 (d, *J* = 5.3 Hz, 1H, OH-3′), 4.94–4.26 (m, 10H, H-1″,2′,Bn), 4.11–4.05 (m, 3H, H-3′,5′,5″), 3.70–3.40 (m, 6H, H-2″,3″,4′,5′,6″), 3.26 (d, *J* = 8.0 Hz, 1H, H-4″); ^{13}C NMR (101 MHz, DMSO): δ 157.9 (C-6), 152.3 (CH-2), 150.8 (C-4), 139.1 (CH), 138.8 (CH), 138.7 (CH), 138.6 (CH), 128.7 (CH), 128.7 (CH), 128.6 (C-Bn), 128.6 (CH), 128.5 (CH), 128.2 (CH), 128.1 (CH), 128.1 (CH), 128.1 (CH), 128.0 (CH), 127.9 (CH), 127.8 (CH), 127.8 (CH), 122.0 (CH-8), 103.3 (CH-1″), 103.2 (C-5), 100.5 (CH-7), 87.6 (CH-1′), 84.1 (CH-4′), 82.6 (CH-5″), 82.1 (CH-4″), 78.1 (CH-3″), 74.9 (CH-2′), 74.4 (CH_2), 74.3 (CH_2), 74.2 (CH_2), 74.1 (CH-3′), 72.8 (CH_2), 71.1 (CH-2″), 70.6 (CH_2-5′), 69.1 (CH_2-6″); HRMS Calcd. For $C_{45}H_{49}N_4O_9{}^+$ [M + H]$^+$: 789.3494, found: 789.3499.

3.14. Synthesis of 5′-O-α-D-Glucopyranosyl tubercidin (**2**)

To the stirred suspension of compound **16** (100 mg, 0.13 mmol) in THF (10 ml) and methanol (10 mL) was added 20% Pd(OH)$_2$/C (50 mg, 0.04 mmol), and the mixture was stirred at 40 °C for 12 h under a continuous hydrogen environment. The mixture was filtered and the filtrate was concentrated to provide the crude product, which was purified by isocratic HPLC (H_2O/CH_3OH, 4:1) to provide 43.5 mg compound **2** (80.1% yield) as a white solid; mp: 120–122 °C; $[\alpha]_D^{25}$ = 10.5 (*c* = 0.08, H_2O); ^{1}H NMR (400 MHz, DMSO): δ 8.09 (s, 1H, H-2), 7.76 (d, *J* = 3.2 Hz, 1H, H-8), 7.20 (s, 2H, NH_2), 6.61 (d, *J* = 3.4 Hz, 1H, H-7), 6.14 (d, *J* = 7.0 Hz, 1H, H-1′), 5.21 (d, *J* = 4.6 Hz, 2H, 2″, 3′-OH), 5.10 (d, *J* = 4.3 Hz, 1H, 2′-OH), 4.92 (d, *J* = 4.9 Hz, 1H, 3″-OH), 4.84 (d, *J* = 3.6 Hz, 1H, 4″-OH), 4.72 (d, *J* = 3.1 Hz, 1H, H-1″), 4.49 (s, 1H, 6″-OH), 4.43 (m, 1H, H-2′), 4.10 (s, 1H, H-4′), 4.08 (d, *J* = 1.7 Hz, 1H, H-3′), 3.78 (dd, *J* = 10.9, 2.8 Hz, 1H, H-5′), 3.66–3.63 (m, 1H, H-6′), 3.47–3.42 (m, 3H, H-5′,4″,6″), 3.40–3.34 (m, 2H, H-2″,5″), 3.13–3.07 (m, 1H, H-3″); ^{13}C NMR (101 MHz, DMSO): δ 157.3 (C-6), 151.5 (CH-2), 150.6 (C-4), 122.4 (CH-8), 102.8 (C-5), 100.0 (CH-7), 98.5 (CH-1″), 85.9 (CH-1′), 83.2 (CH-4′), 74.2 (CH-2′), 73.3 (CH-4″), 72.7 (CH-5″), 71.5 (CH-2″), 70.9 (CH-3″), 70.0 (CH-3′), 67.0 (CH_2-5′), 60.9 (CH_2-6″); HRMS Calcd. For $C_{17}H_{25}N_4O_9{}^+$ [M + H]$^+$: 429.1616, found: 429.1620.

3.15. Synthesis of 5′-O-β-D-Glucopyranosyl Tubercidin (**17**)

Compound **16′** was converted into **17** (79.0% yield) as described for the synthesis of **2**; mp: 156–157 °C; $[\alpha]_D^{25}$ = −37 (*c* = 0.08, H_2O); ^{1}H NMR (400 MHz, DMSO): δ 8.07 (s, 1H, H-2), 7.38 (d, *J* = 3.7 Hz, 1H, H-8), 7.09 (s, 2H, 6-NH_2), 6.61 (d, *J* = 3.6 Hz, 1H, H-7), 6.08 (d, *J* = 5.6 Hz, 1H, H-1′), 5.26 (d, *J* = 5.5 Hz, 1H, 2′-OH), 5.17 (d, *J* = 4.6 Hz, 1H, 3′-OH), 5.03 (d, *J* = 3.7 Hz, 1H, 2″-OH), 4.95 (s, 1H, 3″-OH), 4.95 (s, 1H, 4″-OH), 4.51 (s, 1H, 6″-OH), 4.36 (d, *J* = 5.0 Hz, 1H, H-2′), 4.21 (d, *J* = 7.8 Hz, 1H, H-1″), 4.14 (d, *J* = 4.4 Hz, 1H, H-3′), 4.02–3.95 (m, 2H, H-4′,5′), 3.68 (d, *J* = 9.2 Hz, 1H, H-6″), 3.59 (dd, *J* = 11.9, 5.7 Hz, 1H,

H-5′), 3.44 (d, *J* = 6.7 Hz, 1H, H-6″), 3.15 (d, *J* = 9.9 Hz, 1H, H-3″), 3.12–3.04 (m, 2H, H-4″,5″), 3.00 (d, *J* = 7.8 Hz, 1H, H-2″); ^{13}C NMR (101 MHz, DMSO): δ 157.5 (C-6), 151.7 (CH-2), 150.7 (C-4), 122.2 (CH-8), 103.5 (CH-1″), 103.1 (C-5), 100.5 (CH-7), 87.1 (CH-1′), 83.1 (CH-4′), 77.4 (CH-2″), 77.2 (CH-2′), 74.2 (CH-4″), 74.1 (C-5″), 71.1 (CH-3′), 70.5 (CH-3″), 69.5 (CH$_2$-5′), 61.5 (CH$_2$-6″); HRMS Calcd. For C$_{17}$H$_{25}$N$_4$O$_9{}^+$ [M + H]$^+$: 429.1616, found: 429.1619.

4. Conclusions

In summary, we developed a concise total synthesis of 5′-*O*-α-D-glucopyranosyl tubercidin **2** in eight steps with a 28% overall yield from commercially available 1-*O*-acetyl-2,3,5-tri-*O*-benzoyl-β-D-ribose **7**. This is the first report on the synthesis of disaccharide 7-deazapurine nucleosides.   This synthetic route features two key steps: (1) a one-pot Vorbrüggen glycosylation of ribose **7** with 6-chloro-7-bromo-7-deazapurine **8**, and (2) stereoselective α-*O*-glycosylation of 7-deazapurine nucleoside **11** with 2,3,4,6-tetra-*O*-benzyl-glucopyranosyl trichloroacetimidate **6**.   Additionally, 5′-*O*-β-D-glucopyranosyl tubercidin **17** was also prepared. A comparison with the natural product's spectra further confirmed the reported structure. Applications of this newly developed modular synthetic approach to prepare other glycosylated tubercidin analogs are ongoing in our laboratory.

Supplementary Materials: The following are available online at http://www.mdpi.com/1660-3397/18/8/398/s1, Table S1: ^{13}C NMR chemical shifts of naturally occurring nucleoside **2** and synthetic nucleosides **2** and **17**; Table S2: ^{1}H NMR chemical shifts and coupling constants of naturally occurring nucleoside **2**, synthetic **2** and **17**, and ^{1}H-and ^{13}C-NMR charts of all compounds.

Author Contributions: H.H. and Q.X. conceived and designed this research and analyzed the experimental data; W.O. prepared compounds and collected their spectral data; R.Y. and H.D. checked the experimental data; W.O., H.H. and Q.X. wrote the paper; all authors reviewed and approved the manuscript. All authors have read and agreed to the published version of the manuscript.

Funding: This work is financially supported by the National Natural Science Foundation of China (No. 21676131 and No. 21462019), the Science Foundation of Jiangxi Province (20161BAB213085), Education Department of Jiangxi Province (No. GJJ180625), and Jiangxi Science & Technology Normal University (No. 2018BSQD025).

Conflicts of Interest: The authors declare no conflict of interest.

References

1. Anzai, K.; Nakamura, G.; Suzuki, S. A new antibiotic, tubercidin. *J. Antibiot.* **1957**, *10*, 201–204. [PubMed]
2. Biabani, M.F.; Gunasekera, S.P.; Longley, R.E.; Wright, A.E.; Pomponi, S.A. Tubercidin, a cytotoxic agent from the marine sponge Caulospongia biflabellata. *Pharm. Biol.* **2002**, *40*, 302–303. [CrossRef]
3. Mitchell, S.S.; Pomerantz, S.C.; Concepcion, G.P.; Ireland, C.M.J. Tubercidin analogs from the ascidian Didemnum voeltzkowi. *Nat. Prod.* **1996**, *59*, 1000–1001. [CrossRef] [PubMed]
4. Duvall, L.R. Tubercidin. *Cancer Chemother. Rep.* **1963**, *30*, 61–62. [PubMed]
5. Bisel, H.F.; Ansfield, F.J.; Mason, J.H.; Wilson, W.L. Clinical studies with tubercidin administered by direct intravenous injection. *Cancer Res.* **1970**, *30*, 76–78.
6. Lynch, T.P.; Jakobs, E.S.; Paran, J.H.; Paterson, A.R.P. Treatment of mouse neoplasms with high doses of tubercidin. *Cancer Res.* **1981**, *41*, 3200–3204.
7. Mooberry, S.L.; Stratman, K.; Moore, R.E. Tubercidin stabilizes microtubules against vinblastine-induced depolymerization, a taxol-like effect. *Cancer Lett.* **1995**, *96*, 261–266. [CrossRef]
8. Perlikova, P.; Hocek, M. Pyrrolo[2,3-d]pyrimidine (7-deazapurine) as a privileged scaffold in design of antitumor and antiviral nucleosides. *Med. Res. Rev.* **2017**, *37*, 1429–1460. [CrossRef]
9. Hulpia, F.; Campagnaro, G.D.; Scortichini, M.; Van Hecke, K.; Maes, L.; de Koning, H.P.; Caljon, G.; Van Calenbergh, S. Revisiting tubercidin against kinetoplastid parasites: Aromatic substitutions at position 7 improve activity and reduce toxicity. *Eur. J. Med. Chem.* **2019**, *164*, 689–705. [CrossRef]
10. Hulpia, F.; Mabille, D.; Campagnaro, G.D.; Schumann, G.; Maes, L.; Roditi, I.; Hofer, A.; de Koning, H.P.; Caljon, G.; Van Calenbergh, S. Combining tubercidin and cordycepin scaffolds results in highly active candidates to treat late-stage sleeping sickness. *Nat. Commun.* **2019**, *10*, 5564. [CrossRef]

11. Hulpia, F.; Bouton, J.; Campagnaro, G.D.; Alfayez, I.A.; Mabille, D.; Maes, L.; de Koning, H.P.; Caljon, G.; Van Calenbergh, S. C6–O-alkylated 7-deazainosine nucleoside analogues: Discovery of potent and selective anti-sleeping sickness agents. *Eur. J. Med. Chem.* **2020**, *188*, 112018. [CrossRef] [PubMed]

12. De Coen, L.M.; Heugebaert, T.S.A.; Garcia, D.; Stevens, C.V. Synthetic entries to and biological activity of pyrrolopyrimidines. *Chem. Rev.* **2016**, *116*, 80–139. [CrossRef] [PubMed]

13. Tumkevicius, S.; Dodonova, J. Functionalization of pyrrolo[2,3-d]pyrimidine by palladium-catalyzed cross-coupling reactions. *Chem. Heterocycl. Comp.* **2012**, *48*, 258–279. [CrossRef]

14. Mulamoottil, V.A. Tubercidin and related analogues: An inspiration for 50 years in drug discovery. *Curr. Org. Chem.* **2016**, *20*, 830–838. [CrossRef]

15. Stewart, J.B.; Bornemann, V.; Chen, J.L.; Moore, R.E.; Caplan, F.R.; Karuso, H.; Larsen, L.K.; Patterson, G.M.L.J. Cytotoxic, fungicidal nucleosides from blue green algae belonging to the Scytonemataceae. *J. Antibiot.* **1988**, *41*, 1048–1056. [CrossRef]

16. Efimtseva, E.V.; Kulikova, I.V.; Mikhailov, S.N. Disaccharide nucleosides as an important group of natural compounds. *Mol. Biol.* **2009**, *43*, 301–312. [CrossRef]

17. Sylla, B.; Gauthier, C.; Legault, J.; Fleury, P.-Y.; Lavoie, S.; Mshvildadze, V.; Muzashvili, T.; Kemertelidze, E.; Pichette, A. Isolation of a new disaccharide nucleoside from Helleborus caucasicus: Structure elucidation and total synthesis of hellecaucaside A and its β-anomer. *Carbohy. Res.* **2014**, *398*, 80–89. [CrossRef]

18. Dong, X.; Tang, J.; Hu, C.; Bai, J.; Ding, H.; Xiao, Q. An Expeditious total synthesis of 5′-Deoxy-toyocamycin and 5′-Deoxysangivamycin. *Molecules* **2019**, *24*, 737. [CrossRef]

19. Ding, H.; Ruan, Z.; Kou, P.; Dong, X.; Bai, J.; Xiao, Q. Total synthesis of mycalisine B. *Mar. Drugs* **2019**, *17*, 226. [CrossRef]

20. Huang, H.; Ruan, Z.; Hu, T.; Xiao, Q. An improved total synthesis of tubercidin. *Chin. J. Org. Chem.* **2014**, *34*, 1358–1363. [CrossRef]

21. Sun, J.; Dou, Y.; Ding, H.; Yang, R.; Sun, Q.; Xiao, Q. First total synthesis of a naturally occurring iodinated 5′-deoxyxylofuranosyl marine nucleoside. *Mar. Drugs* **2012**, *10*, 881–889. [CrossRef] [PubMed]

22. Song, Y.; Ding, H.; Dou, Y.; Yang, R.; Sun, Q.; Xiao, Q.; Ju, Y. Efficient and practical synthesis of 5′-deoxytubercidin and its analogues via vorbrüggen glycosylation. *Synthesis* **2011**, *2011*, 1442–1446.

23. Aoki, S.; Fukumoto, T.; Itoh, T.; Kurihara, M.; Saito, S.; Komabiki, S.Y. Synthesis of disaccharide nucleosides by the O-glycosylation of natural nucleosides with thioglycoside donors. *Chem. Asian J.* **2015**, *10*, 740–751. [CrossRef] [PubMed]

24. Someya, H.; Itoh, T.; Kato, M.; Aoki, S.J. Regioselective O-Glycosylation of Nucleosides via the Temporary 2′, 3′-Diol Protection by a boronic ester for the synthesis of Disaccharide Nucleosides. *Vis. Exp.* **2018**, *137*, e57897. [CrossRef]

25. Zhang, Y.; Knapp, S.J. Glycosylation of nucleosides. *Org. Chem.* **2016**, *81*, 2228–2242. [CrossRef]

26. Ingale, S.A.; Leonard, P.; Seela, F.J. Glycosylation of Pyrrolo [2, 3-d] pyrimidines with 1-O-Acetyl-2, 3, 5-tri-O-benzoyl-β-d-ribofuranose: Substituents and protecting groups effecting the synthesis of 7-Deazapurine ribonucleosides. *Org. Chem.* **2018**, *83*, 8589–8595. [CrossRef]

27. Seela, F.; Peng, X.H.J. 7-Functionalized 7-deazapurine ribonucleosides related to 2-aminoadenosine, guanosine, and xanthosine: Glycosylation of pyrrolo [2, 3-d] pyrimidines with 1-O-acetyl-2, 3, 5-tri-O-benzoyl-D-ribofuranose. *Org. Chem.* **2006**, *71*, 81–90. [CrossRef]

28. Seela, F.; Peng, X.H. Progress in 7-deazapurine-pyrrolo [2, 3-d] pyrimidine-ribonucleoside synthesis. *Curr. Top. Med. Chem.* **2006**, *6*, 867–892. [CrossRef]

29. Gupta, S.; Thakur, K.; Khare, N.K. Regio- and stereoselectivity in O-glycosylation reactions. *Trends Carbohydr. Res.* **2019**, *11*, 1–42.

30. Takahashi, D.; Tanaka, M.; Nishi, N.; Toshima, K. Novel 1, 2-cis-stereoselective glycosylations utilizing organoboron reagents and their application to natural products and complex oligosaccharide synthesis. *Carbohydr Res.* **2017**, *452*, 64–77. [CrossRef]

31. Mensink, R.A.; Boltje, T.J. Advances in stereoselective 1,2-cis Glycosylation using C-2 Auxiliaries. *Chem. Eur. J.* **2017**, *23*, 17637–17653. [CrossRef] [PubMed]

32. Demchenko, A.V. 1, 2-cis O-Glycosylation: Methods, strategies, principles. *Curr. Org. Chem.* **2003**, *7*, 35–79. [CrossRef]

33. Zhu, X.; Schmidt, R.R. New principles for glycoside-bond formation. *Angew. Chem. Int. Ed.* **2009**, *48*, 1900–1934. [CrossRef] [PubMed]
34. Schmidt, R.R.; Kinzy, W. Anomeric-oxygen activation for glycoside synthesis: The trichloroacetimidate method. *Adv. Carbohydr. Chem. Biochem.* **1994**, *50*, 21–123. [PubMed]

 *marine drugs*

Article

Unexpected Enhancement of HDACs Inhibition by MeS Substitution at C-2 Position of Fluoro Largazole

Bingbing Zhang [1,†], Zhu-Wei Ruan [1,†], Dongdong Luo [2,†], Yueyue Zhu [1], Tingbo Ding [1], Qiang Sui [3] and Xinsheng Lei [1,4,*]

1   School of Pharmacy, Fudan University, 826 Zhangheng Road, Pudong Zone, Shanghai 201203, China; 14211030012@fudan.edu.cn (B.Z.); 13301030004@fudan.edu.cn (Z.-W.R.); 18211030012@fudan.edu.cn (Y.Z.); tbding@shmu.edu.cn (T.D.)
2   School of Medicine and Pharmacy, Ocean University of China, Qingdao 266003, China; luck_luodong@163.com
3   China State Institute of Pharmaceutical Industry, No. 285 Gebaini Road, Pudong Zone, Shanghai 201203, China; chem_sq@163.com
4   Key Laboratory of Synthetic Chemistry of Natural Substances, Shanghai Institute of Organic Chemistry, Chinese Academy of Sciences, Shanghai 200032, China
*   Correspondence: leixs@fudan.edu.cn; Tel.: +86-021-51980128
†   These authors contributed equally to the work.

Received: 30 May 2020; Accepted: 28 June 2020; Published: 30 June 2020

Abstract: Given our previous finding that fluorination at the C18 position of largazole showed reasonably good tolerance towards inhibitory activity and selectivity of histone deacetylases (HDACs), further modification on the valine residue in the fluoro-largazole's macrocyclic moiety with S-Me L-Cysteine or Glycine residue was performed. While the Glycine-modified fluoro analog showed poor activity, the S-Me L-Cysteine-modified analog emerged to be a very potent HDAC inhibitor. Unlike all previously reported C2-modified compounds in the largazole family (including our recent fluoro-largazole analogs) where replacement of the Val residue has failed to provide any potency improvement, the S-Me L-Cysteine-modified analog displayed significantly enhanced (five–nine-fold) inhibition of all the tested HDACs while maintaining the selectivity of HDAC1 over HDAC6, as compared to largazole thiol. A molecular modeling study provided rational explanation and structural evidence for the enhanced inhibitory activity. This new finding will aid the design of novel potent HDAC inhibitors.

Keywords: marine natural product; largazole; HDAC inhibitors; modification; fluoro olefin

1. Introduction

Both histone acetyltransferases (HATs) and histone deacetylases (HDACs) play a key role in the regulation of histone-tailed lysine acetylation status, which is closely associated with cell processes [1]. Up to now, 18 HDAC isoforms have been identified and divided into four classes, including the Zn^{2+}-dependent Class I (HDAC1, 2, 3, 8), Class II (HDAC4–7, 9–10), Class IV (HDAC11), and NAD+-dependent Class III (namely SIRT1-7) [2]. Although the function of individual HDAC isoform is not fully understood in cells, HDACs have become an appealing target in cancer therapy through balancing histone hypoacetylation and overexpression of HDACs in multiple cancers. Currently, several Zn^{2+}-dependent HDAC inhibitors, such as SAHA (Vorinostat), Belinostat, Panobinostat, Chidamide and Romidepsin (FK228), have been approved for the treatment of cutaneous T-cell lymphoma. However, these inhibitors are pan-selective inhibitors [3,4]. The current trend is to explore potent and selective HDAC inhibitors [5–7].

Largazole, as a potent and selective Class I HDAC inhibitor, was discovered in 2008 by Leusch et al. [8,9]. Due to its superior anticancer properties, this marine natural product has attracted widespread attention in the medicinal chemistry community [10–15]. In order to improve the inhibitory activity and selectivity of HDACs, great efforts have been made to modify the macrocyclic moiety, the hydrophobic linker and the warhead of largazole. Thus far, almost all the modifications on the linker and the warhead cause obvious losses in activity against HDACs, possibly due to preventing the thiol group to efficiently penetrate into the narrow channel containing the Zn^{2+} ion of HDACs and to coordinate it with the optimal geometry. In contrast, the modifications on the macrocyclic moiety seem to be allowable to some extent. Among them, certain variations in valine residue (Val) at the C2 position of largazole may maintain the HDACs inhibition despite with a little sacrifice in its potency or isoenzyme selectivity (Figure 1). For example, Hong and Leusch's group reported that replacement of Val with Ala decreased the activity by three-fold in growth inhibition of HCT-116 [16]. Subsequently, they introduced aromatic (Phe, Tyr), acidic (Asp) or basic (His) amino acid residues at that position, but all the analogs exhibited reduced activity in HDAC1 [17]. Based on the hypothesis of potentially hydrophobic interactions between Val and the surrounding residues of HDAC1 (Tyr196 and Leu263), Jiang et al. replaced Val with Leu, Phe and Tyr, reporting that Tyr had about five-fold higher GI_{50} values of HCT116 and A549 cells compared to largazole despite with an increasing selectivity over normal cells [18]. Ganesan's and Williams' groups reported that replacement with Gly or Pro resulted in a significantly decreased inhibition against HDACs, respectively [19,20]. Recently, Hong's group investigated again the effect of the replacement of Val with Phe, Tyr, Asp or His on both the activity and selectivity of the class I HDACs, and the observed His substitution showed comparable activity, and a slight selectivity towards HDAC1 over HDAC2–3. This slight selectivity was hypothesized to be resulted from a possible hydrogen-bond ability of His. However, when they replaced Val with a set of residues tagged with the terminal amine or amides, the results suggested that the hydrogen bonding interaction did not play an essential role in HDAC inhibition [21]. All these works indicated that the structural optimization at the C2 position appeared not to improve the potency of HDACs inhibition.

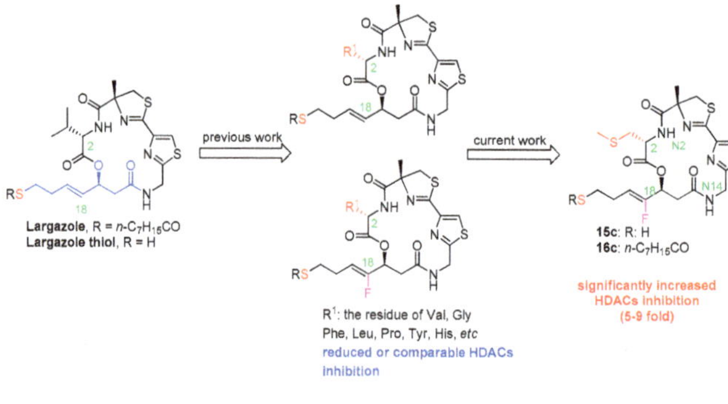

Figure 1. Largazole and its analogs modified at C-2 position.

Recently, we discovered that the modification on the linker of largazole was allowable in some cases (Figure 1) [22–24]. For example, fixing fluorine at C18 position of the linker could keep almost the same inhibitory activity and selectivity of HDACs. Subsequently, the replacement of Val with Phe gave slightly reduced inhibition of HDACs, similar to the previous observations reported for largazole [23]. As our research continued, we discovered to our surprise that the replacement of Val with S-Me Cys significantly enhanced the inhibition of HDACs, whereas the replacement with Gly significantly decreased the activity. With an expanded research program under way based on this observation, we would like to communicate here our preliminary results and provide rational explanation for the unexpected results through a molecular modeling study. To the best of our knowledge, the S-Me Cys modification is the first reported example where replacement of the Val residue has led to significant enhancement in HDACs inhibition. We believe this insight may enable the design of highly potent HDAC inhibitors.

2. Results and Discussion

2.1. Chemistry

In our previous paper [23], we synthesized the two fluoroolefin analogs of largazole with Val and Phe residue at C2 position. The key macrolactamization was chosen at N2 position, but the yields were poor (less than 31% yield). As an alternative, the key macrolactamization for the two new fluoroolefin analogs in this study was arranged at N14 position with an intend to improve the yield of the key step. The evolved synthetic route is depicted in Scheme 1.

The key fluoro fragment (**8**) was prepared from acrolein via four steps according to our previous method [23]. Alcoholysis of **8** with TMSCH$_2$CH$_2$OH in the presence of DMAP provided **9** in good yield (67%). Condensation of **9** with S-Me Fmoc-L-Cystine or Fmoc-Glycine in the presence of EDCI, DMAP and DIEPA afforded ester **10c** and **10d** (yield: 75% for **10c**). Removal of Fmoc group of **10c** or **10d** with Et$_2$NH in CH$_2$Cl$_2$ and subsequent condensation with the acid (**11**) resulted in the linear precursors (**12c**, 84%; **12d**, 41% based on **9**). Deprotection of Boc group of **10c** or **10d** with CF$_3$CO$_2$H in CH$_2$Cl$_2$ and subsequent release of free acid provided the linear depsipeptides (**13c** and **13d**) exposed at the N- and C-terminus, respectively. The depsipeptides were then subjected to the optimal cyclization condition (HATU/HOAT/DIPEA in anhydrous CH$_2$Cl$_2$ solution with a diluted concentration of about 0.001 M). The macrolactamization yields were 48% for **14c** and 40% for **14d** in two steps, respectively. Compared to our previous cyclization at N2 position, cyclizing at N14 position afforded apparently improved yield. Following our previously reported procedure [24], the free thiol **15c** and **15d** were obtained through deprotection of Trt group from **14c** and **14d**, respectively. The subsequent acylation with *n*-C$_7$H$_{15}$COCl under the standard condition led to the final fluoro analogs **16c** and **16d** (yields: 57%, 40%), respectively.

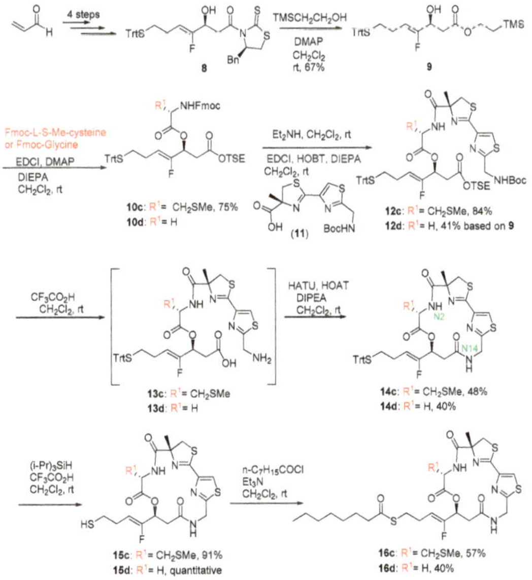

Scheme 1. Synthesis of the analogs of Largazole. The red R^1 is the different substituent and the green number is the atom code in the molecule.

2.2. Biology

It is well-known that largazole is a prodrug species for a beneficial cell permeability and its free thiol is indeed the activated species for the inhibition against HDACs [10]. Given the instability of the free thiol for storage, the in vitro cell assays were used as the initial screening by our well-established cell tests (A549, HCT116, MDA-MB-231 and SK-OV-3 cells), aiming at speedy identification of potent HDAC inhibitors. The preliminary results were shown in Table 1, using largazole as the positive control compound.

Compared with largazole, **16c** displayed the obvious growth inhibition towards A549, HCT116 and SK-OV-3 cells with IC_{50} values ranged with 1.41–3.13 μM, but not towards MDA-MB-231 (IC_{50}: >10 μM). In contrast, **16d** almost lost the growth inhibition towards most of those cells, and only maintained a slight growth inhibition towards HCT116 (IC_{50}: 6.62 μM), indicating that **16d** was a less potent HDAC inhibitor, which was consistent with previous reports [19,20], where both Ganesan's and Williams' groups demonstrated that the replacement of Val with Gly in the case of largazole resulted in a significantly decrease of inhibition against HDACs.

Table 1. The preliminary in vitro growth inhibition of several cells (IC_{50}, μM) with Largazole and its analog **16c–d**.

	IC_{50} [a] (μM)			
	A549	**HCT116**	**MDA-MB-231**	**SK-OV-3**
largazole	0.52 ± 0.17	0.11 ± 0.01	4.75 ± 1.95	0.25 ± 0.17
16c	2.92 ± 0.53	1.41 ± 0.54	>10	3.13 ± 0.50.
16d	>10	6.62 ± 1.86	>10	>10

[a] Compounds were tested in at least seven-dose IC_{50} mode in duplicate with three-fold serial dilution.

To further confirm the growth inhibition of **16c** in cellular assays, we tested **16c** and largazole with other cells. As shown in Table 2, **16c** displayed growth inhibition against Hela, Eca-109, Bel 7402 and U937 cells with high potencies that are quite identical to largazole.

Table 2. The further in vitro growth inhibition of other cells (IC_{50}, μM) with Largazole and **16c–d**.

	IC_{50} [a] (μM)			
	Hela	**Eca-109**	**Bel 7402**	**U937**
largazole	0.17 ± 0.01	0.10 ± 0.00	0.17 ± 0.02	0.02 ± 0.00
16c	0.10 ± 0.01	0.09 ± 0.02	0.25 ± 0.07	0.08 ± 0.02

[a] Compounds were tested in at least seven-dose IC_{50} mode in duplicate with three-fold serial dilution.

Based on the obvious growth inhibition of **16c** in in vitro cell assays, we next performed in vitro enzyme assays with the correspond active species **15c**, and the data were compared with those of our previous compounds. The results are shown in Table 3.

In sharp contrast to the cellular assays where **16c** exhibited identical or slightly inferior activities as compared to largazole, in the enzymatic assays, compound **15c** surprisingly displayed significantly increased inhibition towards HDACs, with the IC_{50} values of HDAC 1, 2, 3, 8 and 6 being 0.27, 1.33, 2.33, 0.44 and 23.91 nM, respectively. It was five–nine-fold more potent towards all the tested HDACs when compared with largazole thiol. Moreover, the selectivity of HDAC1 over HDAC6 remained unchanged (selectivity: 89). Notably, we demonstrated earlier that the fluoro analogs (**15a** and **15b**) exhibited slightly less activities than largazole thiol, which was consistent with all previous observations made with substitutions at C2 for largazole [16–21]. We wanted to compare **15c** with similar largazole compounds. However, a literature search revealed no prior report on modification at C2 position with a sulphur-containing substituent. **15c** appeared to be the first example of such compounds in the largazole family. The surprisingly enhanced activity of **15c** might be attributed to the variation in Val at the C2 position, suggesting that the MeS group of **15c** played an important role in the interaction between **15c** and HDACs. This was confirmed subsequently by the results from a molecular modeling study.

Table 3. In vitro inhibition of human histone deacetylases (HDACs) with the free thiols of largazole and its two analogs.

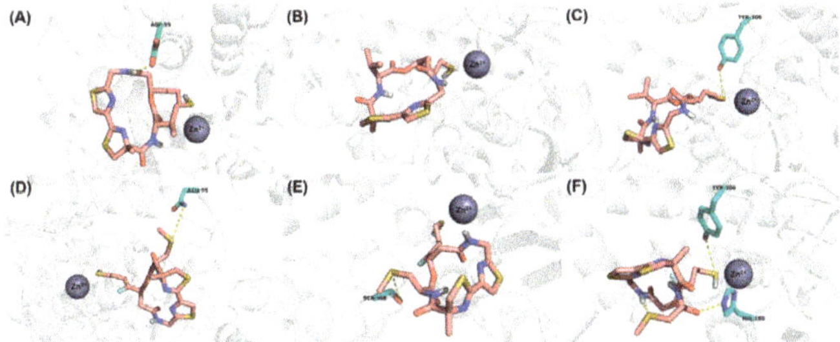

Largazole thiol
R^1 = i-Pr

15a: R^1 = i-Pr
15b: R^1 = Bn
15c: R^1 = CH$_2$SCH$_3$

	IC$_{50}$ (nM) [a]					Selectivity (HDAC1/6)
	HDAC1	**HDAC2**	**HDAC3**	**HDAC8**	**HDAC6**	
15a [b]	4.39 ± 0.17	21.0 ± 0.3	39.2 ± 2.3	8.38 ± 2.46	300 ± 5	68
15b [b]	4.24 ± 0.38	16.3 ± 0.4	37.1 ± 0.8	17.5 ± 4.2	338 ± 17	80
15c	0.27 ± 0.02	1.33 ± 0.02	2.33 ± 0.15	0.44 ± 0.16	23.9 ± 0.2	89
largazole thiol	2.01 ± 0.01	9.49 ± 0.19	14.4 ± 0.2	3.75 ± 0.20	121 ± 5	60

[a] Compounds were tested in 10-dose IC$_{50}$ mode in duplicate with three-fold serial dilution. [b] The data were previously reported by us in [23].

2.3. Molecular Modeling Study

To gain some structural insight on the increased inhibitory effects of **15c** on the tested HDACs, molecular docking was performed by using MOE 2019 with MMF94 force field. The crystal structures of HDACs were downloaded from the Protein DataBank (PDB, http://www.rcsb.org), and were used to investigate the binding modes of **15c** with HDAC1, HDAC6 and HDAC8, respectively (PDB code: HDAC1, 5ICN; HDAC6, 5EDU; HDAC8, 4RN0).

The binding modes of largazole thiol in HDAC1, HDAC6 and HDAC8 indicated that the thiol side chain could coordinate to the catalytic Zn^{2+} ion and the overall metal coordination geometry was nearly perfectly tetrahedral [25]. Additionally, largazole thiol formed hydrogen bond interactions with ASP99 in HDAC1 and TYR306 in HDAC8, respectively (Figure 2A–C). For **15c**, which was derived from largazole thiol, additional hydrogen bonds were formed between sulphur atom in the methylthio group and ASN95 in HDAC1 and SER568 in HDAC6, respectively (Figure 2D,E). Interestingly, the introduction of MeS substitution resulted in orientation change of the ester bond at C1 position, and then made an oxygen atom of the carboxyl group form a new hydrogen bond with HIS180 of HDAC8 (Figure 2F). These additional interactions are likely the factors that have led to significant improvement in **15c**'s potency towards HDACs. These findings from molecular modeling study confirmed our hypothesis about the key role that MeS group played, and provided telling structural evidence for the significantly increased inhibition of **15c** towards HDACs.

Figure 2. Proposed binding mode of Largazole thiol (carmine carbon sticks) with (**A**) HDAC1 (5ICN), (**B**) HDAC6 (5EDU) and (**C**) HDAC8 (4RN0). Proposed binding mode of **15c** (carmine carbon sticks) with (**D**) HDAC1 (5ICN), (**E**) HDAC6 (5EDU) and (**F**) HDAC8 (4RN0). Atom color code: red = oxygen, blue = nitrogen, white = hydrogen, yellow = sulfur, cyan = fluorine. Hydrogen bonds between the ligand and receptor are indicated by yellow dashed lines.

3. Materials and Methods

3.1. Chemistry

The chemicals and reagents were purchased from Acros (Shanghai, China), Alfa Aesar (Shanghai, China), and National Chemical Reagent Group Co. Ltd., P. R. China (Shanghai, China), and used without further purification. Anhydrous solvents (THF, MeOH, DMF, CH_2Cl_2 and CH_3CN) used in the reactions were dried and freshly distilled before use. All the reactions were carried out under Ar atmosphere, otherwise stated else. The progress of the reactions was monitored by TLC (silica-coated glass plates) and visualized under UV light, and by using iodine or phosphomolybdic acid. Melting points were measured on an SGW X-4 microscopy melting point apparatus without correction. ^{1}H NMR and ^{13}C NMR spectra were recorded either on a 400 MHz Varian Instrument at 25 °C or 600 MHz Bruker Instrument at 25 °C, using TMS as an internal standard, respectively. Multiplicity is tabulated as s for singlet, d for doublet, dd for doublet of doublet, t for triplet, and m for multiplet. The original spectra of the relative compounds could be found in Supplementary Materials. HRMS spectra were recorded on Finnigan-Mat-95 mass spectrometer, equipped with ESI source. Largazole, its free thiol and key intermediates (compound **8** and **11**) were prepared according to our previous method [22–24].

3.1.1. 2-(Trimethylsilyl)ethyl (S,Z)-4-fluoro-3-hydroxy-7-(tritylthio)hept-4-enoate (9)

To a solution of **8** (0.595 g, 0.95 mmoL, 1.0 equiv.) and DMAP (0.183 g, 1.5 mmoL, 1.5 equiv.) in CH_2Cl_2 (DCM, 20 mL), $Me_3CH_2CH_2OH$ was added (1.36 mL, 9.5 mmol, 10 equiv.), and the resulting solution was stirred at room temperature (rt) overnight. Concentration in vacuo left the residue, which was purified by column chromatography, giving compound **9** (0.341 mg) with a 67% yield as colorless oil. R_f = 0.17 [petroleum ether (PE)/ethyl acetate (EA) 10/1]. ^{1}H-NMR (600 MHz, $CDCl_3$): δ 7.50–7.13 (m, 15H, Ar*H*), 4.81 (dt, *J* = 37.2, 7.2 Hz, 1H, C*H*=CF), 4.46 (m, 1H, C*H*OH), 4.20 (m, 2H, OC*H*$_2$CH$_2$TMS), 3.14 (brs, 1H, O*H*), 2.63 (dd, *J* = 16.2, 3.6 Hz, 1H, C*H*$_2$CO), 2.58 (dd, *J* = 16.2, 9.0 Hz, 1H, C*H*$_2$CO), 2.20 (t, *J* = 7.2 Hz, 2H, SC*H*$_2$CH$_2$), 2.14 (t, *J* = 7.2 Hz, 2H, SCH$_2$C*H*$_2$), 0.99 (t, *J* = 8.4 Hz, 2H, OCH$_2$C*H*$_2$TMS), 0.03 (s, 9H, Si*Me*$_3$). ^{13}C-NMR (150 MHz, $CDCl_3$): δ 171.4, 157.9 (d, *J* = 255 Hz), 144.2, 128.9, 127.2, 126.0, 104.2, 66.1 (d, *J* = 27 Hz), 62.7, 37.9, 30.8, 22.0, 16.6, 2.1. ^{19}F NMR (376 MHz, $CDCl_3$): δ −123.24 (dd, *J* = 37.6, 11.2 Hz). ESI (*m/z*): [M + H]$^+$ 537.4.

3.1.2. 2-(Trimethylsilyl)ethyl (S,Z)-3-((N-(((9H-fluoren-9-yl)methoxy)carbonyl)-S-methyl-L-cysteinyl) oxy)-4-fluoro-7-(tritylthio)hept-4-enoate (10c)

To a solution of **9** (0.690 g, 1.28 mmol, 1.0 equiv.) in DCM (10 mL) a solution of S-Me Fmoc-L-Cysteine was added (2.38 g, 6.42 mmol, 5.0 equiv.); DMAP (0.016 g, 0.128 mmol, 0.1 equiv.) and EDCI (1.48 g, 7.7 mmol, 6.0 equiv.) was added in DCM (20 mL) at rt, and DIPEA (1.30 mL, 7.7 mmol, 2.5 equiv.) was added to the stirring solution. The reaction was stirred for 18 h. This solution was quenched with a NH_4Cl aqueous solution and separated. The aqueous phase was extracted with DCM (30 mL × 3) and the combined organic phase was washed with H_2O, brine, dried with Na_2SO_4 and filtered. After the removal of the solvent, the resulting crude was purified by flash chromatography on silica gel, eluting with PE/EA (5/1) to afford **10c** (0.856 g, 75%) as a white foamy solid. R_f = 0.20 (PE/EA 5/1). ^{1}H-NMR (600 MHz, $CDCl_3$): δ 7.80–7.10 (m, 23H, Ar*H*), 6.64 (m, 1H, C*H*OC=O), 5.37 (d, *J* = 8.4 Hz, 1H, N*H*), 4.88 (dt, *J* = 36, 7.2 Hz, 1H, C*H*=CF), 4.49 (m, 1H, C*H*CH$_2$SMe), 4.40 (m, 2H, OC*H*$_2$CH[Fmoc]), 4.21 (m, 1H, OCH$_2$C*H*[Fmoc]), 4.16 (m, 2H, OC*H*$_2$CH$_2$TMS), 2.80 (dd, *J* = 16.8, 8.4 Hz, 1H, C*H*$_2$CO), 2.68 (dd, *J* = 16.8, 4.8 Hz, 1H, C*H*$_2$CO), 2.42 (m, 1H, CHC*H*$_2$SMe), 2.20–2.14 (m, 4H, SC*H*$_2$C*H*$_2$), 2.03 (s, 3H, CHCH$_2$S*Me*), 1.91 (m, CHC*H*$_2$SMe), 0.96 (t, *J* = 8.4 Hz, 2H, OCH$_2$C*H*$_2$TMS), 0.01 (s, 9H, Si*Me*$_3$). ^{13}C-NMR (150 MHz, $CDCl_3$): δ 170.2, 168.4, 155.1, 153.9 (d, *J* = 257 Hz), 144.1, 143.2, 143.0, 140.7, 128.9, 127.2, 127.1, 126.4, 126.0, 124.4, 119.3, 108.8 (d, *J* = 14 Hz), 68.8 (d, *J* = 29 Hz), 66.4, 66.2, 62.8, 52.4, 46.5, 35.6, 31.2, 30.4, 22.2, 16.6, 14.7, −2.1. ^{19}F NMR (376 MHz, $CDCl_3$): δ −123.28 (dd, *J* = 37.6, 18.8 Hz). HRMS-ESI (*m/z*): [M + H]$^+$ calcd. for $C_{50}H_{54}FNO_6S_2SiH$: 876.3219, found: 876.3223.

3.1.3. 2-(Trimethylsilyl)ethyl (S,Z)-3-((N-((R)-2′-(((tert-butoxycarbonyl)amino)methyl)-4-methyl-4,5-dihydro-[2,4′-bithiazole]-4-carbonyl)-S-methyl-L-cysteinyl)oxy)-4-fluoro-7-(tritylthio)hept-4-enoate (12c)

To a solution of **10c** (0.890 g, 0.90 mmol, 1.0 equiv.) in DCM (60 mL) was added Et$_2$NH (4.74 mL, 29 mmol, 10 equiv.) was added at rt. After stirring for 2 h and removal of the solvent, the residue was purified by flash chromatography on silica gel, eluting with PE/EA (3/2) to afford the corresponding amine (0.465 g, 77%) as a white foamy solid.

A mixture of compound **11** (0.196 g, 0.55 mmol, 1.0 equiv.), the amine (0.438 g, 0.66 mmol, 1.2 equiv.), HOBT (0.089 g, 0.66 mmol, 1.2 equiv.) and EDCI (0.126 g, 0.66 mmol, 1.2 equiv.) was dissolved in dry DCM (50 ml) and stirred at rt for 3 h. The solution was concentrated under reduced pressure, and diluted with EA (150 mL), washed saturated NaHCO$_3$ aqueous solution and brine successively, dried over Na$_2$SO$_4$ and filtered. After removal of the solvent, the residue was purified by flash chromatography on silica gel, eluting with PE/EA (5/2) to afford **12c** (84% based on **10c**) as a white foamy solid. R_f = 0.30 (PE/EA 5/2). ^{1}H-NMR (600 MHz, CDCl$_3$): δ 7.87 (s, 1H, Ar*H*), 7.5–7.10 (m, 17H, Ar*H*, 2N*H*), 5.60 (ddd, *J* = 19.8, 8.4, 5.4 Hz, 1H, C*H*OC=O), 5.29 (dd, *J* = 36, 7.2 Hz, 1H, C*H*=CF), 4.68 (m,1H, C*H*CH$_2$SMe), 4.62 (d, *J* = 5.4 Hz, 2H, C*H*$_2$NHCOCMe$_3$), 4.18 (m, 2H, OC*H*$_2$CH$_2$SiMe$_3$), 4.04 (d, *J* = 11.4 Hz, 1H, C$_q$C*H*$_2$S), 3.32 (d, *J* = 11.4 Hz, 1H, C$_q$C*H*$_2$S), 2.81 (dd, *J* = 16, 8.4 Hz, 1H, C*H*$_2$CO$_2$), 2.70 (dd, *J* = 16, 8.4 Hz, 1H, C*H*$_2$CO$_2$), 2.39 (m, 2H, CHC*H*$_2$SMe), 2.20–2.00 (m, 4H, Ph$_3$CSC*H*$_2$C*H*$_2$), 1.96 (s, 3H, CHCH$_2$S*Me*), 1.57 (s, 3H, C$_q$C*H*$_3$), 1.46 (s, 9H, CH$_2$NHCOC*Me*$_3$), 0.97 (t, *J* = 8.4 Hz, 3H, OCH$_2$C*H*$_2$SiMe$_3$), 0.03 (s, 9H, OCH$_2$CH$_2$Si*Me*$_3$). ^{13}C-NMR (150 MHz, CHCl$_3$): δ 173.6, 168.9, 168.7, 167.8, 164.5, 155.2 (d, $^1J_{C\text{-}F}$ = 256 Hz), 147.4, 129.5, 127.9, 126.6, 124.3, 109.2 (d, $^2J_{C\text{-}F}$ = 12.4 Hz), 84.3, 69.9 (d, $^3J_{C\text{-}F}$ = 29.6 Hz), 57.5, 43.3, 41.1, 38.6, 37.4, 34.2, 31.1, 29.7, 24.1, 22.8, 18.8, 16.5, 14.2. ^{19}F NMR (376 MHz, CDCl$_3$): δ -124.91 (dd, *J* = 37.6, 18.8 Hz). ESI (*m/z*): [M + Na]$^+$ 1015.4.

3.1.4. 2-(Trimethylsilyl)ethyl (S,Z)-3-((((R)-2′-(((tert-butoxycarbonyl)amino)methyl)-4-methyl-4,5-dihydro-[2,4′-bithiazole]-4-carbonyl)glycyl)oxy)-4-fluoro-7-(tritylthio)hept-4-enoate (12d)

Compound **12d** was prepared from Fmoc-Glycine in 41% yield based on **9** using the same procedure for **12c**. ^{1}H NMR (600 MHz, CDCl$_3$) δ 7.85 (s, 1H, Ar*H*), 7.53–7.09 (m, 16H, Ar*H*, N*H*), 5.68 (ddd, *J* = 19.7, 8.8, 5.2 Hz, 1H, C*H*OC=O), 5.36 (brs, 1H, N*H*), 4.88 (dt, *J* = 36.0, 8.0 Hz, 1H, C*H*=CF), 4.61 (d, *J* = 6.3 Hz, 2H, C*H*$_2$NHCOCMe$_3$), 4.25–4.08 (m, 2H, OC*H*$_2$CH$_2$SiMe$_3$), 4.12–3.94 (m, 2H, NHC*H*$_2$CO$_2$), 3.74 (d, *J* = 11.7 Hz, 1H, C$_q$C*H*$_2$S), 3.32 (d, *J* = 11.5 Hz, 1H, C$_q$C*H*$_2$S), 2.77 (dd, *J* = 16.6, 8.7 Hz, 1H, C*H*$_2$CO$_2$), 2.66 (dd, *J* = 16.5, 5.0 Hz, 1H, C*H*$_2$CO$_2$), 2.12 (dd, *J* = 25.5, 7.1 Hz, 4H, Ph$_3$CSC*H*$_2$C*H*$_2$), 1.58 (s, 3H, C$_q$C*H*$_3$), 1.46 (s, 9H, CH$_2$NHCOC*Me*$_3$), 1.04–0.88 (m, 2H, OCH$_2$C*H*$_2$SiMe$_3$), 0.02 (s, 9H, OCH$_2$CH$_2$Si*Me*$_3$). ^{13}C NMR (150 MHz, CDCl$_3$) δ 174.7, 169.7, 169.1, 168.2, 163.3, 155.6, 154.5 (d, *J* = 258.2 Hz), 148.5, 144.7, 129.5, 127.8, 126.6, 121.9, 108.9 (d, *J* = 13.7 Hz), 84.9, 80.4, 69.1 (d, *J* = 30.2 Hz), 66.6, 63.4, 42.3, 41.2, 41.0, 36.3, 31.0, 28.3, 24.7, 22.8, 17.2, -1.5. ^{19}F NMR (376 MHz, CDCl$_3$): 124.96 (m). ESI (*m/z*): [M + Na]$^+$ 955.5.

3.1.5. (1^{2}Z,2^{2}Z,2^{4}R,5R,8S)-8-((Z)-1-fluoro-4-(tritylthio)but-1-en-1-yl)-2^4-methyl-5-((methylthio)methyl)-2^4,2^5-dihydro-7-oxa-4,11-diaza-1(4,2),2(2,4)-dithiazolacyclododecaphane-3,6,10-trione (14c)

12c (0.446 g, 0.44 mmol, 1.0 equiv.) was dissolved in a solution of CF$_3$CO$_2$H (5 mL) and dry DCM (20 mL), and stirred at rt overnight. After removal of the solvent under reduced pressure, and the crude amino acid (about 0.408 g) was directly used for the next step.

The amine acid was dissolved in anhydrous DMF (40 mL) and then DIEPA (0.50 mL, 2.66 mmol, 6.0 equiv.) was added to the solution at rt in order to prepare the amine solution. Next, the amine solution was added dropwise to a stirring solution of HATU (0.337 g, 0.89 mmol, 2.0 equiv.) and HOBT (0.120 g, 0.89 mmol, 2.0 equiv.) in anhydrous DMF (80 mL) at rt over 3 h. The resulting solution continued to stir at rt overnight. The mixture was diluted with water (150 mL), and extracted with MeOBu^t (80 mL). The combined organic phase was washed with a saturated NaHCO$_3$ aqueous solution (50 mL × 3), water (50 mL) and brine (50 mL). After drying by NaSO$_4$, filtration and evaporation, the residue was purified by flash chromatography on silica gel, eluting with PE/EA (1/1) to afford **14c** (48% in 2 steps) as a white foamy solid. R_f = 0.20 (PE/EA 1/1). ^{1}H-NMR (600 MHz, CDCl$_3$): δ 7.75 (s, 1H, Ar*H*), 7.5–7.10 (m, 16H, Ar*H*, N*H*), 6.39 (m, 1H, N*H*), 5.70 (m, 1H, C*H*OC=O), 5.52 (dd, *J* = 17.4, 9.6 Hz, 1H, ArC*H*$_2$NH), 4.96 (dd, *J* = 36, 7.2 Hz, 1H, C*H*=CF), 4.68 (dd, *J* = 12, 4.2 Hz, 1H, C*H*CH$_2$SMe), 4.22 (dd, *J* = 17.4, 3.0 Hz, 1H, ArC*H*$_2$NH), 4.08 (d, *J* = 11.4 Hz, 1H, C$_q$C*H*$_2$S), 3.20 (d, *J* = 11.4 Hz, 1H, C$_q$C*H*$_2$S), 3.13 (dd, *J* = 16, 11.4 Hz, 1H, C*H*$_2$CO$_2$), 2.70 (d, *J* = 16, 1H, C*H*$_2$CO$_2$), 2.30–1.90 (m, 6H, Ph$_3$CSC*H*$_2$C*H*$_2$, CHC*H*$_2$SMe), 1.96 (s, 3H, CHCH$_2$S*Me*), 1.70 (s, 3H, C$_q$C*H*$_3$). ^{13}C-NMR (150 MHz, CHCl$_3$): δ 172.8, 168.3, 168.2, 167.2, 163.6, 154.1 (d, $^1J_{C\text{-}F}$ = 256 Hz), 146.8, 144.1, 128.9, 127.2, 126.0, 123.8, 108.8 (d, $^2J_{C\text{-}F}$ = 12 Hz), 83.6, 69.4 (d, $^3J_{C\text{-}F}$ = 30 Hz), 66.0, 51.8, 42.2, 40.6, 36.5, 30.9, 30.4, 27.8, 23.6, 22.2, 14.2. ^{19}F NMR (376 MHz, CDCl$_3$): δ −125.10 (dd, *J* = 37.6, 18.8 Hz). HRMS-ESI (*m/z*): [M + H]$^+$ calcd. for C$_{39}$H$_{39}$FN$_4$O$_4$S$_4$H: 775.1911, found: 775.1918.

3.1.6. (1^{2}Z,2^{2}Z,2^{4}R,8S)-8-((Z)-1-fluoro-4-(tritylthio)but-1-en-1-yl)-2^4-methyl-2^4,2^5-dihydro-7-oxa-4,11-diaza-1(4,2),2(2,4)-dithiazolacyclododecaphane-3,6,10-trione (14d)

Compound **12d** was prepared in 40% yield using the same procedure for **12c**. ^{1}H NMR (600 MHz, CDCl$_3$) δ 7.72 (s, 1H, Ar*H*), 7.44–7.17 (m, 15H, Ar*H*), 7.03 (brs, 1H, N*H*), 6.55–6.46 (m, 1H, N*H*), 5.76 (dd, *J* = 22.6, 11.1 Hz, 1H, C*H*OC=O), 5.19 (dd, *J* = 17.4, 8.9 Hz, 1H, ArC*H*$_2$NH), 4.94 (dt, *J* = 36.0, 8.0 Hz, 1H, C*H*=CF), 4.24 (dd, *J* = 17.4, 4.0 Hz, 1H, ArC*H*$_2$NH), 4.13 (m, 1H, C*H*$_2$CO$_2$), 4.12 (d, *J* = 12.0 Hz, 1H, C$_q$C*H*$_2$S), 3.88 (dd, *J* = 18.7, 4.0 Hz, 1H, C*H*$_2$CO$_2$), 3.22 (d, *J* = 12.0 Hz, 1H, C*H*$_2$CO$_2$), 3.13 (dd, *J* = 16.9, 11.3 Hz, 1H, C*H*$_2$CONH), 2.70 (d, *J* = 16.8 Hz, 1H, C*H*$_2$CONH), 2.16–2.14 (m, 4H, Ph$_3$CSC*H*$_2$C*H*$_2$), 1.82 (s, 3H, C$_q$C*H*$_3$). ^{13}C NMR (150 MHz, CDCl$_3$) δ 173.8, 169.0, 167.5, 165.9, 163.7, 154.7 (d, *J* = 258.2 Hz), 147.4, 144.7, 129.5, 127.8, 126.6, 124.6, 109.6 (t, *J* = 12.1 Hz), 84.3, 70.0 (d, *J* = 28.7 Hz), 66.6, 43.6, 42.0, 41.4, 37.1, 31.0, 25.2, 22.8. ^{19}F NMR (376 MHz, CDCl$_3$): δ −129.82 (dd, *J* = 37.6, 22.5 Hz). HRMS-ESI (*m/z*): [M + Na]$^+$ calcd. for C$_{37}$H$_{35}$FN$_4$O$_4$S$_3$Na: 737.1697, found: 737.1675.

3.1.7. (1^{2}Z,2^{2}Z,2^{4}R,8S)-8-((Z)-1-fluoro-4-mercaptobut-1-en-1-yl)-2^4-methyl-5-((methylthio)methyl)-2^4,2^5-dihydro-7-oxa-4,11-diaza-1(4,2),2(2,4)- dithiazolacyclododecaphane-3,6,10-trione (15c)

14c (0.050 mg, 0.63 mmol, 1.0 equiv.) was dissolved in dry DCM (10 mL) and cooled to 0 °C. The mixture was successively treated with *i*-Pr$_3$SiH (26 μL, 0.13 mmol, 2.0 equiv.) and TFA (0.30 mL, 4.0 mmol, 6.7 equiv.). The reaction mixture was allowed to warm to rt and stirred for 1.5 h. The reaction was quenched with a saturated NaHCO$_3$ solution (10 mL) and separated. The aqueous phase was extracted with DCM (10 mL × 3), and the combined layers was washed with brine, dried over Na$_2$SO$_4$ and filtered. After removal of the solvent, the residue was purified by flash chromatography on silica gel, eluting with DCM/MeOH (60/1) to afford **15c** in 91% yield as a white foamy solid. R_f = 0.30 (PE/EA 2/3). ^{1}H-NMR (400 MHz, CDCl$_3$): δ ^{1}H-NMR (600 MHz, CDCl$_3$): δ 7.78 (s, 1H, Ar*H*), 7.26 (d, *J* = 7.6 Hz, 1H, N*H*), 6.44 (dd, *J* = 10, 3.6 Hz, 1H, N*H*), 5.77 (ddd, *J* = 21.2, 11.2, 2.0 Hz, 1H, C*H*OC=O), 5.28 (dd, *J* = 17.6, 9.6 Hz, 1H, ArC*H*$_2$NH), 5.11 (dt, *J* = 36, 7.6 Hz, 1H, C*H*=CF), 4.70 (m, 1H, C*H*CH$_2$SMe), 4.27 (dd, *J* = 17.6, 3.6 Hz, 1H, ArC*H*$_2$NH), 4.08 (d, *J* = 11.2 Hz, 1H, C$_q$C*H*$_2$S), 3.25 (d, *J* = 11.2 Hz, 1H, C$_q$C*H*$_2$S), 3.19 (dd, *J* = 16.8, 11.2 Hz, 1H, C*H*$_2$CONH), 2.76 (dd, *J* = 16.8, 2.4 Hz, 1H, C*H*$_2$CONH), 2.41 (m, 2H, C*H*$_2$CH=CF), 2.30–1.90 (m, 4H, C*H*$_2$SH, CHC*H*$_2$SMe), 1.87 (s, 3H, CHCH$_2$S*Me*), 1.85 (s, 3H, C$_q$C*H*$_3$). ^{13}C-NMR (150 MHz, CHCl$_3$): δ 174.1, 169.5, 169.4, 164.8, 155.9 (d, $^1J_{C\text{-}F}$ = 256 Hz), 147.9, 125.0, 109.3 (d, $^2J_{C\text{-}F}$ = 12 Hz), 84.8, 70.9 (d, $^3J_{C\text{-}F}$ = 30 Hz), 53.0, 43.4, 41.8, 37.8, 28.4, 24.8, 24.3, 15.6. ^{19}F NMR (376 MHz, CDCl$_3$): δ −124.79 (dd, *J* = 37.6, 18 Hz). HRMS-ESI (*m/z*): [M + H]$^+$ calcd. for C$_{20}$H$_{25}$FN$_4$O$_4$S$_4$H: 533.0815, found: 533.0819.

3.1.8. S-((Z)-4-fluoro-4-((1^{2}Z,2^{2}Z,2^{4}R,5R,8S)-2^4-methyl-5-((methylthio)methyl)-3,6,10-trioxo-2^4,2^5-dihydro-7-oxa-4,11-diaza-1(4,2),2(2,4)-dithiazolacyclododecaphane-8-yl)but-3-en-1-yl) octanethioate (16c)

The free thiol **15c** (32 mg, 0.05mmol, equiv.) was dissolved in dry DCM (10 mL) and cooled to 0 °C. The mixture was successively treated with pyridine (23 μL, 0.29 mmol, 5.0 equiv.) and octanoyl chloride (29 μL, 0.17 mmol, 3.0 equiv.). The reaction was allowed to warm to rt and stirred for 2 h, and then quenched with a saturated $NaHCO_3$ solution (10 mL) and separated. The aqueous phase was extracted with DCM (10 mL × 3), and the combined layers was washed with brine, dried over Na_2SO_4 and filtered. After removal of the solvent, the residue was purified by flash chromatography on silica gel, eluting with DCM/EA (1/1) to afford **16c** (57%) as a white foamy solid. R_f = 0.42 (DCM/EA 3/1). $[\alpha]_{20}^{D}$ 26.4 (c 0.34, $CHCl_3$). ^{1}H-NMR (600 MHz, $CDCl_3$): δ 7.78 (s, 1H, Ar*H*), 7.24 (d, *J* = 11.4 Hz, 1H, N*H*), 6.54 (dd, *J* = 9.6, 3.6 Hz, 1H, N*H*), 5.76 (ddd, *J* = 21, 11.4, 2.4 Hz, 1H, C*H*OC=O), 5.30 (dd, *J* = 17.4, 9.6 Hz, 1H, ArC*H*₂NH), 5.09 (dt, *J* = 36, 7.2 Hz, 1H, C*H*=CF), 4.71 (m, 1H, C*H*CH₂SMe), 4.27 (dd, *J* = 17.5, 3.0 Hz, 1H, ArC*H*₂NH), 4.10 (d, *J* = 11.4 Hz, 1H, C$_q$C*H*₂S), 3.28 (d, *J* = 11.4 Hz, 1H, C$_q$C*H*₂S), 3.19 (dd, *J* = 16.8, 11.4 Hz, 1H, C*H*₂CONH), 2.90 (t, *J* = 7.2 Hz, 2H, SC*H*₂CH₂), 2.75 (dd, *J* = 16.8, 2.4 Hz, 1H, C*H*₂CONH), 2.54 (t, *J* = 7.5 Hz, 2H, C*H*₂COS), 2.40–2.30 (m, 2H, SC*H*₂CH₂), 2.30–1.90 (m, 2H, CHC*H*₂SMe), 1.87 (s, 3H, CHCH₂S*Me*), 1.86 (s, 3H, C$_q$C*H*₃), 1.70–1.6 (m, 2H, C*H*₂), 1.40–1.10 (m, 8H, C*H*₂), 0.89 (t, *J* = 6.0 Hz, 3H, C*H*₃CH₂). ^{13}C-NMR (150 MHz, $CDCl_3$): δ 198.6, 172.9, 168.3, 168.2, 167.3, 163.8, 154.7 (d, *J* = 256 Hz), 146.7, 123.9, 108.1 (d, *J* = 12 Hz), 83.6, 69. 5 (d, *J* = 30 Hz), 51.8, 43.5, 42.2, 40.6, 36.5, 30.9, 28.4, 27.2, 24.9, 24.1, 23. 6, 23.2, 21.9, 14.4, 13.4. ^{19}F NMR (376 MHz, $CDCl_3$): δ −124.69 (dd, *J* = 35.9, 20.9 Hz). HRMS-ESI (*m/z*): [M + Na]⁺ calcd. for $C_{28}H_{39}FN_4O_5S_4Na$: 681.1680, found: 681.1693.

3.1.9. S-((Z)-4-fluoro-4-((1²Z,2²Z,2⁴R,8S)-2⁴-methyl-3,6,10-trioxo-2⁴,2⁵-dihydro-7-oxa-4,11-diaza-1(4,2),2(2,4)-dithiazolacyclododecaphane-8-yl)but-3-en-1-yl) octanethioate (**16d**)

Compound **16d** was prepared in 40% yield from **14d** using the same procedure for **16c**. $[\alpha]_{20}^{D}$ 37.9 (c 0.14, $CHCl_3$). ^{1}H NMR (600 MHz, $CDCl_3$) δ 7.71 (s, 1H, Ar*H*), 7.00 (m, 1H, N*H*), 6.60 (dd, *J* = 9.1, 4.1 Hz, 1H, N*H*), 5.78 (dd, *J* = 22.3, 11.1 Hz, 1H, C*H*OC=O), 5.18 (dd, *J* = 17.4, 8.8 Hz, 1H, ArC*H*₂NH), 5.04 (dt, *J* = 36.1, 7.5 Hz, 1H, C*H*=CF), 4.24 (dd, *J* = 17.5, 4.0 Hz, 1H, ArC*H*₂NH), 4.15 (m, 1H, C*H*₂CO₂), 4.10 (d, *J* = 11.7 Hz, 1H, C$_q$C*H*₂S), 3.87 (dd, *J* = 18.9, 3.1 Hz, 1H), 3.20 (d, *J* = 11.2 Hz, 1H, C$_q$C*H*₂S), 3.14 (dd, *J* = 16.0, 12.0 Hz, 1H, C*H*₂CONH), 2.86 (t, *J* = 7.1 Hz, 2H, SC*H*₂CH₂), 2.71 (d, *J* = 16.9 Hz, 1H, C*H*₂CO), 2.50 (t, *J* = 7.6 Hz, 2H, C₆H₁₃C*H*₂CO), 2.41–2.27 (m, 2H, SC*H*₂CH₂), 2.11–1.87 (m, 2H), 1.79 (s, 3H, C$_q$C*H*₃), 1.60–1.15 (m, 10H), 0.84 (t, *J* = 6.6 Hz, 3H, C*H*₃-C₅H₁₁-CH₂CO). ^{19}F NMR (376 MHz, $CDCl_3$): δ −123.29 (dd, *J* = 35.9, 20.9 Hz). HRMS-ESI (*m/z*): [M + Na]⁺ calcd. for $C_{26}H_{35}FN_4O_5S_3Na$: 621.1646, found: 621.1650.

3.2. Biology

3.2.1. Recombinant Human HDAC1, HDAC2, HDAC3, HDAC6 and HDAC8 Enzymatic Assays

The assays were carried out by Shanghai ChemPartner Co., Ltd (Shanghai, China), according to our previous method [23,24]. Briefly, different concentrations of compounds were incubated with recombinant human HDAC1, HDAC2, HDAC3, HDAC6 and HDAC8 (BPS Biosciences, San Diego, CA, USA) at room temperature for 15 min, which was followed by adding Ac-peptide-AMC substrates to initiate the reaction in Tris-based assay buffer. Reaction mixtures were incubated at room temperature for 60 min in HDAC1, HDAC2, HDAC3 and HDAC6 assays, and were incubated for 240 min in HDAC8 assay. Then, the stop solution containing trypsin was added. The coupled reaction was incubated for another 90 min at 37 °C. Fluorescent AMC released from substrate was measured in SynergyMx (BioTek, Winooski, VT, US) using filter sets as excitation = 355 nm and emission = 460 nm. IC_{50} values were calculated by GraphPad Prism software (7.0 version., GraphPad Software, San Diego, CA, USA).

3.2.2. Cellular Assay

Eca-109, Bel 7402 and U937 cell lines were purchased from ATCC (Manassas, VA, USA), and the other tested cell lines were obtained from the Shanghai Institute for Biological Sciences, Chinese Academy of Sciences (Shanghai, China). MDA-MB-231, Hela, Eca-109 and Bel 7402 cells were cultured in DMEM medium supplemented with 10% heat-inactivated fetal bovine serum, 100 U/mL penicillin and 100 µg/mL streptomycin (Beyotime Biotech, Shanghai, China). A549 cells were cultured in F-12K medium supplemented with 10% heat-inactivated fetal bovine serum, 100 U/mL penicillin and 100 µg/mL streptomycin. HCT116 and SK-OV-3 cells were cultured in McCoy's 5A medium supplemented with 10% heat-inactivated fetal bovine serum, 100 U/mL penicillin and 100 µg/mL streptomycin. U937 cells were cultured in RPMI-1640 medium supplemented with 10% heat inactivated fetal bovine serum, 100 U/mL penicillin and 100 µg/mL streptomycin. All cells were cultured at 37 °C in humidified air containing 5% CO_2. According to our previous method [23,24], A549, HCT116, MDA-MB-231, SK-OV-3, Hela, Eca-109, Bel 7403 and U937 (600,000 cells/well) were seeded in 96-well plates and incubated for 24 h before being treated with various concentration of compounds or solvent control. Cells were further incubated for 72 h and then treated with MTT and incubated for another 3 h. The media was then removed and 100 µL DMSO was added to each well. The absorbance at 550 nm was measured by a SpectraMAX340 microplate reader (Molecular Devices, San Jose, CA, USA) with a reference wavelength at 690 nm. Adriamycin was used as a positive control in the assay.

3.3. Molecular Modeling Study

The known crystal complexes of HDAC1, HDAC6, HDAC8 and their ligand (PDB code: 5ICN, 5EDU, 4RN0) were obtained from PDB (http://www.rcsb.org). Molecular docking simulations in the HDACs were run using the MOE 2019 (Molecular Operating Environment, Chemical Computing Group, Montreal, Quebec, Canada) due to its universality and very fast speed. Ligands were prepared with the ChemBio3D Ultra 14.0 (PerkinElmer, Waltham, MA, USA), followed by MM2 energy minimization. Protein structures were also prepared with the MOE, which could automatically add hydrogen atoms to proteins by explicitly considering the protonation state of histidine and optimize the force field. All crystal water, small ligands and cofactors except HEM were removed. After this step, the binding sites were deduced from the known crystal complexes and the ligands were docked to the prepared proteins through flexible docking mode. Top scoring function poses were selected as representative of the simulations and were displayed with Open-Source PyMOLTM 1.8X software (Schrödinger, Ltd, New York, NY, USA).

4. Conclusions

Given our previous finding that fluorination at the C18 position of largazole showed good tolerance towards the inhibitory activity and selectivity of HDACs, the current study investigated further modifications on the valine residue in its macrocyclic moiety with S-Me Cysteine or Glycine residue. While the Glycine-modified fluoro analog showed poor activity, the S-Me L-Cysteine-modified analog emerged to be a very potent HDAC inhibitor. Unlike all previously reported C2-modified compounds in the largazole family (including our recent fluoro-largazole analogs) where replacement of the Val residue has failed to provide any potency improvement, the S-Me L-Cysteine-modified analog displayed significantly enhanced (five–nine-fold) inhibition of all the tested HDACs while maintaining the selectivity of HDAC1 over HDAC6, as compared to largazole thiol. Molecular modeling study provided rational explanation and structural evidence for the enhanced activity. This new finding will aid the design of novel potent HDAC inhibitors. An expanded research program is currently under way to investigate largazole analogs bearing similar structural characteristics.

Supplementary Materials: The supporting information is available online at http://www.mdpi.com/1660-3397/18/7/344/s1. Figures S1–S3: ^{1}H, ^{13}C and ^{19}F NMR spectra of compound 8, Figure S4: ^{1}H NMR spectra of compound 11, Figures S5–S7: ^{1}H, ^{13}C and ^{19}F NMR spectra of compound 9, Figures S8 and S9: ^{1}H and ^{13}C NMR spectra of compound 10c, Figures S10–S12: ^{1}H, ^{13}C and ^{19}F NMR spectra of compound 12c; Figures S13–S15: ^{1}H, ^{13}C and ^{19}F NMR spectra of compound 12d, Figures S16–S18: ^{1}H, ^{13}C and ^{19}F NMR spectra of compound 14c, Figures S19–S21: ^{1}H, ^{13}C and ^{19}F NMR spectra of compound 14d, Figures S22–S25: ^{1}H, COSY, ^{13}C and ^{19}F NMR spectra of compound 15c, Figures S26–S28: ^{1}H, ^{13}C and ^{19}F NMR spectra of compound 16c, Figures S29–S31: ^{1}H, ^{13}C and ^{19}F NMR spectra of compound 16d, Figures S32–S39: The concentration-response curves of Largazole and 16c in the cellular assays, Figures S40–S42: The concentration-response curves of Largazole thiol in HDACs assays, Figures S43–S45: The concentration-response curves of 15c in HDACs assays.

Author Contributions: B.Z. and Z.-W.R. performed the chemical synthesis and analyzed the experimental data. D.L. performed the molecular modeling research. Y.Z. and T.D. participated in the analysis of the experimental data. X.L. was responsible for the funding of project and the design of the research. Q.S. was responsible for the writing of the manuscript. All authors have read and agreed to the published version of the manuscript.

Funding: This research was funded by the National Key Research and Development Program of China (No. 2018YFC0310900) and the National Natural Science Foundation of China (No. 21472024 and No. 21242008).

Acknowledgments: The authors acknowledge the National Key Research and Development Program of China and the National Natural Science Foundation of China for their financial support. The authors also acknowledge Prof. Tan Wen-Fu and Prof. Bing Cai for the cellular assays and Shanghai ChemPartner Co., Ltd. for the HDAC inhibition assays.

Conflicts of Interest: The authors declare no conflict of interest.

References

1. Delcuve, G.P.; Khan, D.H.; Davie, J.R. Roles of histone deacetylases in epigenetic regulation: Emerging paradigms from studies with inhibitors. *Clin. Epigenetics* **2012**, *4*, 5. [CrossRef]
2. De Ruijter, A.J.M.; Van Gennip, A.H.; Caron, H.N.; Kemp, S.; Van Kuilenburg, A.B.P. Histone deacetylases (HDACs): Characterization of the classical HDAC family. *Biochem. J.* **2003**, *370*, 737–749. [CrossRef]
3. Johnstone, R.W. Histone-deacetylase inhibitors: Novel drugs for the treatment of cancer. *Nat. Rev. Drug Discovery* **2002**, *1*, 287–299. [CrossRef]
4. Zagni, C.; Floresta, G.; Monciino, G.; Rescifina, A. The search for potent, small-molecule HDACIs in cancer treatment: A decade after vorinostat. *Med. Res. Rev.* **2017**, *37*, 1373–1428.
5. Roche, J.; Bertrand, P. Inside HDACs with more selective HDAC inhibitors. *Eur. J. Med. Chem.* **2016**, *121*, 451–483. [CrossRef]
6. Qin, H.-T.; Li, H.-Q.; Liu, F. Selective histone deacetylase small molecule inhibitors: Recent progress and perspectives. *Expert Opin. Therap. Pat.* **2017**, *27*, 621–636. [CrossRef]
7. Wang, X.X.; Wan, R.Z.; Liu, Z.P. Recent advances in the discovery of potent and selective HDAC6 inhibitors. *Eur. J. Med. Chem.* **2018**, *143*, 1406–1418. [CrossRef] [PubMed]
8. Taori, K.; Paul, V.J.; Luesch, H. Structure and activity of largazole, a potent antiproliferative agent from the Floridian marine cyanobacterium *Symploca* sp. *J. Am. Chem. Soc.* **2008**, *130*, 1806–1807. [CrossRef] [PubMed]
9. Ying, Y.; Taori, K.; Kim, H.; Hong, J.; Hendrik Luesch, H. Total synthesis and molecular target of largazole, a histone deacetylase inhibitor. *J. Am. Chem. Soc.* **2008**, *130*, 8455–8459. [CrossRef] [PubMed]
10. Bowers, A.; West, N.; Taunton, J.; Schreiber, S.L.; Bradner, J.E.; Williams, R.M. Total synthesis and biological mode of action of largazole: A potent class I histone deacetylase inhibitorc. *J. Am. Chem. Soc.* **2008**, *130*, 11219–11222. [CrossRef]
11. Maolanon, A.R.; Kristensen, H.M.E.; Leman, L.J.; Ghadiri, M.R.; Olsen, C.A. Natural and synthetic macrocyclic inhibitors of the histone deacetylase enzymes. *ChemBioChem* **2017**, *18*, 5–49. [CrossRef]
12. Reddy, D.N.; Ballante, F.; Chuang, T.; Pirolli, A.; Marrocco, B.; Marshall, G.R. Design and synthesis of simplified largazole analogues as isoform-selective human lysine deacetylase inhibitors. *J. Med. Chem.* **2016**, *59*, 1613–1633. [CrossRef] [PubMed]
13. Yao, Y.; Tu, Z.; Liao, C.; Wang, Z.; Li, S.; Yao, H.; Li, Z.; Jiang, S. Discovery of novel class I histone deacetylase inhibitors with promising in vitro and in vivo antitumor activities. *J. Med. Chem.* **2015**, *58*, 7672–7680. [CrossRef]
14. Poli, G.; Fabio, R.D.; Ferrante, L.; Summa, V.; Maurizio Botta, M. Largazole Analogues as Histone Deacetylase Inhibitors and Anticancer Agents: An Overview of Structure–Activity Relationships. *ChemMedChem* **2017**, *12*, 1917–1926. [CrossRef]
15. Kim, B.; Ratnayake, R.; Lee, H.; Shi, G.; Zeller, S.L.; Li, C.; Luesch, H.; Hong, J. Synthesis and biological evaluation of largazole zinc-binding group analogs. *Bioorg. Med. Chem.* **2017**, *25*, 3077–3086. [CrossRef] [PubMed]
16. Ying, Y.; Liu, Y.X.; Byeon, S.R.; Kim, H.; Luesch, H.; Hong, J. Synthesis and activity of largazole analogues with linker and macrocycle modification. *Org. Lett.* **2008**, *10*, 4021–4024. [CrossRef] [PubMed]
17. Liu, Y.X.; Salvador, L.A.; Byeon, S.; Ying, Y.; Kwan, J.C.; Law, B.K.; Hong, J.; Luesch, H. Anticolon cancer activity of largazole, a marine-derived tunable histone deacetylase inhibitor. *J. Pharmacol. Exp. Ther.* **2010**, *335*, 351–361. [CrossRef] [PubMed]

18. Zeng, X.; Yin, B.L.; Hu, Z.; Liao, C.Z.; Liu, J.L.; Li, S.; Li, Z.; Nicklaus, M.C.; Zhou, G.B.; Jiang, S. Total synthesis and biological evaluation of largazole and derivatives with promising selectivity for cancers cells. *Org. Lett.* **2010**, *12*, 1368–1371. [CrossRef] [PubMed]

19. Benelkebir, H.; Marie, S.; Hayden, A.L.; Lyle, J.; Loadman, P.M.; Crabb, S.J.; Packham, G.; Ganesan, A. Total synthesis of largazole and analogues: HDAC inhibition, antiproliferative activity and metabolic stability. *Bioorg. Med. Chem.* **2011**, *19*, 3650–3658. [CrossRef]

20. Bowers, A.A.; West, N.; Newkirk, T.L.; Troutman-Youngman, A.E.; Schreiber, S.L.; Wiest, O.; Bradner, J.E.; Williams, R.M. Synthesis and histone deacetylase inhibitory activity of largazole analogs: Alteration of the zinc-binding domain and macrocyclic scaffold. *Org. Lett.* **2009**, *11*, 1301–1304. [CrossRef]

21. Kim, B.; Park, H.; Salvador, L.A.; Serrano, P.E.; Kwan, J.C.; Zeller, S.L.; Chen, Q.Y.; Ryu, S.; Liu, Y.; Byeon, S.; et al. Evaluation of class I HDAC isoform selectivity of largazole analogues. *Bioorg. Med. Chem. Lett.* **2014**, *24*, 3728–3731. [CrossRef] [PubMed]

22. Yu, X.; Zhang, B.; Shan, G.; Wu, Y.; Yang, F.-L.; Lei, X. Synthesis of the molecular hybrid inspired by Largazole and Psammaplin A. *Tetrahedron* **2018**, *74*, 549–555. [CrossRef]

23. Zhang, B.; Shan, G.; Zheng, Y.; Yu, X.; Ruan, Z.-W.; Li, Y.; Lei, X. Synthesis and preliminary biological evaluation of two fluoroolefin analogs of Largazole inspired by the structural similarity of the side chain unit in Psammaplin A. *Mar. Drugs* **2019**, *17*, 333. [CrossRef]

24. Zhang, B.; Liu, J.; Gao, D.; Yu, X.; Wang, J.; Lei, X. A fluorine scan on the Zn2+-binding thiolate side chain of HDAC inhibitor largazole: Synthesis, biological evaluation, and molecular modeling. *Eur. J. Med. Chem.* **2019**, *182*. [CrossRef]

25. Cole, K.E.; Dowling, D.P.; Boone, M.A.; Phillips, A.J.; Christianson, D.W. Structural basis of the antiproliferative activity of largazole, a depsipeptide inhibitor of the histone deacetylases. *J. Am. Chem. Soc.* **2011**, *133*, 12474–12477. [CrossRef] [PubMed]

 *marine drugs* 

Review

(Semi)-Synthetic Fucosylated Chondroitin Sulfate Oligo- and Polysaccharides

Giulia Vessella [1], Serena Traboni [1], Antonio Laezza [2], Alfonso Iadonisi [1] and Emiliano Bedini [1,*]

[1] Department of Chemical Sciences, University of Naples Federico II, Complesso Universitario Monte
     S. Angelo, via Cintia 4, I-80126 Napoli, Italy; giulia.vessella@unina.it (G.V.); serena.traboni@unina.it (S.T.);
     iadonisi@unina.it (A.I.)
[2] Department of Sciences, University of Basilicata, viale dell'Ateneo Lucano 10, I-85100 Potenza, Italy;
     antonio.laezza@unibas.it
* Correspondence: ebedini@unina.it; Tel.: +39-081-674153

Received: 12 May 2020; Accepted: 28 May 2020; Published: 1 June 2020

Abstract: Fucosylated chondroitin sulfate (fCS) is a glycosaminoglycan (GAG) polysaccharide with a unique structure, displaying a backbone composed of alternating *N*-acetyl-D-galactosamine (GalNAc) and D-glucuronic acid (GlcA) units on which L-fucose (Fuc) branches are installed. fCS shows several potential biomedical applications, with the anticoagulant activity standing as the most promising and widely investigated one. Natural fCS polysaccharides extracted from marine organisms (*Echinoidea*, *Holothuroidea*) present some advantages over a largely employed antithrombotic drug such as heparin, but some adverse effects as well as a frequently found structural heterogeneity hamper its development as a new drug. To circumvent these drawbacks, several efforts have been made in the last decade to obtain synthetic and semi-synthetic fCS oligosaccharides and low molecular weight polysaccharides. In this Review we have for the first time collected these reports together, dividing them in two topics: (i) total syntheses of fCS oligosaccharides and (ii) semi-synthetic approaches to fCS oligosaccharides and low molecular weight polysaccharides as well as glycoclusters displaying multiple copies of fCS species.

Keywords: carbohydrates; polysaccharides; semi-synthesis; sulfation; glycosylation; fucose; fucosylated chondroitin sulfate

---

1. Introduction

Glycosaminoglycans (GAGs) are highly negatively charged polysaccharides ubiquitously distributed in the animal kingdom. They are usually found covalently linked to proteins to form proteoglycans (PGs), which are one of the major and most important components of the extracellular matrix. GAGs are involved in a myriad of biological events in both physiological and pathological processes [1]. From a structural point of view GAG polysaccharides are constituted of a linear sequence of disaccharide units, each consisting of an aminosugar and a hexose or an uronic acid, very often decorated with one or more sulfate groups on their structure.

Some marine invertebrates display GAGs with unique, non-linear structures, characterized by the presence of monosaccharide or short oligosaccharides as branches [2,3]. Among these branched GAGs, most attention is currently focused on fCS, a glycosaminoglycan found up to now exclusively in the body wall of sea cucumbers (*Echinoidea*, *Holothuroidea*) and showing several potential biomedical applications related to inflammation, hyperglycemia, atherosclerosis, cellular growth, cancer metastasis, and angiogenesis [4]. However, its most interesting feature is the potential use as an antithrombotic agent alternative to heparin [5], compared to which fCS displays two very important advantages:

(i) unlike heparin, the activity is retained also on antithrombin (AT)- and heparin cofactor II (HC-II)-free plasmas, because the mechanism of action of fCS on the blood coagulation cascade has some differences with respect to heparin [6,7]; (ii) it is orally deliverable because it can be digested neither during its adsorption in the gastrointestinal tract nor by intestinal bacterial enzymes [8].

The unique structure of fCS consists of a linear backbone of GalNAc and GlcA units linked together through alternating β-1→3 and β-1→4 glycosidic bonds, on which a single Fuc unit or, more rarely, di- to nonasaccharide Fuc chains are inserted as branches. Sulfate groups are also present to a various extent both on the backbone and on Fuc branches [9]. The latter, which are essential for the biological activities of fCS [10–12], are usually linked at position *O*-3 of GlcA residues, even if different fucosylation sites have been sometimes found too [13–16]. The structural diversity of fCS from different sea cucumbers species is mainly investigated by two-dimensional nuclear magnetic resonance (2D–NMR) techniques [17] and achievements in this field have been reviewed [9,18,19]. Since the structural elucidation of fCS polysaccharides from more and more species is constantly published, upgraded summarizing tables can be found in very recent papers [20,21]. A schematic representation of fCS structural variability found up to now is depicted in Figure 1.

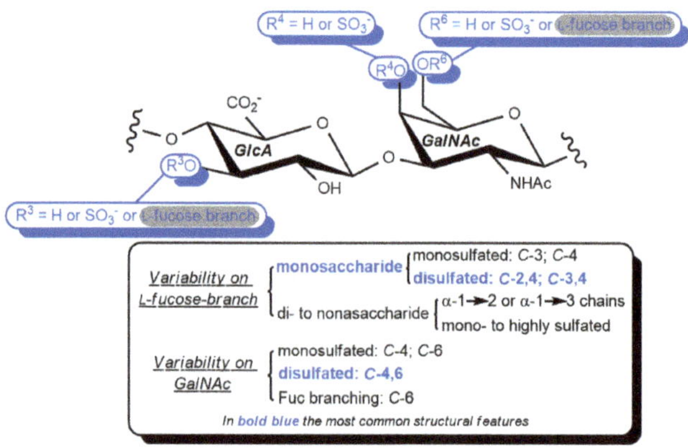

Figure 1. Structural variability of fCS from sea cucumbers.

Although fCSs have shown very interesting anticoagulant and antithrombotic activities, the intravenous administration of the native polysaccharides can cause some adverse effects such as platelet aggregation, hypotension, and bleeding [22]. These drawbacks are typically associated with high molecular weight (M_w) sulfated species. Indeed, M_w values of fCS polysaccharides typically span from 20 to 100 kDa [23]. To develop fCS-based anticoagulant and antithrombotic lead compounds while avoiding these side effects, in the last decade several research groups have focused their attention on the production of fCS oligosaccharides and low molecular weight polysaccharides through total synthetic or semi-synthetic approaches. This has also helped detailed structure-activity relationship studies that are instead hampered by the usually heterogeneous structure of native fCS polysaccharides. Indeed, only in a few cases a homogeneous structure with a single sulfation pattern on Fuc and GalNAc units has been found [21,24–26]. Even if the synthesis of GAG oligo- and polysaccharides has been recently reviewed [27–29], to the best of our knowledge no comprehensive account selectively focusing on the synthetic and semi-synthetic efforts to fCS species that increased more and more in the last decade, has been published. Here we fill this gap, including in this review also the very recent reports on the obtainment of fCS-based multivalent structures. The content is divided into two chapters: the former reviews the total synthetic approaches for building fCS oligosaccharides starting from commercially available monosaccharides, the latter concerns the semi-synthetic strategies to obtain fCS oligosaccharides and low molecular weight polysaccharides starting from natural fCS itself or related GAGs.

2. Total Synthetic Approaches

The first total synthesis of a fCS oligosaccharide was reported by Tamura and co-workers in 2013 [30]. They synthesized the trisaccharide unit most commonly found in natural fCS—4,6-di-*O*-sulfated-β-GalNAc-(1→4)-[2,4-di-*O*-sulfated-α-Fuc-(1→3)]-GlcA—as β-*p*-methoxyphenyl glycoside (**1**, Scheme 1), starting from monosaccharide building blocks **2–4** that were carefully designed with respect to their orthogonal pattern of protecting groups.

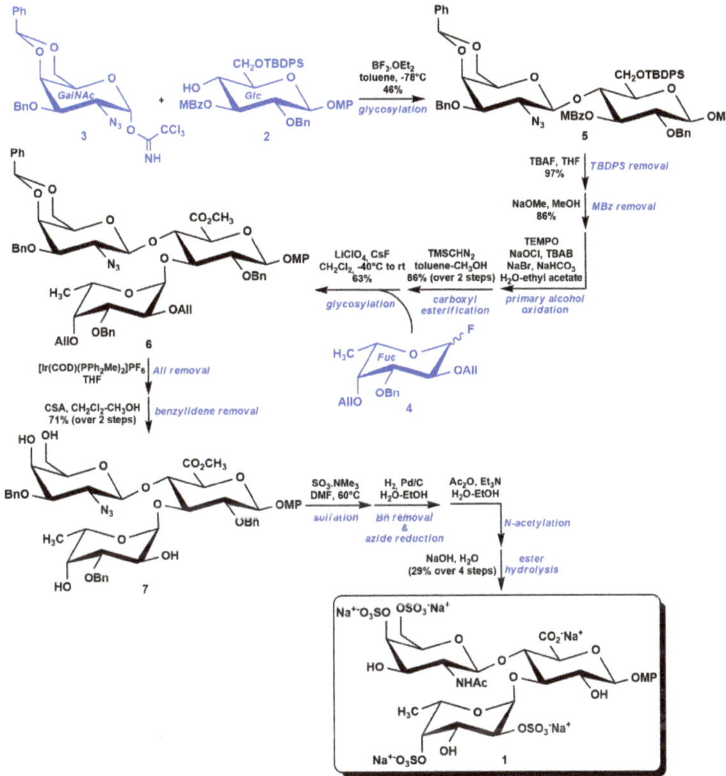

Scheme 1. First synthesis of fCS trisaccharide by Tamura and co-workers [30]; abbreviations: acetic anhydride (Ac₂O), (1S)-(+)-10-camphorsulfonic acid (CSA), 1,5-cyclooctadiene (COD), benzyl (Bn), *p*-methoxyphenyl (MP), tetra-*n*-butylammonium bromide (TBAB), tetra-*n*-butylammonium fluoride (TBAF), 2,2,6,6-tetramethyl-1-piperidinyloxy free-radical (TEMPO), tetrahydrofuran (THF), (trimethylsilyl)diazomethane (TMSCHN₂).

In particular, glucose (Glc) derivative **2** displayed a free alcohol at *C*-4 site together with a *p*-methoxybenzoyl (MBz) ester and *t*-butyldiphenylsilyl (TBDPS) ether installed at *C*-3 and *C*-6 position, respectively. These two protecting groups could be orthogonally cleaved to liberate the corresponding hydroxyls that could be then fucosylated or oxidized at the desired synthetic stage. Similarly, trichloroacetimidate donor **3** served as building block for the GalNAc unit, with an azide as acetamide masking group and a benzylidene as orthogonally cleavable protecting group for *C*-4,6 diol, which had to be sulfated at a late stage of the synthesis. Finally, glycosyl fluoride **4** was chosen as Fuc donor, displaying allyl (All) ethers as temporary protecting groups at *C*-2 and *C*-4 positions that should be sulfated as well. The assembly of the building blocks started with the glycosylation between acceptor **2** and donor **3** to give disaccharide **5**, that after TBDPS and MBz cleavage, primary alcohol oxidation, carboxyl esterification and α-fucosylation, afforded the fully protected trisaccharide **6** in 21% overall yield. Then, selective removal of allyl and benzylidene protecting groups liberated the alcohol moieties at the positions selected for sulfation that was conducted with $SO_3 \cdot NMe_3$ complex in *N,N*-dimethylformamide (DMF) before full deprotection in three steps to afford the target fCS trisaccharide **1** with a 4.3% global yield, calculated from the Glc, GalNAc and Fuc monosaccharide building blocks.

Most recently, Qin and co-workers reported the total synthesis of fCS trisaccharide **8**, displaying the same sulfation pattern as **1** but with the GalNAc unit instead of GlcA at the pseudo-reducing position [31] (Scheme 2).

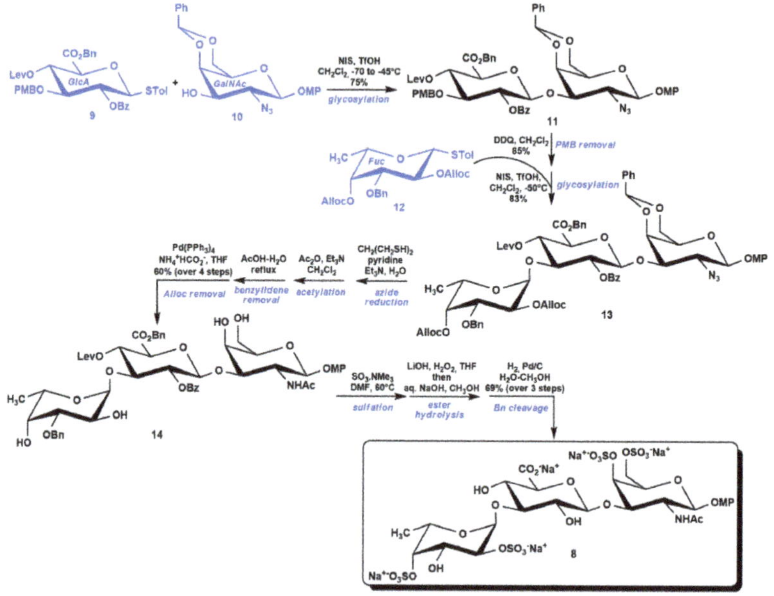

Scheme 2. Synthesis of fCS trisaccharide with GalNAc unit at pseudo-reducing position by Qi and co-workers [31]; abbreviations: benzoate (Bz), 2,3-dichloro-5,6-dicyano-*p*-benzoquinone (DDQ), levulinoyl (Lev), *N*-iodosuccinimide (NIS), trifluoromethanesulfonic acid (TfOH), toluene (Tol).

In their synthetic strategy the *C*-6 oxidation on Glc to give a GlcA unit was performed at a monosaccharide level, therefore the first glycosylation was conducted between the suitably protected GlcA thioglycoside donor **9**, carrying a *p*-methoxybenzyl (PMB) ether as orthogonally cleavable protecting group at *C*-3 site, and GalNAc acceptor **10**, again with an azide as acetamide masking group (also several *N*-protecting groups were tested but with much worse results). The obtained disaccharide **11** was subjected to a selective cleavage of the PMB ether to liberate a single hydroxyl that was in turn glycosylated in the presence of Fuc thioglycoside donor **12**, carrying an allyloxycarbonyl (Alloc) as orthogonal protecting group at *C*-2 and *C*-4 positions. Interestingly, an α-linked glycosidic bond was formed, although Alloc protecting group at *C*-2 site would direct to a 1,2-*trans* β-stereochemistry due to a neighboring participating effect in the glycosylation mechanism. The authors suggested that the obtainment of the 1,2-*cis* α-product was due to an in situ anomerization from the firstly formed β-trisaccharide to the more stable α-anomer. Fully protected trisaccharide **13** was then subjected to azide-acetamide conversion and the alcohol moieties were liberated at the positions selected for sulfation by Alloc and benzylidene cleavage. Tetraol **14** was sulfated and then globally deprotected under conditions similar to the synthesis of **1**. Target fCS trisaccharide **8** was obtained in 26% overall yield from the GlcA, GalNAc and Fuc monosaccharide building blocks.

Nifantiev and co-workers reported the synthesis and conformational analysis of a library of di- and trisaccharide fragments carrying Fuc branches differing by length, sulfation pattern and site of linkage [32–34]. These serve as simple model compounds covering the structural variability of native fCS, including the less frequent cases of Fuc units linked at GalNAc *O*-6 instead of GlcA *O*-3 position [13–16] and of Fuc oligosaccharides instead of monosaccharides as branches [21,35,36].

A divergent approach was employed to afford 12 different oligosaccharides from only four monosaccharide building blocks (**15–17** and **30**, Schemes 3 and 4) that were designed with a suitable pattern of permanent and orthogonally cleavable protecting groups. In particular, GlcA acceptor **15** and Fuc trichloroacetimidate donor **16** and acceptor **17** were used to prepare five disaccharides and three trisaccharides (**19–23** and **27–29**, respectively, Scheme 3), all displaying Fuc branches at GlcA *O*-3 site and differentiated for the length of the branch and/or the sulfation pattern [32]. The five disaccharides were all obtained from the completely protected precursor **18** that was prepared in turn by a glycosylation reaction between **15** and **16**. The very high yield in the α-anomer was ascribed to a remote participation of the chloroacetyl (ClAc) groups at Fuc donor *O*-3 and *O*-4 positions, with the formation of a stabilized glycosyl cation during the glycosylation mechanism. Then, different sequences of few synthetic steps on **18**, including orthogonal protecting group – ClAc or Bn – cleavage, regioselective installation of Bz ester protecting group, sulfation and global deprotection, allowed the obtainment of differently sulfated disaccharides **19–23**.

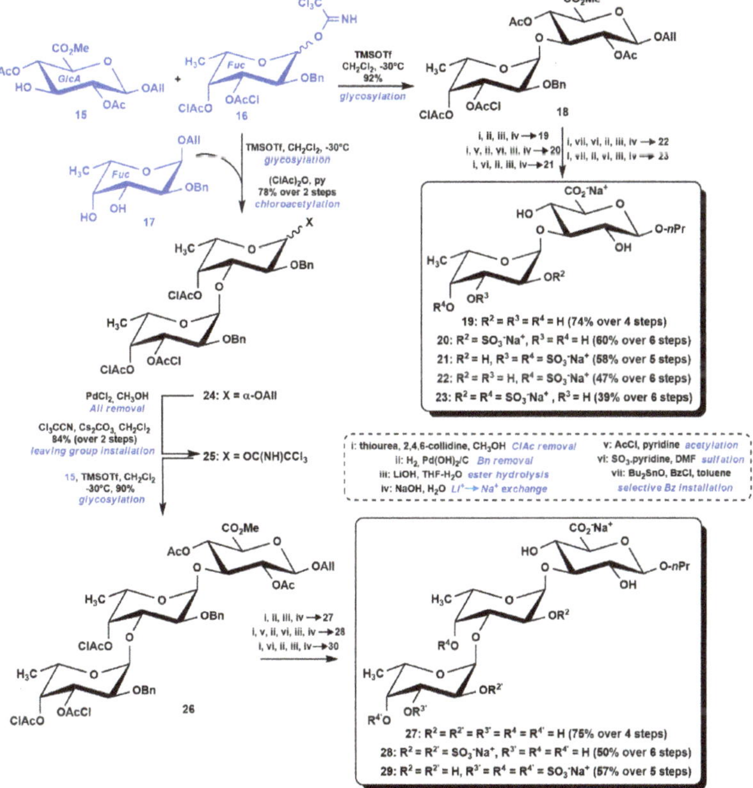

Scheme 3. Nifantiev synthesis of di- and trisaccharides with Fuc branches at GlcA *O*-3 site [32]; abbreviations: acetyl chloride (AcCl), benzoyl chloride (BzCl), chloroacetic anhydride ((ClAc)₂O), *n*-propyl (*n*Pr), trimethylsilyl trifluoromethanesulfonate (TMSOTf).

The access to trisaccharides **27–29** required firstly the synthesis of Fuc-Fuc disaccharide **24** by a regio- and α-stereoselective glycosylation between donor **16** and acceptor **17**. Further manipulation of **24** by chloroacetylation, de-*O*-allylation and installation of a trichloroacetimidate leaving group at the anomeric position furnished disaccharide donor **25** that was in turn glycosylated with GlcA acceptor **15**. The obtained trisaccharide **26** was finally subjected to three different reaction sequences to afford differently sulfated trisaccharides **27–29**.

The synthesis of four disaccharides with Fuc branch linked at GalNAc *O*-6 site was accomplished by using GalNAc acceptor **30** and a Fuc donor [34]. Interestingly, Fuc trichloroacetimidate **16** gave in this case a moderate yield (51%) and no stereoselectivity (α/β 1:1), whereas Fuc bromide **31** afforded α-linked disaccharide **32** exclusively (Scheme 4), even if a much longer reaction time (7 days vs. 30 min) was necessary. As with di- and trisaccharides of Scheme 3, proper sequences of further synthetic steps furnished differently sulfated disaccharides **33–36** (Scheme 4).

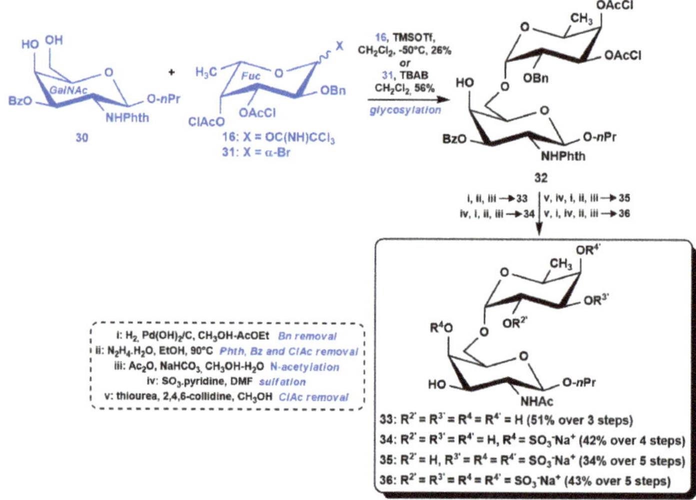

Scheme 4. Nifantiev synthesis of disaccharides with Fuc branches at GalNAc *O*-6 site [34]; abbreviations: ethyl acetate (AcOEt), phthaloyl (Phth).

3. Semi-Synthetic Strategies

3.1. Semi-Synthesis of fCS Oligosaccharides

In addition to the total synthesis of fCS oligosaccharides, the research on fCS species production has concomitantly developed also semi-synthetic strategies based on GAGs or similar natural polysaccharides as starting material. Since chondroitin sulfate (CS) and fCS have very similar molecular structures, the former sharing the same polysaccharide backbone with the latter but without any Fuc branch on it, the use of CS polysaccharides, ubiquitously widespread in the animal kingdom, as starting material for the access to fCS species was pursued. Fifteen years ago, it was already demonstrated that a selective hydrolytic breakage of β-1→4 vs. β-1→3 glycosidic linkages in CS polysaccharides can be carried out under controlled acid conditions, thus obtaining the β-GlcA-(1→3)-GalNAc disaccharide in multi-decagram quantities and in a much shorter time than with any known total synthesis [37–39]. Very recently, this approach has been employed for the obtainment of fCS trisaccharides and glycoclusters derived from them [40,41]. By adding two further steps after the controlled acid hydrolysis, peracylated disaccharide **37** could be obtained in 31% yield over three steps from commercially available CS-A polysaccharide (Scheme 5) [40]. It was then subjected to a two-step reaction sequence for the insertion of an aglycone carrying an azide moiety useful for fCS glycocluster synthesis. The obtained disaccharide

38 was further derivatized in order to liberate a single hydroxyl at GlcA *C*-3 site, where Fuc branch had to be attached. This was accomplished by de-*O*-acetylation followed by installation of a benzylidene ring on GalNAc 4,6-diol and of Bz esters at GlcA 2,4-hydroxyls. The latter regioselective reaction was possible through a one-pot 3,6-lactonization/benzoylation/lactone methanolysis sequence [42]. Disaccharide acceptor **39** was obtained in 39% overall yield over eight steps from **37**. Its coupling with three differently protected Fuc thioglycoside donors **40–42** gave trisaccharides **43–45** in high yield and α-stereoselectivity, as expected for the remote participation of Lev or ClAc esters protecting position *O*-3 and/or *O*-4 of Fuc donors. Fully protected derivatives **43–45** were then subjected to different sequences of orthogonal protecting groups cleavage to give trisaccharides **46–48** with some liberated hydroxyls that could be in turn sulfated and subjected to a final global ester hydrolysis to afford target fCS trisaccharides **49–51** carrying a 2,4-*O*-, 3,4-*O*- or 4-*O*-sulfate decoration, respectively, on Fuc unit.

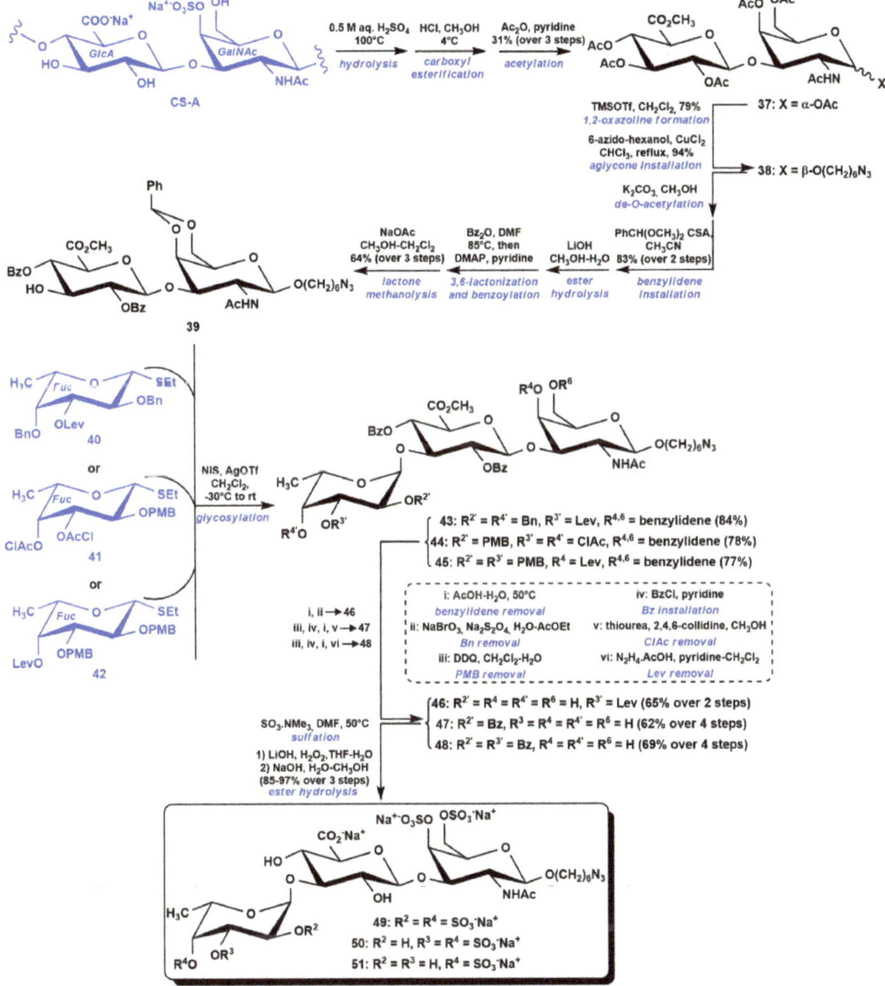

Scheme 5. Semi-synthesis of three fCS trisaccharides from CS-A [40]; abbreviations: silver trifluoromethanesulfonate (AgOTf), benzoic anhydride (Bz₂O).

The three partially protected, semi-synthetic trisaccharides **46–48** were also employed for the construction of a library of fCS glycoclusters displaying three different Fuc sulfation pattern and nine different multivalent scaffold architectures (**52–54, 56, 58–62**, Figure 2). The multiple decoration of the scaffolds with the fCS trisaccharides was accomplished by exploiting the azide moiety of **46–48** aglycone in a Cu(I)-catalyzed azide-alkyne cycloaddition (CuAAC) "click" reaction in a ternary solvent mixture ($CHCl_3$–CH_3OH–H_2O), followed by sulfation and ester hydrolysis under the conditions already employed for **49–51**.

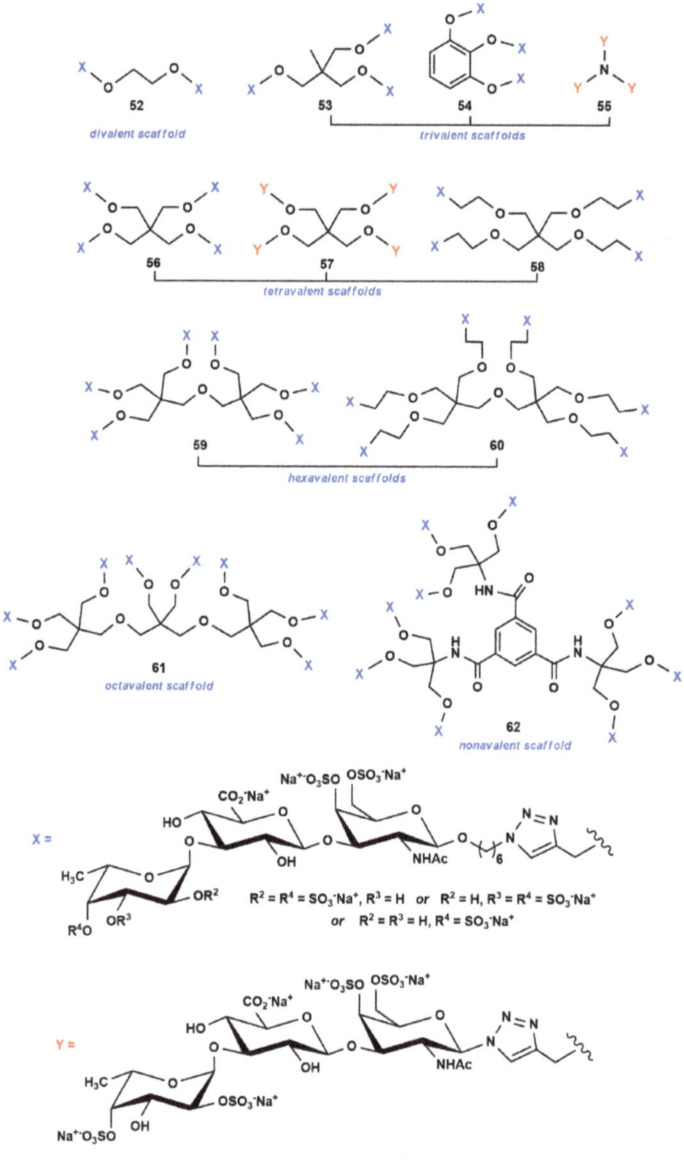

Figure 2. fCS glycoclusters semi-synthesized from CS-A [40,41]

Two additional glycoclusters **55** and **57**, carrying a shorter linker between the 2,4-di-*O*-sulfated fCS trisaccharide moieties and the core scaffolds, were reported too. A very similar semi-synthetic approach from CS-A was employed in this case, with the only relevant difference lying in conducting the CuAAC reactions directly on the final fCS trisaccharide under aqueous conditions that were suitably optimized to avoid the interference of the anionic sulfates with the copper catalyst [41]. The activity of glycoclusters **52–62** in blocking the intrinsic, extrinsic, and common coagulation cascade pathways was evaluated by measuring the activated partial thromboplastin time (APTT), prothrombin time (PT) and thrombine time (TT), respectively. Hexa-, octa- and nonavalent scaffolds all showed a significant intrinsic pathway inhibition. The octavalent glycocluster **61** carrying 2,4-*O*-disulfated Fuc moieties was the most active one [40], with an APTT value only one order of magnitude higher than natural fCS polysaccharide from *Stichopus monotuberculatus* possessing 92% of Fuc units with the same sulfation pattern [43]. Interestingly, "long-armed" glycoclusters **60** showed a significantly lower inhibition activity than "short-armed" **59**, although the two scaffolds are both hexavalent. It is also worth noting that trivalent and tetravalent glycoclusters **55** and **57**, displaying a shorter linker connecting the fCS trisaccharide to the core moiety, gave an intrinsic pathway inhibition activity comparable or even higher than hexa-, octa- and nonavalent glycoclusters **59–62** designed with a longer linker between the oligosaccharides and the core region [41]. These findings clearly indicated that the arrangement of the fCS repeating units in the glycoclusters makes great difference on the bioactivity.

Natural CS polysaccharides were employed for the access to fCS species not only as starting material of β-GlcA-(1→3)-GalNAc disaccharide through a controlled chemical hydrolysis, but also for the production of longer oligosaccharides by enzymatic degradation [44]. Indeed, bovine testicular hyaluronidase (BTH) is known to catalyze the breakage of hyaluronic acid and chondroitin polysaccharide chains to the respective tetra- and hexasaccharide species [45]. A decagram scale BTH-catalyzed degradation of desulfated CS afforded [β-GlcA-(1→3)-GalNAc]$_2$ tetrasaccharide **63** and [β-GlcA-(1→3)-GalNAc]$_3$ hexasaccharide **64** in 38% and 35% yields, respectively (Scheme 6) [42]. On the two oligosaccharides, an azide aglycone was firstly installed selectively at the reducing anomeric position with sodium azide and *N*-methylmorpholine (NMM) in the presence of Shoda reagent (2-chloro-*N*,*N'*-1,3-dimethylimidazolium chloride, DMC) [46]. Then, in order to protect all the functionalities but the hydroxyls at *O*-3 position of GlcA units, a sequence of protection and deprotection steps following the already reported manipulation of disaccharide **38** into **39** (Scheme 5) was demonstrated to be efficient also on tetra- and hexasaccharides. With oligosaccharide acceptors **65** and **66** in hands, challenging double and triple fucosylations with suitably protected Fuc thioglycoside **67** were attempted (Scheme 6). Optimized conditions included a post-glycosylation treatment with Ac$_2$O in hot acetic acid to rearrange glycosyl imidate byproducts, formed by acetamide competition as nucleophile [47], into the desired hexa- and nonasaccharide **68** and **69** in 70% and 62% yield, calculated from tetra- and hexasaccharide acceptors **65** and **66**, respectively. Final steps of the synthesis were the cleavage of PMB and benzylidene protecting groups to liberate hydroxyls at Fuc *O*-2,4 and GalNAc *O*-4,6 sites that were then sulfated, and the hydrolysis of the ester protecting groups. Target fCS hexa- and nonasaccharide **70** and **71** were obtained in 5.0% and 3.0% global yield, respectively, from CS-A. It is worth noting that only nonasaccharide **71** showed a significant intrinsic pathway anticoagulation activity, in agreement with the concomitant finding of fCS octasaccharide **79** (see Scheme 7 and discussion below) as the minimum structural unit able to confer anticoagulant activity [48]. Nonasaccharide **71** was also further reacted with a biotin-appended alkyne in a CuAAC reaction involving azide aglycone. The obtained glycoconjugate and a soluble human factor IXa were found by biolayer interferometry studies to fit well to a 2:1 binding model [44].

With the aim to avoid too many chemical reactions on complex oligosaccharides, including the challenging multiple fucosylations discussed above, an alternative approach employing natural fCS polysaccharides as starting material of controlled, partial depolymerizations to fCS oligosaccharides was widely pursued. Several different degradation methods have been investigated. Mild acid hydrolysis afforded a selective cleavage of Fuc branches together with sulfate groups, thus resulting in a valuable method for unveiling the details of natural fCS structures through a bottom-up approach [23,49–51] but not applicable for the production of pure fCS oligosaccharides. fCS backbone depolymerizations without a significant loss of sulfate and Fuc branches were reported under hydrothermal [52], ^{60}Co γ-rays irradiation [53] and free-radical oxidative conditions. The last method typically employs a Cu(II)-catalyzed Fenton system in a H_2O_2 aqueous solution [54–57], and some careful studies of the influence of several reaction parameters on the rate and extent of depolymerization have been reported [58,59]. Nonetheless, all these methods resulted in no or only slight selectivity in the breakage of GlcA vs. GalNAc glycosidic bonds (or *vice versa*) of the backbone, thus giving very complex mixtures of fCS oligosaccharides for which no fractionation to pure species was reported but in a single report [60].

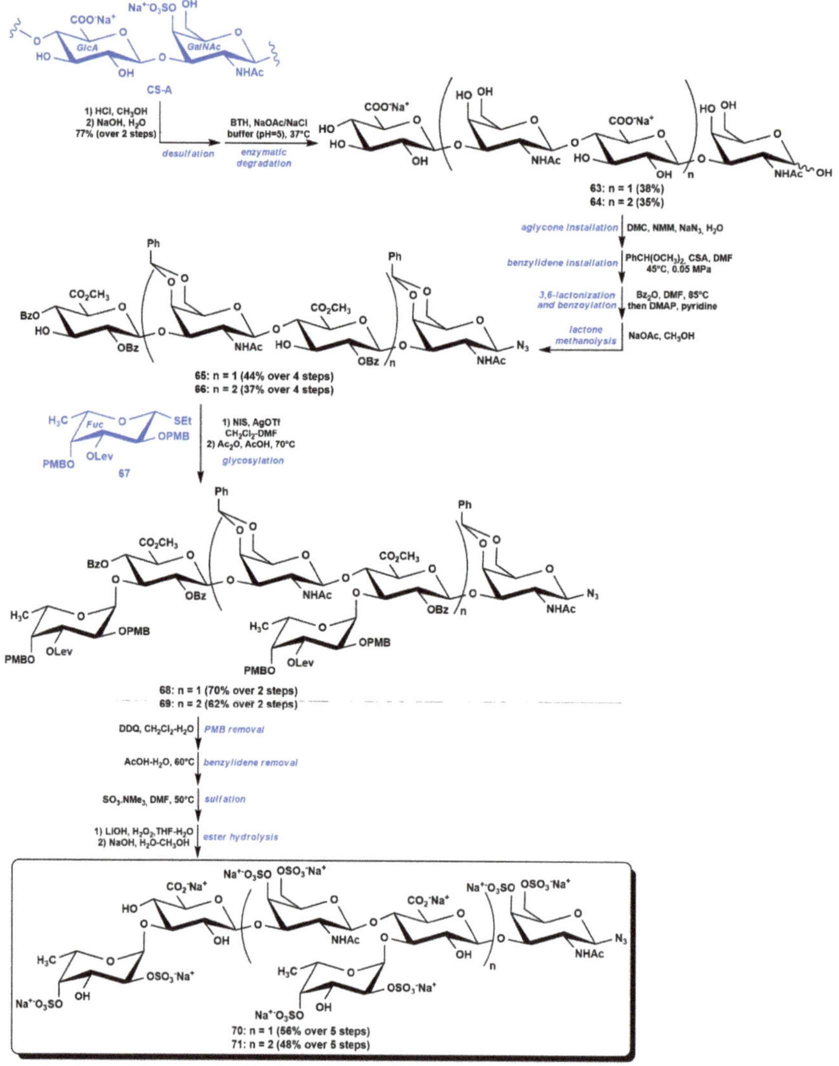

Scheme 6. Semi-synthesis of fCS hexa- and nonasaccharide from CS-A [42]; abbreviations: 4-(dimethylamino)pyridine (DMAP), sodium acetate (NaOAc).

Instead, this was possible by applying two different multi-step protocols for the highly selective breakage of the glycosidic linkage at the anomeric site of only GalNAc units. One method is based on a deaminative cleavage with nitrous acid, as developed three decades ago firstly for heparin and then for other GAGs [61]. Firstly, fCS was subjected to hydrazinolysis to have a partially de-*N*-acetylated polysaccharide. The degree of de-*N*-acetylation could be varied from 1% to 78% in dependence of several reaction parameters [62]. A subsequent treatment with diluted nitrous acid gave a very fast, highly selective cleavage at *N*-deacetylated GalNAc sites through diazotization of the free amine moieties, and furnished fCS fragments with an unnatural 2,5-anhydro-D-talose unit as reducing end (**72–77**, Scheme 7). By modulating the degree of de-*N*-acetylation, fCS oligosaccharides with different length distributions could be obtained in high, overall mass yield (71–80%) [63,64] and then purified by gel-permeation chromatography (GPC) techniques. Pure tri-, hexa-, nona-, dodeca-, pentadeca- and octadecasaccharides with different sulfation patterns have been obtained up to now through this protocol from fCSs extracted from seven different sea cucumber species [62–66]. Among these fCS fragments, nonasaccharide **74** with 2,4-disulfated Fuc branches on GlcA units, 4,6-disulfated GalNAc residues and a 2,5-anhydro-D-talitol as pseudoreducing end revealed to be the minimum fragment retaining the potent selective inhibition of the intrinsic coagulation pathway shown by natural fCS polysaccharide but displaying no side effects [63].

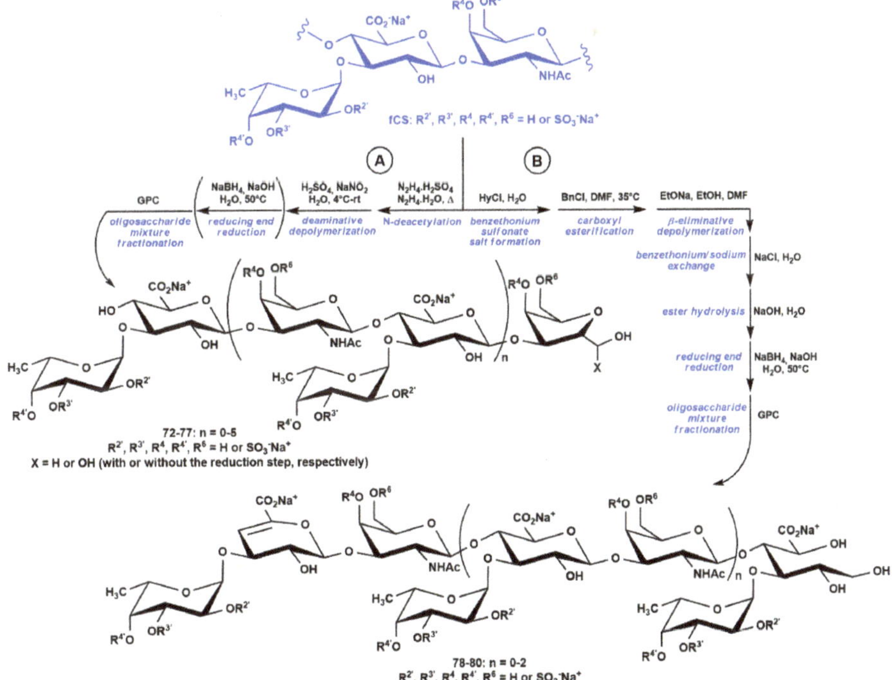

Scheme 7. Selective depolymerization (A: deamination; B: β-elimination) protocols for the obtainment of fCS oligosaccharides from natural fCS polysaccharides; abbreviations: benzyl chloride (BnCl), sodium ethoxide (EtONa), benzethonium chloride (HyCl).

A second protocol for the selective cleavage of fCS backbone employs the well-known β-eliminative degradation for polysaccharides containing uronic acid residues [67]. In particular, the method optimized for fCS polysaccharides relies upon a six-step procedure with the key depolymerization step performed on a fCS benzethonium salt derivative with some of the GlcA residues converted into benzyl esters. By treatment of this polysaccharide derivative with sodium ethoxide as base promoting the β-elimination, the breakage of some of the β-1→4 glycosidic linkages involving GlcA esters could be observed (Scheme 7) [68]. Analogously to the hydrazinolysis-deamination protocol discussed above, the modulation of the degree of GlcA esterification allowed the obtainment of fCS fragments with different length distributions in 48-65% overall mass yield [24,25,48]. The obtained mixture could be then fractioned by GPC to furnish pure fCS oligosaccharides. Penta-, octa- and undecasaccharides with different sulfation patterns (**78–80**, Scheme 7) were obtained in pure form from fCSs extracted from seven different sea cucumber species [25,26,48,68,69]. Interestingly, these fCS oligosaccharides showed a different structure with respect to those obtained by depolymerization under deamination conditions. Indeed, it is well-known that a peeling reaction can easily occur under alkaline conditions on a 1→3-linked reducing end [70], such as the GalNAc residue released after fCS β-elimination. This resulted in fCS oligosaccharides with a single GalNAc unit missing with respect to oligosaccharides obtained by deamination, and a glucuronitol residue instead of 2,5-anhydro-D-talitol as pseudo-reducing end (Scheme 7). Noteworthy, in this group of fCS oligosaccharides, octasaccharide **79** revealed to be the minimum fragment showing a significant activity as intrinsic coagulation pathway inhibitor [48].

3.2. Semi-Synthesis of Low Molecular Weight fCS Polysaccharides

Semi-synthetic approaches were pursued for accessing not only fCS oligosaccharides but also low molecular weight polysaccharides resembling the structure of natural fCS. Most of these works employed unsulfated chondroitin (CS-0) as starting material that could be obtained by fed-batch fermentation of *Escherichia coli* O5:K4:H4, followed by an expedite downstream purification [71]. This polysaccharide shares with fCS the same polymeric backbone, but is devoid of any decoration of sulfate groups and Fuc branches that could be installed through suitably developed, semi-synthetic sequences [72–74]. Indeed, a small library of fCS polysaccharides with different Fuc branching and/or sulfation pattern was obtained [73], by combining two chondroitin polysaccharide acceptors (**81** and **83**, Scheme 8) with three differently protected Fuc donors in challenging glycosylation reactions conducted in a CH$_2$Cl$_2$-DMF solvent mixture to ensure a high α-stereoselectivity [75]. In particular, chondroitin acceptor **81** resulted from carboxyl esterification followed by protection of GalNAc diol of CS-0, with a nearly quantitative degree of substitution (DS) in both steps. By GlcA diol acetylation and subsequent oxidative cleavage of the benzylidene protecting group [76] on GalNAc units, polysaccharide acceptor **83** could be obtained. Fucosylation of both acceptors was conducted with differently protected donors **85–87**. In particular, after a preliminary screening, thioglycosides were selected as better donors than *N*-phenyl-trifluoroacetimidates for both a longer shelf life and a shorter sequence of steps for their preparation, in spite the latter are known to be highly performing in fucosylations [77] and actually gave comparable results to thioglycosides [72]. Fuc donors **85–87** displayed benzyl ethers and benzoyl esters as temporary and permanent protecting groups, respectively. Indeed, benzyl ethers could be orthogonally cleaved under oxidative conditions to liberate hydroxyls at specific positions of Fuc units that were then sulfated. This allowed the obtainment of six differently sulfated and/or fucosylated fCS polysaccharides (**88**, **90**, **91**, **93**, **95**, **96**, Scheme 8) with a lower molecular weight (6.9–10.9 kDa) with respect to natural fCS species (20–100 kDa, [23]). Furthermore, by skipping the sulfation step, the two unprecedented non-sulfated, fucosylated chondroitin polysaccharides **89** and **94** could be accessed. Preliminary in vitro anticoagulant assays on the semi-synthesized low molecular weight fCS polysaccharides pointed out some structure-activity relationships, with **93**, **95** and **96** carrying Fuc branches on GlcA units as well as **88** with trisulfated Fuc units on GalNAc residues showing a behavior similar to low molecular weight species obtained by partial depolymerization of natural fCSs [73]. Nonetheless, the random distribution of Fuc branches between positions 2 and 3 of GlcA units in **93–96** as well as between positions 4 and 6 of GalNAc residues in **88–91** hampered more detailed structure-activity relationship investigations. To overcome these limitations, a new chondroitin acceptor (**84**, Scheme 8), obtained by mild hydrolysis of benzylidene residues of polysaccharide **82** without a significant cleavage of glycosidic bonds, was very recently employed for the regioselective insertion of Fuc branches at GalNAc *O*-6 site [74]. Furthermore, the semi-synthesis of low molecular weight fCS polysaccharides with Fuc units exclusively linked at GlcA *O*-3 positions—as in most of natural fCSs—has been very recently accessed [78]. This result was achieved by applying the one-pot 3,6-lactonization/benzoylation/lactone methanolysis sequence—already developed on di-, tetra- and hexasaccharide chondroitin derivatives (Schemes 5 and 6)—also on microbial sourced CS-0 polysaccharide in order to obtain a polymeric acceptor with a single, free hydroxyl per repeating unit exclusively placed at GlcA *C*-3 site [79].

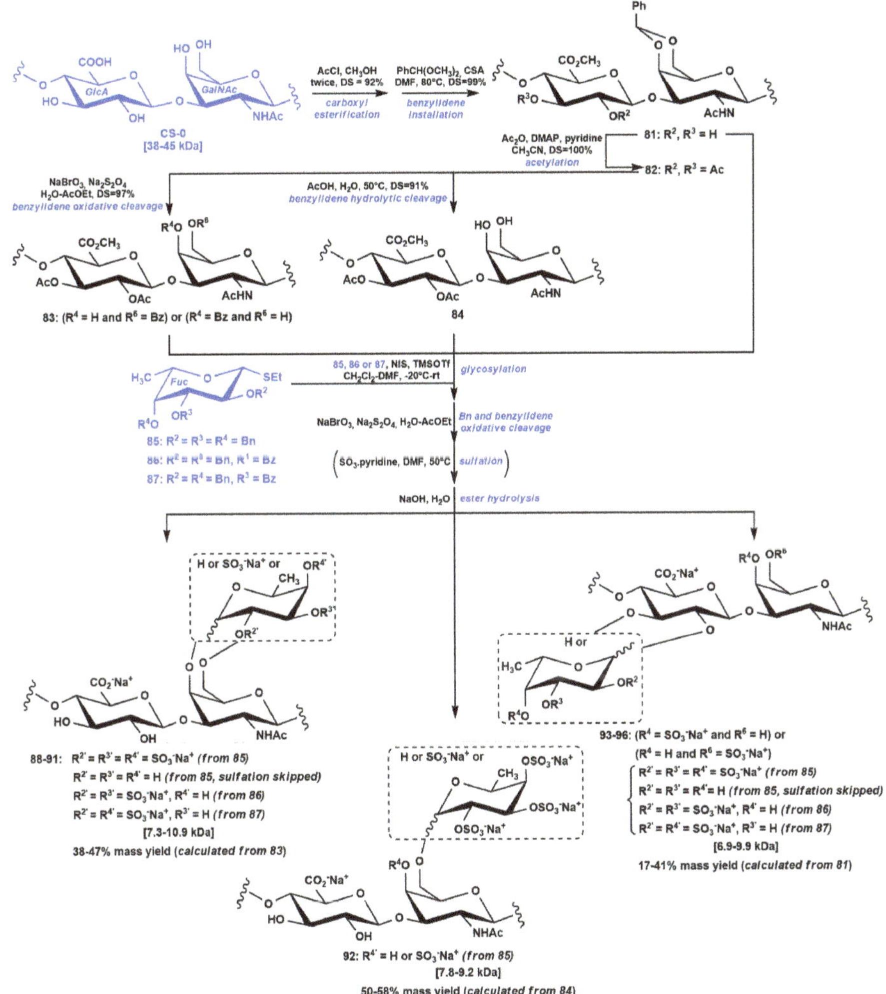

Scheme 8. Semi-synthesis of low molecular weight fCS polysaccharides from *E. coli* sourced CS-0.

A small library of low molecular weight polysaccharides resembling fCS structure, but with Fuc branches linked to the polymeric backbone through amide instead of glycosidic linkages, was prepared starting from two differently sulfated CS polysaccharides (CS-A and CS-E) [80]. The semi-synthesis relied upon a CuAAC "click" reaction between chondroitin polysaccharides derivatives **97** and **98** carrying *N*-propargyl amides with different degrees of substitution on GlcA units and fucosides **99** and **100** displaying a 2-azidoethyl aglycone (Scheme 9). The obtained grafted polysaccharides **101–104**, showing different sulfation patterns on Fuc and/or GalNAc units as well as a different degree of fucosylation, were assayed for their anticoagulant activity. Only polysaccharides **103** and **104** with the lowest degree of fucosylation (14%) clearly displayed an inhibition of the intrinsic coagulation pathway. This suggests a role of both GlcA carboxylic acid moieties and GalNAc sulfation degree for the anticoagulant activity of these fCS mimetics.

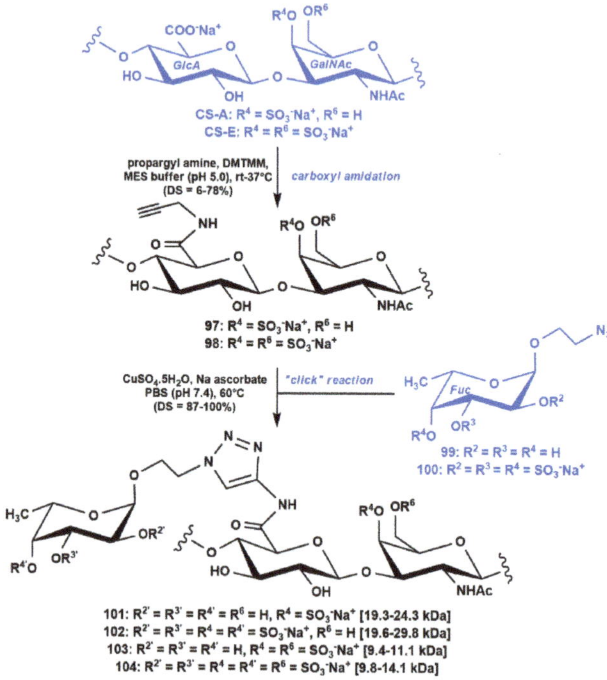

Scheme 9. Semi-synthesis of polysaccharides with amide-linked Fuc branches from CS-A and CS-E [80]; abbreviations: 4-(4,6-dimethoxy-1,3,5-triazin-2-yl)-4-methylmorpholinium chloride (DMTMM), 4-morpholineethanesulfonic acid (MES), phosphate buffered saline (PBS).

Semi-synthetic, low molecular weight polysaccharides with a non-natural fCS structure were also obtained by derivatization of fCSs themselves. In particular, a free-radical, partially depolymerized fCS from *Thelenata ananas* that shows 3-*O*-, 4-*O*- and 2,4-di-*O*-sulfated-Fuc branches [54], was derivatized through five different kinds of semi-synthetic modifications: random *O*-acylation (acetylation, propionylation or succinoylation) [81], GlcA carboxylic acid esterification (ethylation, benzylation or 1-butenylation) or reduction to alcohol, GalNAc *N*-deacetylation or Fuc de-branching [43]. All the derivatizations—except acylations with low degrees of substitution (up to 40%)—caused a reduction of the anticoagulant activity, with defucosylation with the greatest effect, as expected [10–12].

4. Conclusions and Perspectives

The highly promising activity of fCS polysaccharides extracted from sea cucumbers in blocking the intrinsic coagulation cascade has prompted several studies on this topic in the last decade. Reviews focused on structural diversity [9,18,19], characterization methods [17,23] and bioactivities [4] of fCSs have been reported, whereas to the best of our knowledge no review paper on (semi)-syntheses of (macro)molecules resembling the structure of natural fCS polysaccharides has appeared in the literature yet. Here we have filled this gap, by including in this review not only all the total synthetic and semi-synthetic strategies to fCS oligosaccharides and low molecular weight polysaccharides reported up to now, but also very recent achievements on semi-synthetic glycoclusters displaying multiple copies of fCS species. For all the targets, the total synthetic and/or semi-synthetic strategy has been discussed, underlining for each approach advantages and drawbacks and also reporting the main results on structure-bioactivity relationships.

Despite the several achievements obtained up to now by studies of fCS chemistry and biology, there are still some key points that should be addressed in a near future. Specifically, extensive structure-activity relationships of fCS have yet to be reported. Therefore, fast, easily scalable and/or cost effective (semi)-synthetic methods to fCS oligosaccharides, plausibly higher than an octasaccharide (the minimum structural unit able to confer a remarkable anticoagulant activity), should be developed. Furthermore, the chemical space around the natural fCS structure should be explored further and more intensively, in order to access non-natural oligosaccharides, low molecular weight polysaccharides or multivalent compounds that could display similar or even more powerful and interesting biological activities than all the fCS-related species investigated up to now. Therefore, we foresee that the research targeting the (semi)-synthesis of fCS oligo- and polysaccharides and analogs thereof will attract a growing interest in the next years and bring even more advances than in the last decade in the field of synthetic chemistry inspired to marine biomolecules.

Author Contributions: All authors contributed substantially to the preparation of this review. All authors approved the manuscript in its current form. All authors have read and agreed to the published version of the manuscript.

Funding: This research was funded by MIUR, Ministero dell'Istruzione, dell'Università e della Ricerca, Rome, Italy, FFABR 2017 grant.

Conflicts of Interest: The authors declare no conflict of interest.

References

1. Lindahl, U.; Couchman, J.; Kimata, K.; Esko, J.D. Proteoglycans and sulfated glycosaminoglycans. In *Essential of Glycobiology*, 3rd ed.; Cold Spring Harbor Laboratory Press: Cold Spring Harbor, NY, USA, 2017; Chapter 17.
2. Kinoshita-Toyoda, A.; Yamada, S.; Haslam, S.M.; Khoo, K.H.; Sugiura, M.; Morris, H.R.; Dell, A.; Sugahara, K. Structural determination of five novel tetrasaccharides containing 3-*O*-sulfated D-glucuronic acid and two rare oligosaccharides containing a β-D-glucose branch isolated from squid cartilage chondroitin sulfate. *Biochemistry* **2004**, *43*, 11063–11074. [CrossRef] [PubMed]
3. Higashi, K.; Takeda, K.; Mukuno, A.; Okamoto, Y.; Masuko, S.; Linhardt, R.J.; Toida, T. Identification of keratan sulfate disaccharide at C-3 position of glucuronate of chondroitin sulfate from *Mactra chinensis*. *Biochem. J.* **2016**, *473*, 4145–4148. [CrossRef] [PubMed]
4. Pomin, V.H. Holothurian fucosylated chondroitin sulfate. *Mar. Drugs* **2014**, *12*, 232–254. [CrossRef] [PubMed]
5. Mourão, P.A.S. Perspective on the use of sulfated polysaccharides from marine organisms as a source of new antithrombotic drugs. *Mar. Drugs* **2015**, *13*, 2770–2784. [CrossRef] [PubMed]
6. Glauser, B.F.; Pereira, M.S.; Monteiro, R.Q.; Mourão, P.A.S. Serpin-independent anticoagulant activity of a fucosylated chondroitin sulfate. *Thromb. Haemost.* **2018**, *100*, 420–428. [CrossRef]
7. Buyue, Y.; Sheehan, J.P. Fucosylated chondroitin sulfate inhibits plasma thrombin generation via targeting of the factor IXa heparin-binding exosite. *Blood* **2009**, *114*, 3092–3100. [CrossRef] [PubMed]
8. Fonseca, R.J.C.; Sucupira, I.D.; Oliveira, S.-N.M.C.G.; Santos, G.R.C.; Mourão, P.A.S. Improved anticoagulant effect of fucosylated chondroitin sulfate orally administered as gastroresistant tablets. *Thromb. Haemost.* **2017**, *117*, 662–670. [PubMed]
9. Myron, P.; Siddiquee, S.; Al Azad, S. Fucosylated chondroitin sulfate diversity in sea cucumbers: A review. *Carbohydr. Polym.* **2014**, *112*, 173–178. [CrossRef]
10. Monteiro-Machado, M.; Tomaz, M.A.; Fonseca, R.J.C.; Strauch, M.A.; Cons, B.L.; Borges, P.A.; Patrão-Neto, F.C.; Tavares-Henriques, M.S.; Teixeira-Cruz, J.M.; Calil-Elias, S.; et al. Occurrence of sulfated fucose branches in fucosylated chondroitin sulfate are essential for the polysaccharide effect preventing muscle damage induced by toxins and crude venom from *Bothrops jararacussu* snake. *Toxicon* **2015**, *98*, 20–33. [CrossRef]

11. Mourão, P.A.S.; Guimarães, M.A.M.; Mulloy, B.; Thomas, S.; Gray, E. Antithrombotic activity of a fucosylated chondroitin sulphate from echinoderm: Sulphated fucose branches on the polysaccharide account for its antithrombotic action. *Br. J. Haematol.* **1998**, *101*, 647–652. [CrossRef]

12. Mourão, P.A.S.; Pereira, M.S.; Pavão, M.S.; Mulloy, B.; Tollefsen, D.M.; Mowinckel, M.C.; Abildgaard, U. Structure and anticoagulant activity of a fucosylated chondroitin sulfate from echinoderm. Sulfated fucose branches on the polysaccharide account for its high anticoagulant action. *J. Biol. Chem.* **1996**, *271*, 23973–23984. [CrossRef] [PubMed]

13. Li, Q.; Cai, C.; Chang, Y.; Zhang, F.; Linhardt, R.J.; Xue, C.; Li, G.; Yu, G. A novel structural fucosylated chondroitin sulfate from *Holothuria mexicana* and its effects on growth factors binding and anticoagulation. *Carbohydr. Polym.* **2018**, *181*, 1160–1168. [CrossRef] [PubMed]

14. Ustyuzhanina, N.E.; Bilan, M.I.; Dmitrenok, A.S.; Nifantiev, N.E.; Usov, A.I. The structure of a fucosylated chondroitin sulfate from the sea cucumber *Cucumaria frondosa*. *Carbohydr. Polym.* **2017**, *165*, 7–12. [CrossRef] [PubMed]

15. Ustyuzhanina, N.E.; Bilan, M.I.; Dmitrenok, A.S.; Tsvetkova, E.A.; Shashkov, A.S.; Stonik, V.A.; Nifantiev, N.E.; Usov, A.I. Structural characterization of fucosylated chondroitin sulfates from sea cucumbers *Apostichopus japonicus* and *Actinopyga mauritiana*. *Carbohydr. Polym.* **2016**, *153*, 399–405. [CrossRef] [PubMed]

16. Yang, L.; Wang, Y.; Yang, S.; Lv, Z. Separation, purification, structures and anticoagulant activities of fucosylated chondroitin sulfates from *Holothuria scabra*. *Int. J. Biol. Macromol.* **2018**, *108*, 710–718. [CrossRef] [PubMed]

17. Mourão, P.A.S.; Vilanova, E.; Soares, P.A.G. Unveiling the structure of sulfated fucose-rich polysaccharides via nuclear magnetic resonance spectroscopy. *Curr. Opin. Struct. Biol.* **2018**, *50*, 33–41. [CrossRef]

18. Ustyuzhanina, N.E.; Bilan, M.I.; Nifantiev, N.E.; Usov, A.I. New insight on the structural diversity of holothurian fucosylated chondroitin sulfates. *Pure Appl. Chem.* **2019**, *91*, 1065–1071. [CrossRef]

19. Khotimchenko, Y. Pharmacological potential of sea cucumbers. *Int. J. Mol. Sci.* **2018**, *19*, 1342. [CrossRef]

20. Chahed, L.; Balti, R.; Elhiss, S.; Bouchemal, N.; Ajzenberg, N.; Ollivier, V.; Chaubet, F.; Mounir Maaroufi, R.; Ben Mansour, M. Anticoagulant activity of fucosylated chondroitin sulfate isolated from *Cucumaria syracusana*. *Process Biochem.* **2020**, *91*, 149–157. [CrossRef]

21. Soares, P.A.G.; Ribeiro, K.A.; Valente, A.P.; Capillé, N.V.; Oliveira, S.-N.M.C.G.; Tovar, A.M.F.; Pereira, M.S.; Vilanova, E.; Mourão, P.A.S. A unique fucosylated chondroitin sulfate type II with strikingly homogeneous and neatly distributed α-fucose branches. *Glycobiology* **2018**, *28*, 565–579. [CrossRef]

22. Fonseca, R.J.C.; Oliveira, S.-N.M.C.G.; Pomin, V.H.; Mecawi, A.S.; Araujo, I.G.; Mourão, P.A.S. Effects of oversulfated and fucosylated chondroitin sulfates on coagulation. *Thromb. Haemost.* **2010**, *103*, 994–1004. [CrossRef] [PubMed]

23. Ustyuzhanhina, N.E.; Bilan, M.I.; Nifantiev, N.E.; Usov, A.I. Structural analysis of holothurian fucosylated chondroitin sulfates: Degradation *versus* non-degradative approach. *Carbohydr. Res.* **2019**, *476*, 8–11. [CrossRef] [PubMed]

24. Ustyuzhanina, N.E.; Bilan, M.I.; Dmitrenok, A.S.; Borodina, E.Y.; Stonik, V.A.; Nifantiev, N.E.; Usov, A.I. A highly regular fucosylated chondroitin sulfate from the sea cucumber *Massinium magnum*: Structure and effects on coagulation. *Carbohydr. Polym.* **2017**, *167*, 20–26. [CrossRef] [PubMed]

25. Shang, F.; Gao, N.; Yin, R.; Lin, L.; Xiao, C.; Zhou, L.; Li, Z.; Purcell, S.W.; Wu, M.; Zhao, J. Precise structures of fucosylated glycosaminoglycan and its oligosaccharides as novel intrinsic factor Xase inhibitors. *Eur. J. Med. Chem.* **2018**, *148*, 423–435. [CrossRef] [PubMed]

26. Cai, Y.; Yang, W.; Li, X.; Zhou, L.; Wang, Z.; Lin, L.; Chen, D.; Zhao, L.; Li, Z.; Liu, S.; et al. Precise structure and anti-intrinsic tenase complex activity of three fucosylated glycosaminoglycans and their fragments. *Carbohydr. Polym.* **2019**, *246*, 115146. [CrossRef] [PubMed]

27. Zhang, X.; Lin, L.; Huang, H.; Linhardt, R.J. Chemoenzymatic synthesis of glycosaminoglycans. *Acc. Chem. Res.* **2020**, *53*, 335–346. [CrossRef]

28. Mende, M.; Bednarek, C.; Wawryszyn, M.; Sauter, P.; Biskup, M.B.; Schepers, U.; Bräse, S. Chemical synthesis of glycosaminoglycans. *Chem. Rev.* **2016**, *116*, 8193–8255. [CrossRef]

29. Bedini, E.; Laezza, A.; Iadonisi, A. Chemical derivatization of sulfated glycosaminoglycans. *Eur. J. Org. Chem.* **2016**, 3018–3042. [CrossRef]

30. Tamura, J.; Tanaka, H.; Nakamura, A.; Takeda, N. Synthesis of β-ᴅ-GalNAc(4,6-diS)(1–4)[α-ʟ-Fuc(2,4-diS)(1–3)]-β-ᴅ-GlcA, a novel trisaccharide unit of chondroitin sulfate with a fucose branch. *Tetrahedron Lett.* **2013**, *54*, 3940–3943. [CrossRef]

31. He, H.; Chen, D.; Li, X.; Li, C.; Zhao, J.-H.; Qin, H.-B. Synthesis of trisaccharide repeating unit of fucosylated chondroitin sulfate. *Org. Biomol. Chem.* **2019**, *17*, 2877–2882. [CrossRef]

32. Ustyuzhanina, N.E.; Fomitskaya, P.A.; Gerbst, A.G.; Dmitrenok, A.S.; Nifantiev, N.E. Synthesis of the oligosaccharides related to branching sites of fucosylated chondroitin sulfates from sea cucumbers. *Mar. Drugs* **2015**, *13*, 770–787. [CrossRef] [PubMed]

33. Gerbst, A.G.; Dmitrenok, A.S.; Ustyuzhanina, N.E.; Nifantiev, N.E. Conformational analysis of the oligosaccharides related to side chains of holothurian fucosylated chondroitin sulfates. *Mar. Drugs* **2015**, *13*, 936–947. [CrossRef] [PubMed]

34. Vinnitskiy, D.Z.; Ustyuzhanina, N.E.; Dmitrenok, A.S.; Shashkov, A.S.; Nifantiev, N.E. Synthesis and NMR analysis of model compounds related to fucosylated chondroitin sulfates: GalNAc and Fuc(1→6)GalNAc derivatives. *Carbohydr. Res.* **2017**, *438*, 9–17. [CrossRef] [PubMed]

35. Ustyuzhanina, N.E.; Bilan, M.I.; Dmitrenok, A.S.; Nifantiev, N.E.; Usov, A.I. Two fucosylated chondroitin sulfates from the sea cucumber *Eupentacta fraudatrix*. *Carbohydr. Polym.* **2017**, *164*, 8–12. [CrossRef]

36. Santos, G.R.C.; Glauser, B.F.; Parreiras, L.A.; Vilanova, E.; Mourão, P.A.S. Distinct structures of the α-fucose branches in fucosylated chondroitin sulfates do not affect their anticoagulant activity. *Glycobiology* **2015**, *25*, 1043–1052. [CrossRef] [PubMed]

37. Lopin-Bon, C.; Jacquinet, J.-C. From polymer to size-defined oligomers: An expeditious route for the preparation of chondroitin oligosaccharides. *Angew. Chem. Int. Ed.* **2006**, *45*, 2574–2578. [CrossRef] [PubMed]

38. Vibert, A.; Lopin-Bon, C.; Jacquinet, J.-C. From polymer to size-defined oligomers: A step economy process for the efficient and stereocontrolled construction of chondroitin oligosaccharides and biotinylated conjugates thereof: Part 1. *Chem. Eur. J.* **2009**, *15*, 9561–9578. [CrossRef] [PubMed]

39. Jacquinet, J.-C.; Lopin-Bon, C.; Vibert, A. A Highly Divergent and stereocontrolled construction of chondroitin sulfate A, C, D, E, K, L, and M oligomers from a single precursor: Part 2. *Chem. Eur. J.* **2009**, *15*, 9579–9595. [CrossRef] [PubMed]

40. Zhang, X.; Yao, W.; Xu, X.; Sun, H.; Zhao, J.; Meng, X.; Wu, M.; Li, Z. Synthesis of fucosylated chondroitin sulfate glycoclusters: A robust route to new anticoagulant agents. *Chem. Eur. J.* **2018**, *24*, 1694–1700. [CrossRef] [PubMed]

41. Liu, H.; Zhang, X.; Wu, M.; Li, Z. Synthesis and anticoagulation studies of "short-armed" fucosylated chondroitin sulfate glycoclusters. *Carbohydr. Res.* **2018**, *467*, 45–51. [CrossRef]

42. Kornilov, A.V.; Sukhova, E.V.; Nifantiev, N.E. Preparative route to glucuronyl donors bearing temporary protecting group at *O*-3 via 6,3-lactonisation by Bz₂O or Piv₂O. *Carbohydr. Res.* **2001**, *336*, 309–313. [CrossRef]

43. Wu, M.; Wen, D.; Gao, N.; Xiao, L.; Yang, L.; Xu, W.; Lian, W.; Peng, J.; Jiang, J.; Zhao, J. Anticoagulant and antithrombotic evaluation of native fucosylated chondroitin sulfates and their derivatives as selective inhibitors of intrinsic factor Xase. *Eur. J. Med. Chem.* **2015**, *92*, 257–269. [CrossRef] [PubMed]

44. Zhang, X.; Liu, H.; Lin, L.; Yao, W.; Zhao, J.; Wu, M.; Li, Z. Synthesis of fucosylated chondroitin sulfate nonasaccharide as a novel anticoagulant targeting intrinsic factor Xase complex. *Angew. Chem. Int. Ed.* **2018**, *57*, 12880–12885. [CrossRef] [PubMed]

45. Kakizaki, I.; Koizumi, H.; Chen, F.; Endo, M. Inhibitory effect of chondroitin sulfate oligosaccharides on bovine testicular hyaluronidase. *Carbohydr. Polym.* **2015**, *121*, 362–371. [CrossRef]

46. Tanaka, T.; Nagai, H.; Noguchi, M.; Kobayashi, A.; Shoda, S.-I. One-step conversion of unprotected sugars to β-glycosyl azides using 2-chloroimidazolinium salt in aqueous solution. *Chem. Commun.* **2009**, *23*, 3378–3379. [CrossRef]

47. Cheng, A.; Hendel, J.L.; Colangelo, K.; Bonin, M.; Auzanneau, F.I. Convenient temporary methyl imidate protection of *N*-acetylglucosamine and glycosylation at O-4. *J. Org. Chem.* **2008**, *73*, 7574–7579. [CrossRef]

48. Yin, R.; Zhou, L.; Gao, N.; Li, Z.; Zhao, L.; Shang, F.; Wu, M.; Zhao, J. Oligosaccharides from depolymerized fucosylated glycosaminoglycan: Structures and minimum size for intrinsic factor Xase complex inhibition. *J. Biol. Chem.* **2018**, *293*, 14089–14099. [CrossRef]

49. Santos, G.R.C.; Porto, A.C.O.; Soares, P.A.G.; Vilanova, E.; Mourão, P.A.S. Exploring the structure of fucosylated chondroitin sulfate through bottom-up nuclear magnetic resonance and electrospray ionization-high-resolution mass spectrometry approaches. *Glycobiology* **2017**, 625–634. [CrossRef]

50. Xu, L.; Gao, N.; Xiao, C.; Lin, L.; Purcell, S.W.; Wu, M.; Zhao, J. Modulating the degree of fucosylation of fucosylated chondroitin sulfate enhances heparin cofactor II-dependent thrombin inhibition. *Eur. J. Med. Chem.* **2018**, *154*, 133–143. [CrossRef] [PubMed]

51. Qiu, P.; Wu, F.; Yi, L.; Chen, L.; Jin, Y.; Ding, X.; Ouyang, Y.; Yao, Y.; Jiang, Y.; Zhang, Z. Structure characterization of a heavily fucosylated chondroitin sulfate from sea cucumber (*H. leucospilota*) with bottom-up strategies. *Carbohydr. Polym.* **2020**, *240*, 116337. [CrossRef]

52. Shi, D.; Qi, J.; Zhang, H.; Yang, H.; Yang, Y.; Zhao, X. Comparison of hydrothermal depolymerization and oligosaccharide profile of fucoidan and fucosylated chondroitin sulfate from *Holothuria floridana*. *Int. J. Biol. Macromol.* **2019**, *132*, 738–747. [CrossRef] [PubMed]

53. Wu, N.; Ye, X.; Guo, X.; Liao, N.; Yin, X.; Hu, Y.; Sun, Y.; Liu, D.; Chen, S. Depolymerization of fucosylated chondroitin sulfate from sea cucumber, *Pearsonothuria graeffei*, via ^{60}Co irradiation. *Carbohydr. Polym.* **2013**, *93*, 604–614. [CrossRef]

54. Wu, M.; Xu, S.; Zhao, J.; Kang, H.; Ding, H. Free-radical depolymerization of glycosaminoglycan from sea cucumber *Thelenata ananas* by hydrogen peroxide and copper ions. *Carbohydr. Polym.* **2010**, *80*, 1116–1124. [CrossRef]

55. Yang, J.; Wang, Y.; Jiang, T.; Lv, L.; Zhang, B.; Lv, Z. Depolymerized glycosaminoglycan and its anticoagulant activities from sea cucumber *Apostichopus japonicus*. *Int. J. Biol. Macromol.* **2015**, *72*, 699–705. [CrossRef] [PubMed]

56. Liu, X.; Hao, J.; Shan, X.; Zhang, X.; Zhao, X.; Li, Q.; Wang, X.; Cai, C.; Li, G.; Yu, G. Antithrombotic activities of fucosylated chondroitin sulfates and their depolymerized fragments from two sea cucumbers. *Carbohydr. Polym.* **2016**, *152*, 343–350. [CrossRef] [PubMed]

57. Niu, Q.; Li, G.; Li, C.; Li, Q.; Li, J.; Liu, C.; Pan, L.; Li, S.; Cai, C.; Hao, J.; et al. Two different fucosylated chondroitin sulfates: Structural elucidation, stimulating hematopoiesis and immune-enhancing effects. *Carbohydr. Polym.* **2020**, *230*, 115698. [CrossRef] [PubMed]

58. Li, J.; Li, S.; Zhi, Z.; Yan, L.; Ye, X.; Ding, T.; Yan, L.; Linhardt, R.J.; Chen, S. Depolymerization of fucosylated chondroitin sulfate with a modified Fenton-system and anticoagulant activity of the resulting fragments. *Mar. Drugs* **2016**, *14*, 170. [CrossRef]

59. Li, J.; Li, S.; Wu, L.; Yang, H.; Wei, C.; Ding, T.; Linhardt, R.J.; Zheng, X.; Ye, X.; Chen, S. Ultrasound-assisted fast preparation of low molecular weight fucosylated chondroitin sulfate with antitumor activity. *Carbohydr. Polym.* **2019**, *209*, 82–91. [CrossRef]

60. Li, J.; Li, S.; Yan, L.; Ding, T.; Linhardt, R.J.; Yu, Y.; Liu, X.; Liu, D.; Ye, X.; Chen, S. Fucosylated chondroitin sulfate oligosaccharides exert anticoagulant activity by targeting at intrinsic tenase complex with low FXII activation: Importance of sulfation pattern and molecular size. *Eur. J. Med. Chem.* **2017**, *139*, 191–200. [CrossRef]

61. Guo, Y.; Conrad, H.E. The disaccharide composition of heparins and heparan sulfates. *Anal. Biochem.* **1989**, *176*, 96–104. [CrossRef]

62. Zhao, L.; Lai, S.; Huang, R.; Wu, M.; Gao, N.; Xu, L.; Qin, H.; Peng, W.; Zhao, J. Structure and anticoagulant activity of fucosylated glycosaminoglycan degraded by deaminative cleavage. *Carbohydr. Polym.* **2013**, *98*, 1514–1523. [CrossRef] [PubMed]

63. Zhao, L.; Wu, M.; Xiao, C.; Yang, L.; Zhou, L.; Gao, N.; Li, Z.; Chen, J.; Chen, J.; Liu, J.; et al. Discovery of an intrinsic tenase complex inhibitor: Pure nonasaccharide from fucosylated glycosaminoglycan. *Proc. Natl. Acad. Sci. USA* **2015**, *112*, 8284–8289. [CrossRef] [PubMed]

64. Guan, R.; Peng, Y.; Zhou, L.; Zheng, W.; Liu, X.; Wang, P.; Yuan, Q.; Gao, N.; Zhao, L.; Zhao, J. Precise structure and anticoagulant activity of fucosylated glycosaminoglycans from *Apostichopus japonicas*: Analysis of its depolymerized fragments. *Mar. Drugs* **2019**, *17*, 195. [CrossRef] [PubMed]

65. Yan, L.; Li, J.; Wang, D.; Ding, T.; Hu, Y.; Ye, X.; Linhardt, R.J.; Chen, S. Molecular size is important for safety and selective inhibition of intrinsic factor Xase for fucosylated chondroitin sulfate. *Carbohydr. Polym.* **2017**, *178*, 180–189. [CrossRef] [PubMed]

66. Yan, L.; Li, L.; Li, J.; Yu, Y.; Liu, X.; Ye, X.; Linhardt, R.J.; Chen, S. Bottom-up analysis using liquid chromatography-Fourier transform mass spectrometry to characterize fucosylated chondroitin sulfates from sea cucumbers. *Glycobiology* **2019**, *29*, 755–764. [CrossRef] [PubMed]

67. Kiss, J. β-Eliminative degradation of carbohydrates containing uronic acid residues. *Adv. Carbohydr. Chem. Biochem.* **1974**, *29*, 229–303.

68. Gao, N.; Lu, F.; Xiao, C.; Yang, L.; Chen, J.; Zhou, K.; Wen, D.; Li, Z.; Wu, M.; Jiang, J.; et al. β-Eliminative depolymerization of the fucosylated chondroitin sulfate and anticoagulant activities of resulting fragments. *Carbohydr. Polym.* **2015**, *127*, 427–437. [CrossRef]

69. Yang, W.; Chen, D.; He, Z.; Zhou, L.; Cai, Y.; Mao, H.; Gao, N.; Zuo, Z.; Yin, R.; Zhao, J. NMR characterization and anticoagulant activity of the oligosaccharides from the fucosylated glycosaminoglycan isolated from *Holothuria coluber*. *Carbohydr. Polym.* **2020**, *233*, 115844. [CrossRef]

70. Huang, Y.; Mao, Y.; Zong, C.; Lin, C.; Boons, G.J.; Zaia, J. Discovery of a heparan sulfate 3-*O*-sulfation specific peeling reaction. *Anal. Chem.* **2015**, *87*, 592–600. [CrossRef] [PubMed]

71. Schiraldi, C.; Cimini, D.; De Rosa, M. Production of chondroitin sulfate and chondroitin. *Appl. Microbiol. Biotechnol.* **2010**, *87*, 1779–1787. [CrossRef] [PubMed]

72. Laezza, A.; Iadonisi, A.; De Castro, C.; De Rosa, M.; Schiraldi, C.; Parrilli, M.; Bedini, E. Chemical fucosylation of a polysaccharide: A semisynthetic access to fucosylated chondroitin sulfate. *Biomacromolecules* **2015**, *16*, 2237–2245. [CrossRef] [PubMed]

73. Laezza, A.; Iadonisi, A.; Pirozzi, A.V.A.; Diana, P.; De Rosa, M.; Schiraldi, C.; Parrilli, M.; Bedini, E. A modular approach to a library of semi-synthetic fucosylated chondroitin sulfate polysaccharides with different sulfation and fucosylation patterns. *Chem. Eur. J.* **2016**, *22*, 18215–18226. [CrossRef] [PubMed]

74. Vessella, G.; Traboni, S.; Pirozzi, A.V.A.; Laezza, A.; Iadonisi, A.; Schiraldi, C.; Bedini, E. A study for the access to a semi-synthetic regioisomer of natural fucosylated chondroitin sulfate with fucosyl branches on *N*-acetyl-galactosamine units. *Mar. Drugs* **2019**, *17*, 655. [CrossRef] [PubMed]

75. Lu, S.-R.; Lai, Y.-H.; Chen, J.-H.; Liu, C.-Y.; Mong, K.-K.T. Dimethylformamide: An unusual glycosylation modulator. *Angew. Chem. Int. Ed.* **2011**, *50*, 7315–7320. [CrossRef] [PubMed]

76. Adinolfi, M.; Barone, G.; Guariniello, L.; Iadonisi, A. Facile cleavage of carbohydrate benzyl ethers and benzylidene acetals using the NaBrO$_3$-Na$_2$S$_2$O$_4$ reagent under two-phase conditions. *Tetrahedron Lett.* **1999**, *40*, 8439–8441. [CrossRef]

77. Comegna, D.; Bedini, E.; Di Nola, A.; Iadonisi, A.; Parrilli, M. The behaviour of deoxyhexose trihaloacetimidates in selected glycosylations. *Carbohydr. Res.* **2007**, *342*, 1021–1029. [CrossRef] [PubMed]

78. Vessella, G.; Traboni, S.; Iadonisi, A.; Schiraldi, C.; Bedini, E. *manuscript in preparation*.

79. Vessella, G.; Traboni, S.; Cimini, D.; Iadonisi, A.; Schiraldi, C.; Bedini, E. Development of semisynthetic, regioselective pathways for accessing the missing sulfation patterns of chondroitin sulfate. *Biomacromolecules* **2019**, *20*, 3021–3030. [CrossRef]

80. Fan, F.; Zhang, P.; Wang, L.; Sun, T.; Cai, C.; Yu, G. Synthesis and properties of functional glycomimetics through click grafting of fucose onto chondroitin sulfates. *Biomacromolecules* **2019**, *20*, 3798–3808. [CrossRef]

81. Gao, N.; Wu, M.; Liu, S.; Lian, W.; Li, Z.; Zhao, J. Preparation and characterization of *O*-acylated fucosylated chondroitin sulfate from sea cucumber. *Mar. Drugs* **2012**, *10*, 1647–1661. [CrossRef]

 *marine drugs*

Article

Marine-Inspired Bis-indoles Possessing Antiproliferative Activity against Breast Cancer; Design, Synthesis, and Biological Evaluation

Wagdy M. Eldehna [1], Ghada S. Hassan [2], Sara T. Al-Rashood [3,*], Hamad M. Alkahtani [3], Abdulrahman A. Almehizia [3] and Ghada H. Al-Ansary [4,5]

1   Department of Pharmaceutical Chemistry, Faculty of Pharmacy, Kafrelsheikh University, Kafr El-Sheikh 33516, Egypt; wagdy2000@gmail.com
2   Department of Medicinal Chemistry, Faculty of Pharmacy, Mansoura University, Mansoura 35516, Egypt; Ghadak25@yahoo.com
3   Department of Pharmaceutical Chemistry, College of Pharmacy, King Saud University, P.O. Box 2457, Riyadh 11451, Saudi Arabia; haalkahtani@ksu.edu.sa (H.M.A.); mehizia@ksu.edu.sa (A.A.A.)
4   Department of Pharmaceutical Chemistry, Faculty of Pharmacy, Ain Shams University, Cairo P.O. Box 11566, Egypt; ghada.mohamed@pharma.asu.edu.eg
5   Department of Pharmaceutical Chemistry, Pharmacy Program, Batterejee Medical College, Jeddah P.O. Box 6231, Saudi Arabia
*   Correspondence: salrashood@ksu.edu.sa; Tel.: +966-555-295-929

Received: 6 March 2020; Accepted: 29 March 2020; Published: 2 April 2020

Abstract: Diverse indoles and bis-indoles extracted from marine sources have been identified as promising anticancer leads. Herein, we designed and synthesized novel bis-indole series **7a–f** and **9a–h** as Topsentin and Nortopsentin analogs. Our design is based on replacing the heterocyclic spacer in the natural leads by a more flexible hydrazide linker while sparing the two peripheral indole rings. All the synthesized bis-indoles were examined for their antiproliferative action against human breast cancer (MCF-7 and MDA-MB-231) cell lines. The most potent congeners **7e** and **9a** against MCF-7 cells (IC$_{50}$ = 0.44 ± 0.01 and 1.28 ± 0.04 µM, respectively) induced apoptosis in MCF-7 cells (23.7-, and 16.8-fold increase in the total apoptosis percentage) as evident by the externalization of plasma membrane phosphatidylserine detected by Annexin V-FITC/PI assay. This evidence was supported by the Bax/Bcl-2 ratio augmentation (18.65- and 11.1-fold compared to control) with a concomitant increase in the level of caspase-3 (11.7- and 9.5-fold) and p53 (15.4- and 11.75-fold). Both compounds arrested the cell cycle mainly in the G2/M phase. Furthermore, **7e** and **9a** displayed good selectivity toward tumor cells (*S.I.* = 38.7 and 18.3), upon testing of their cytotoxicity toward non-tumorigenic breast MCF-10A cells. Finally, compounds **7a, 7b, 7d, 7e**, and **9a** were examined for their plausible CDK2 inhibitory action. The obtained results (% inhibition range: 16%–58%) unveiled incompetence of the target bis-indoles to inhibit CDK2 significantly. Collectively, these results suggested that herein reported bis-indoles are good lead compounds for further optimization and development as potential efficient anti-breast cancer drugs.

Keywords: marine-inspired; breast cancer; bis-indoles; synthesis; apoptosis

---

1. Introduction

Drug discovery from marine sources is a prehistoric praxis. Recently, the identification and development of novel molecules based on natural heterocyclic scaffold have been an area of growing focus. Surveying the literature reveals that the anticancer activity of diverse bis-indole compounds extracted from marine sources including plants, fungi, algae, and marine mollusks was broadly discussed in diverse manuscripts [1,2].

Diverse substituted indole and bis-indole derivatives extracted from marine sources have been shown to exhibit significant antiproliferative activity [3,4]. Recently, Edwards et al. [5] conducted a study on purified 6-bromoisatin (Figure 1) extracted from the Australian marine mollusk *Dicathais orbita* known for its antineoplastic activity. The study revealed that 6-bromoisatin markedly reduced the proliferation and concomitantly induced apoptosis in human colon cancer cell lines HT29 and Caco2 cells [5]. A latter study was conducted on the same isatin derivative by Esmaeelian et al. [6] which supported the efficacy of 6-bromoisatin at a concentration of 0.05 mg/g to induce apoptosis in colorectal cancer cells leading ultimately to inhibition of cancer proliferation [6].

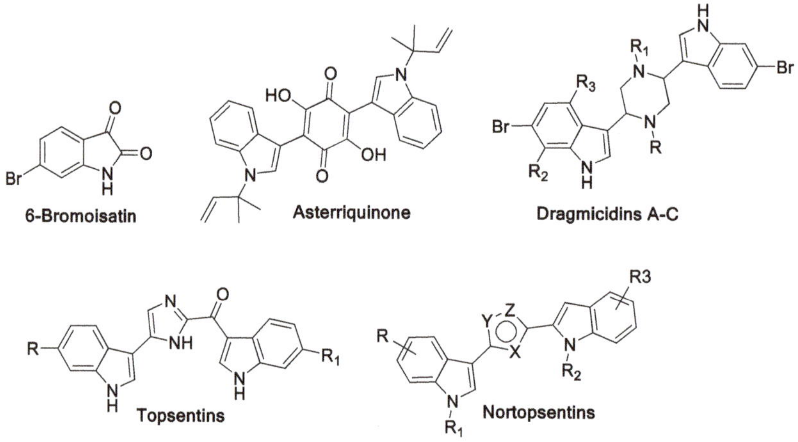

Figure 1. Indole and bis-indole marine products that have reported anticancer activity.

Moreover, many bis-indole derivatives isolated from marine sources manifest anticancer activity. Asterriquinone (Figure 1), isolated from Aspergillus fungi, possesses symmetrical bis-indole moieties separated by quinone spacer and showed in vivo activity against Ehrlich carcinoma, ascites hepatoma AH13, and mouse P388 leukemia [7]. In addition, Dragmacidins A–C, isolated from a large number of deepwater sponges showed modest cytotoxic activity, Figure 1 [8–10]. Topsentins (Figure 1), extracted from the Mediterranean sponge *Topsentia genitrix*, exhibited antitumor and antiviral activities [11,12]. Nortopsentins A–C (Figure 1), that feature imidazole ring spacer, were isolated from *Spongosorites ruetzleri* and showed in vitro cytotoxicity against P388 cells [13,14].

As the reservoir of living organisms is inevitably limited, it became an urgent necessity for medicinal chemists to synthesize biologically active natural molecules and their derivatives to meet the expanding need of such medicinal agents.

Inspired by the aforementioned discoveries and in connection with our research work concerning the development of effective anti-breast cancer agents [15–18], we were endeavored to design novel indole derivatives that promisingly possess antiproliferative activity. Perceiving the significance of these facts and based on our growing research interest in marine natural products, we were persuaded to tackle this study to design and synthesize marine-inspired bis-indole derivatives that have potential in vitro antitumor activity against breast cancer. Herein, we designed and synthesized three novel bis-indole sets **7a–f**, **9a–h**, and **11** as Topsentin and Nortopsentin analogs. Our design was based on replacing the rigid heterocyclic spacer in the natural products by a more flexible hydrazide linker while sparing the two peripheral indole rings to furnish the first set of target compounds **7a–f** (Figure 2). Thereafter, the oxindole moiety was decorated with different *N*-alkyl (allyl, n-propyl, iso-butyl; compounds **9a–c**) and *N*-benzyl (compounds **9d–h**) substituents to fulfill further elaboration for the target bis-indoles and to probe a worthy structure-activity relationship (SAR). Furthermore, a bioisosteric replacement approach was adopted to replace the oxindole ring with carbocyclic tetralin ring (compound **11**), to explore the significance of the bis-indole scaffold, Figure 2.

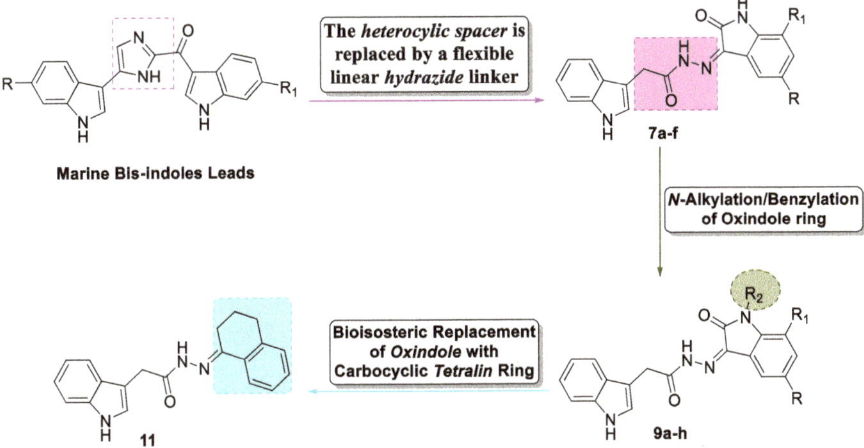

Figure 2. Structure-based design of target bis-indole derivatives (**7a–f** and **9a–h**), and **11**.

In this study, all of the synthesized compounds **7**, **9** and **11** were evaluated for their antiproliferative activity against MCF-7 cells and MDA-MB-231 cancer cell lines. Three of the most potent compounds induced apoptosis in MCF-7 cells as evidenced by the externalization of plasma membrane phosphatidylserine detected by Annexin V-FITC/PI dual staining assay. This evidence was supported by the Bax/Bcl-2 ratio augmentation with a concomitant increase in the level of caspase-3 and p53. Moreover, scrutinizing the results of cell cycle analysis unraveled that these compounds arrest the cell cycle in the G0/G1 phase.

2. Results

2.1. Chemistry

The synthetic pathways proposed to obtain the target bis-indole derivatives (**7a–f** and **9a–h**), and **11** were depicted in Schemes 1 and 2. In Scheme 1, the Fischer esterification procedure was applied to 3-indoleacetic acid **3** [19] to afford methyl 1*H*-indole-2-carboxylate **4**, which subsequently undergone hydrazinolysis through reaction with 99% hydrazine hydrate in ethyl alcohol under reflux temperature to furnish key intermediate 1*H*-indole-2-carbohydrazide **5** in 82% yield.

Thereafter, 1*H*-indole-2-carbohydrazide **5** was condensed with different *N*-unsubstituted 1*H*-indole-2,3-diones **6a–f**, *N*-substituted 1*H*-indole-2,3-diones **8a–h**, or 1-tetralone **10** in absolute ethyl alcohol with catalytic drops of acetic acid to produce target indole derivatives **7a–f**, **9a–h**, and **11**, respectively (Schemes 1 and 2).

Postulated structures of the herein reported indole derivatives **7a–f**, **9a–h**, and **11** are in full agreement with the spectral and elemental analyses data.

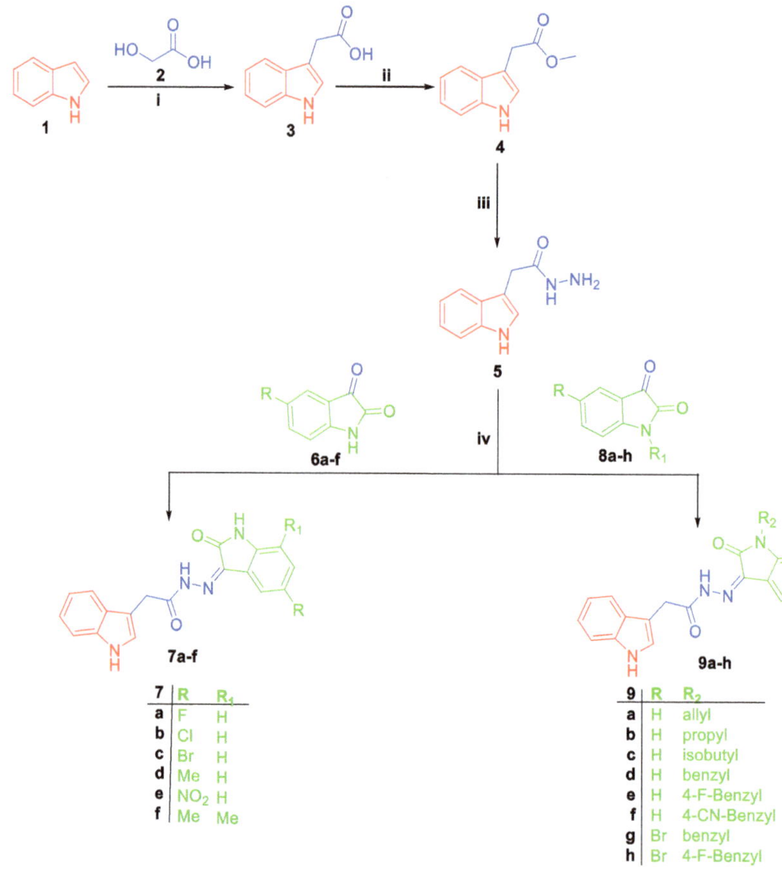

Scheme 1. Synthesis of target bis-indole derivatives **7a–f** and **9a–h**; Reagents and conditions: (i) (a) KOH / heating at 250 °C 18 h, (b) H_2O, cooling to 10 °C, HCl; (ii) $MeOH/H_2SO_4$ (catalytic)/reflux 8 h; (iii) 99% $NH_2NH_2.H_2O/EtOH/reflux$ 3 h; (iv) EtOH/AcOH (catalytic)/reflux 2 h.

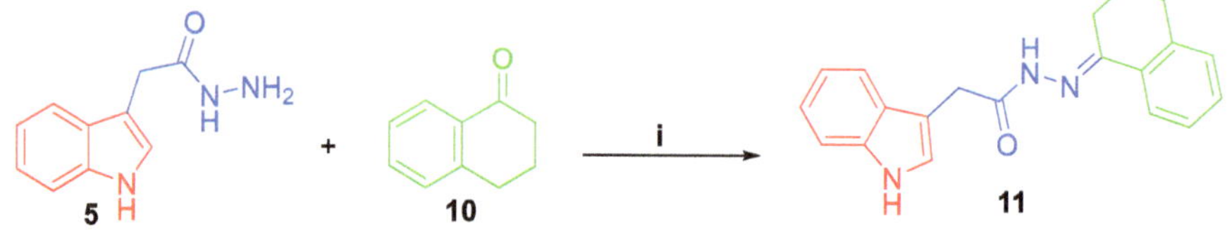

Scheme 2. Synthesis of target compound **11**; Reagents and conditions: (i) EtOH/AcOH (catalytic)/reflux 2 h.

2.2. Biological Evaluation

2.2.1. Antiproliferative Activity against Breast Cancer MCF-7 and MDA-MB-231

The biological evaluation journey started by exploring the antiproliferative activity of the pursed indole derivatives (**7a–f**, **9a–h**, and **11**) against breast cancer cell line; MCF-7 and triple-negative breast cancer cell line; MDA-MB-231, adopting procedures of the sulforhodamine B colorimetric (SRB) assay [20]. Staurosporine was utilized as the reference drug for its well-known broad anticancer activity against diverse tumors.

All of the tested indole derivatives exhibited gradual cellular log kill with IC_{50} values ranging from 0.44 µM to 47.1 µM against breast cancer cell line; MCF-7, while they exerted a much wider range of antiproliferative activity against MDA-MB-231 cell line with IC_{50} values ranging from 0.34 µM up to 77.30 µM, aside from compound **11** which showed very weak antiproliferative activity against MCF-7 cell line (IC_{50} = 84.70 µM) and no antiproliferative activity against MDA-MB-231 cell line (IC_{50} > 100 µM) in a proof of concept of the importance of oxindole moiety for boosting the antiproliferative activity.

Scrutinizing the IC_{50} values of series **7a–f** against MCF-7 cell line revealed that grafting a halide atom on the isatin moiety interestingly influences the antiproliferative activity, where the activity significantly decreased by increasing the size of the halide detected by the IC_{50} values of the floro (**5a**), chloro (**5b**), and bromo (**5c**) derivatives (IC_{50} = 1.53 µM, 8.87 µM, and 36.19 µM, respectively). This suggested that a floro substitution on the isatin group is advantageous for the antiproliferative activity where compound **5a** (IC_{50} = 1.53 µM) is 4.45 times more potent than the reference drug (IC_{50} = 6.81 µM), Table 1.

Table 1. In vitro antiproliferative activity of **7a–f**, **9a–h**, and **11** against breast MCF-7 and MDA-MB-231 cancer cell lines.

Cpd.	R	R_1	R_2	IC_{50} (µM) [a]	
				MCF-7	**MDA-MB-231**
7a	F	H	-	1.53 ± 0.02	9.04 ± 0.32
7b	Cl	H	-	8.87 ± 0.43	2.88 ± 0.08
7c	Br	H	-	36.19 ± 2.78	36.57 ± 1.81
7d	CH_3	H	-	10.95 ± 0.81	0.34 ± 0.02
7e	NO_2	H	-	0.44 ± 0.01	1.32 ± 0.03
7f	CH_3	CH_3	-	NA [b]	77.30 ± 6.21
9a	H	-	$-CH_2CH=CH_2$	1.28 ± 0.04	18.24 ± 0.62
9b	H	-	$-CH_2CH_2CH_3$	28.24 ± 1.53	51.27 ± 3.59
9c	H	-	$-CH_2CH(CH_3)_2$	47.10 ± 3.65	NA [b]
9d	H	-	$-CH_2C_6H_5$	1.51 ± 0.03	4.14 ± 0.19
9e	H	-	$-CH_2C_6H_4$-4-F	10.43 ± 0.81	17.66 ± 0.55
9f	H	-	$-CH_2C_6H_4$-4-CN	8.72 ± 0.39	25.41 ± 1.56
9g	Br	-	$-CH_2C_6H_5$	2.76 ± 0.14	2.85 ± 0.07
9h	Br	-	$-CH_2C_6H_4$-4F	20.89 ± 0.04	2.29 ± 0.09
11	-	-	-	84.70 ± 4.02	NA [b]
Staurosporine	-	-	-	6.81 ± 0.22	10.29 ± 0.72

[a] IC_{50} values are the mean ± S.D. of three separate experiments. [b] NA: Compounds having IC_{50} value > 100 µM.

Furthermore, the influence of grafting electron-donating and electron-withdrawing groups within the oxindole moiety on the antiproliferative activity of the MCF-7 cell line was closely investigated. Interestingly, substitution with the electron-donating (CH_3) group, compound **5d**, decreased the activity in comparison to Staurosporine (IC_{50} = 10.95 µM vs. 6.81 µM), whereas, grafting an electron-withdrawing nitro group on the oxindole moiety, compound **5e,** markedly enhanced the antiproliferative potency (IC_{50} = 0.44 µM), which is, fortunately, 15.5-times the potency of Staurosporine. Noteworthy, decoration of the oxindole moiety with 5,7-dimethyl substitution, compound **7f**, resulted in the abolishment of the growth inhibitory action toward MCF-7 cells (IC_{50} > 100 µM), Table 1. Conclusively, grafting a fluorine atom or a nitro group on C-5 of the oxindole moiety significantly boosts the activity against MCF-7 cells with a more pronounced effect for the nitro group.

As part of our work, we extended our investigation to explore the effect of substituting the nitrogen atom of the oxindole moiety by different alkyl and benzyl moieties in series **9**. Exploring the IC_{50} values unraveled that substitution of the nitrogen atom with an allyl group in **9a** significantly increased the activity 5.3-times compared to the reference drug (IC_{50} = 1.28 µM vs. 6.81 µM, respectively). Conversely,

N-substitution with propyl group in **9b** (IC_{50} = 28.24 µM) and isobutyl group in **9c** (IC_{50} = 47.1 µM) markedly dwindled the growth inhibitory activity toward MCF-7 cell line.

Alternatively, substitution of the nitrogen atom with a benzyl moiety in **9d** (IC_{50} = 1.51 µM) significantly increased the activity 4.5-times in comparison to Staurosporine. Conversely, utilizing substituted benzyl groups in **9e** (4-F-benzyl) and **9f** (4-cyanobenzyl) was not advantageous for the activity as their IC_{50} values were less than the reference drug (IC_{50} = 10.43 µM and 8.72 µM, respectively). Moreover, two compounds (**9g** and **9h**) were synthesized as analogues of compounds **9d** and **9e**, where the oxindole moiety was further substituted with a bromo group in the 5-position. Investigation of the IC_{50} values of **9g** and **9h** (IC_{50} = 2.76 and 20.89 µM, respectively) clearly depicts a deterioration of the activity to the half as compared to **9d** and **9e**. This suggested that bromination of oxindole ring is not advantageous for the antiproliferative activity, an observation that is in accordance with the structure activity relationship extracted from series **7** (compound **7c**, IC_{50} = 36.19 µM).

Triple negative breast cancer (TNBC) is a stubborn type of cancer resistant to many chemotherapeutic agents, thus it represents a powerful challenge for medicinal chemists. Accordingly, we evaluated the potential antiproliferative activity for our compounds against TNBC cell line; MDA-MB-231 (Table 1). Analyzing the IC_{50} values of series **7a–f** and **9a–h** reveals very interesting results as many of the synthesized derivatives (**7a**, **7b**, **7d**, **7f**, **9d**, **9g** and **9h**) exhibited superior potencies compared to Staurosporine (IC_{50} = 9.04 µM, 2.88 µM, 0.34 µM, 1.32 µM, 4.14 µM, 2.85 µM, 2.29 µM and 10.29 µM, respectively). The IC_{50} values of series **7** unraveled that grafting a fluoro (**7a**) or a chloro (**7b**) group on the oxindole ring results in activity enhancement (IC_{50} = 9.04 µM and 2.88 µM, respectively), while grafting of a bromo group (**7c**) markedly decreased the activity (IC_{50} = 36.57 µM) in comparison to Staurosporine (IC_{50} = 10.29 µM). Moreover, substitution with a methyl group (**7d**) and a nitro group (**7e**) interestingly resulted in boosting the activity by 30.3- and 7.8-times, respectively. The effect of the N-substituent of the oxindole ring was further investigated in series **9**. Substitution of the nitrogen atom with an allyl group (**9a**) did not result in enhancement of the antiproliferative activity compared to Staurosporine (IC_{50} = 18.24 µM vs. 10.29 µM). Extending the substituent to propyl or isobutyl even worsens the case producing much less potent derivatives (**9b**) (IC_{50} = 51.27 µM) and (**9c**) which failed to produce any marked cytotoxic effect up to 100 µM. These results are in accordance with the observed results for series **7** where the addition of a larger or branched alkyl group proved to be detrimental to the antiproliferative action against MDA-MB-231 cell line (Table 1).

In addition, the impact of substitution of the nitrogen atom by un/substituted benzyl moieties was explored, revealing that unsubstituted benzyl moiety (**9d**) is advantageous for activity as it enhanced the activity by 2.5-times while utilizing 4-F-benzyl group (**9e**) or 4-CN-benzyl group (**9f**) resulted in a marked decrease of the activity compared to compound **9d** (IC_{50} = 17.66 µM and 25.41 µM vs. 4.14 µM). The antiproliferative activity for the 5-bromo substituted analogs (**9g** and **9h**) against the MDA-MB-231 cell line was compared to that of **9d** and **9e**. The results revealed that, in contrast to the results observed for the antiproliferative activity against MCF-7 cell line, the bromo substitution of the oxindole ring was advantageous for the activity as compound **9g** (IC_{50} = 2.85 µM) proved to be 1.5-times more potent than its unsubstituted bioisostere **9d**, also, compound **9h** (IC_{50} = 2.29 µM) proved even to be 7.7-times more potent than **9e** counterpart.

2.2.2. In Vitro Cytotoxic Activity against Non-Tumorigenic Human Breast Cell Line

To investigate the selectivity and safety profile for the here reported bis-indoles toward the normal cells, compounds that displayed good activity towards MCF-7 and/or MDA-MB-231 cells were examined for their cytotoxic activity against non-tumorigenic human breast epithelial (MCF-10A) cell line (Table 2).

Table 2. Cytotoxic activity toward non-tumorigenic human breast MCF-10A cell line, and selectivity index (MCF-10A/MCF-7).

Comp.	IC$_{50}$ (µM)		Selectivity Index
	MCF-10A	**MCF-7**	**MCF-10A/MCF-7**
7a	14.06	1.53	9.2
7b	39.54	8.87	4.5
7d	54.92	10.95	5.0
7e	17.06	0.44	38.7
9a	23.47	1.28	18.3
9d	19.12	1.51	12.7
9e	48.39	10.43	4.7
9f	42.01	8.72	4.8
9g	17.16	2.76	6.2
9h	26.09	20.89	1.2

The examined bis-indoles exerted non-significant or modest cytotoxic action against non-tumorigenic MCF-10A cells with IC$_{50}$ range: 14.06–54.92 µM, respectively. Bis-indoles **7e** and **9a** showed excellent selectivity indexes (SIs) equal to 38.7 and 18.3, respectively, whereas the remaining compounds, except **9h**, displayed good SIs spanning in the range 4.5–12.7 (Table 2).

2.2.3. Cell Cycle Analysis

Anticancer agents exert their cytotoxic action by aborting cellular proliferation at certain checkpoints. These checkpoints are distinguishable phases in the cell cycle, whose suppression results in termination of the cell proliferation. To deeply comprehend the antiproliferative activity of our tested compounds, the most active two compounds (**7e** and **9a**) toward MCF-7 cells were further investigated for their effect on the different phases of the cell cycle in MCF-7 cell line. MCF-7 cells were treated with IC$_{50}$ concentrations of the two compounds and their effect on the cell population in different cell phases was recorded and displayed in Table 3. Interestingly, exposure of MCF-7 cells to **7e** and **9a** resulted in marked augmentation in the proportion of cells in the G2/M phase by 3- and 2.21-fold, and in the Sub-G1 phase by 18.77- and 13.42-fold, respectively, in comparison to the control. This clearly indicates that the target bis-indoles arrested the cell cycle proliferation of MCF-7 cells in the G2/M phase.

Table 3. Effect of compounds **7e** and **9a** on the phases of the cell cycle of MCF-7 cells.

Comp.	%G0-G1	%S	%G2/M	%Sub-G1
7e	31.66	25.44	42.9	33.61
9a	43.82	24.91	31.27	24.02
Control	57.26	28.59	14.15	1.79

2.2.4. Effect of **5e**, **5f**, and **8a** on the Level of the Apoptotic Markers (Bax, Bcl-2, caspase-3, and p53)

Synchronization of the mitochondrial pathway is headed by the Bcl-2 family of proteins. These proteins are classified into two groups: anti-apoptotic proteins exemplified by Bcl-2 protein and the counteracting pro-apoptotic proteins including Bax protein [21]. As induction of the apoptotic machinery is one of the most useful strategies in cancer therapy [22,23], we investigated the effect of our compounds to boost the pro-apoptotic protein; Bax and reduce the anti-apoptotic protein; Bcl-2 in an attempt to explore the underlying mechanism for their cytotoxic activity (Table 1). As **7e** and **9a** proved to be the most active compounds, their effect on the level of Bax and Bcl-2 was investigated. Fortunately, both compounds markedly boosted the level of Bax by 8.3- and 6.4-fold, respectively (Table 4, Figure 3). Conformingly, they decreased the level of Bcl-2 by 2.25- and 1.74-fold, respectively. A more indicative parameter is the Bax/Bcl-2 ratio [24], which proved to be augmented by **7e** and **9a**

18.65- and 11.1-fold, respectively (Table 4, Figure 3). This further emphasizes that target bis-indoles trigger apoptosis by significantly boosting the Bax/Bcl-2 ratio.

Table 4. Effect of bis-indoles **7e** and **9a** on the expression levels of Bcl-2 and Bax in MCF-7 cancer cells.

Compound	Bax (pg/mg of Total Protein)	Bcl-2 (ng/mg of Total Protein)	Bax/Bcl-2
7e	318.0 ± 10.5	2.07 ± 0.14	153.6
9a	243.6 ± 12.4	2.67 ± 0.16	91.2
Control	38.3 ± 2.2	4.65 ± 0.23	8.2

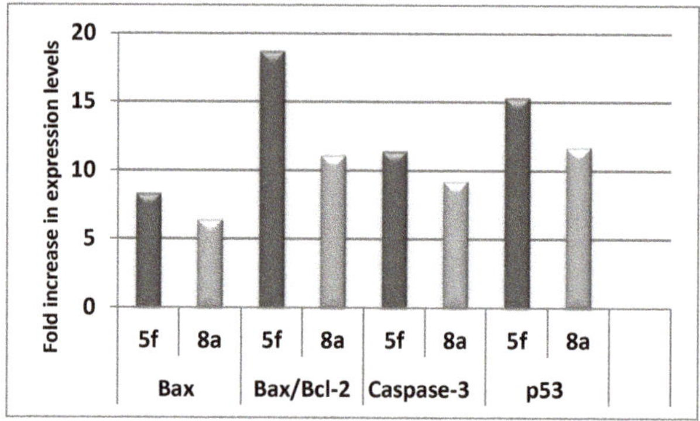

Figure 3. The numbers of fold increase in Bax/Bcl-2 ratio and expression levels of Bax, caspase-3, and p53 in MCF-7 cancer cells upon treatment with compounds **7e** and **9a** in comparison to the control.

Moreover, their effect on the level of caspase-3; the executioner caspase and p53 was evaluated in the MCF-7 cell line. Results revealed that **7e** and **9a** up-regulated the level of caspase-3 by 11.7- and 9.5-fold, respectively as compared to the control (Table 5). In addition, they augmented the level of p53 by 15.4- and 11.75-fold, respectively in comparison to the control (Table 5, Figure 3).

Table 5. Effect of compounds **7e** and **9a** on the expression levels of active caspase-3 and p53 in MCF-7 cancer cells.

Compound	Caspase-3 (pg/mg)	p53 (pg/mg)
7e	409.2 ± 17.2	631.8 ± 35.8
9a	331.0 ± 12.5	482.3 ± 27.4
Control	35.92 ± 1.8	41.26 ± 2.7

2.2.5. Annexin V-FITC Apoptosis Assay

Annexin V-based flow cytometry analysis is a useful tool to investigate whether cell death is pertaining to physiological apoptosis or nonspecific necrosis. Evaluation of the apoptotic effect of bis-indoles **7e** and **9a** was carried out using AnxV-FITC/DAPI dual staining assay (Figure 4).

Treatment of MCF-7 cells with IC_{50} concentration of **7e** and **9a** exerted marked increase in the AnxV-FITC apoptotic cells percentage in both early (from 1.03% to 9.56% and 7.01%, respectively) and late apoptosis (from 0.29% to 21.76% and 15.20%, respectively) phases, Table 6. This corresponds to an increase in the total apoptosis percentage by 23.7-, and 16.8-fold, respectively compared to the control. This proved that the antiproliferative activity of here reported bis-indoles is due to physiological apoptosis, not nonspecific necrosis.

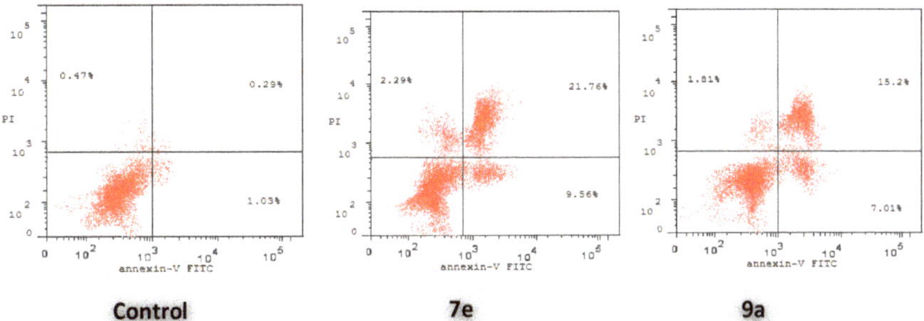

Figure 4. Influence of bis-indoles **7e** and **9a** on the percentage of annexin V-FITC-positive staining in MCF-7 cells. (Lower right: early apoptotic; upper right: late apoptotic; lower left: viable; upper left: necrotic).

Table 6. Distribution of apoptotic cells in the AnnexinV-FITC/PI dual staining assay in MCF-7 cells after treatment with bis-indoles **7e** and **9a**.

Compound	Early Apoptosis (Lower Right %)	Late Apoptosis (Upper Right %)	Total (L.R % + U.R %)
7e	9.56	21.76	31.32
9a	7.01	15.20	22.21
Control	1.03	0.29	1.32

2.2.6. CDK2 Inhibitory Activity

The efficient cell cycle disturbance influence of the herein reported bis-indoles (Table 7) prompted a more examination for their plausible inhibitory action toward the cell cycle regulator CDK2 protein kinase, in an attempt to gain further mechanistic insights for their promising growth inhibitory effect. The potent antiproliferative agents **7a**, **7b**, **7d**, **7e**, and **9a** were examined for their % inhibition of CDK2 at a single dose of 10 µM, (Table 7).

Table 7. Inhibitory effect of bis-indoles **7a**, **7b**, **7d**, **7e**, and **9a** against CDK2 kinase activity at a single dose of 10 µM.

Compound	% Enzyme Inhibitory Activity
7a	41
7b	41
7d	58
7e	16
9a	23
Staurosporine	99

As displayed in Table 7, the examined bis-indoles exerted moderate to weak CDK2 inhibition with a % inhibition range of 16%–58%. Compound **7d** displayed the best % inhibition against CDK2 equals 58, whereas, both **7a** and **7b** showed % inhibition equals 44, Table 7.

These results unveiled incompetence of the target bis-indoles to inhibit CDK2 significantly, highlighting that the cell growth inhibitory and cell cycle arrest capabilities of the target bis-indoles toward the examined human breast cancer cell lines is attributable to another target rather than CDK. Accordingly, further optimization for the herein reported bis-indoles including many mechanistic investigations are in progress and will be reported upon in the future.

3. Experimental

3.1. Chemistry

3.1.1. General

Melting points were measured with a Stuart melting point apparatus and were uncorrected. Infrared spectra were recorded on Schimadzu FT-IR 8400S spectrophotometer. The NMR spectra were obtained on JEOL ECA-500 II spectrophotometer (500 MHz ^{1}H and 125 MHz ^{13}C NMR), in deuterated dimethylsulfoxide (DMSO-d_6). Chemical shifts (δ_H) are reported relative to TMS as the internal standard. All coupling constant (*J*) values are given in hertz. Chemical shifts (δ_C) are reported relative to DMSO-d_6 as internal standards. Elemental analyses were carried out at the Regional Center for Microbiology and Biotechnology, Al-Azhar University. Compounds methyl 1*H*-indole-2-carboxylate **4** and 1*H*-indole-2-carbohydrazide **5** were prepared as reported earlier [25].

3.1.2. General Procedure for Synthesis of the Target Bis-indoles (**7a–f** and **9a–h**), and **11**

To a hot stirred solution of key intermediate 1*H*-indole-2-carbohydrazide **5** (0.18 gm, 1 mmoL) in absolute ethyl alcohol (7 mL) and glacial acetic acid (catalytic amount), the appropriate *N*-unsubstituted 1*H*-indole-2,3-dione **6a–f**, *N*-substituted 1*H*-indole-2,3-dione **8a–h**, or 1-tetralone **10** (1 mmoL) was added. The resulting mixture was refluxed for 2 hours, and then the formed solid was filtered off while hot, washed with cold isopropyl alcohol, dried and recrystallized from DMF to afford target bis-indoles (**7a–f** and **9a–h**), and **11**, respectively.

N′-(5-Fluoro-2-oxoindolin-3-ylidene)-2-(1H-indol-3-yl)acetohydrazide (**7a**)

Red powder, m.p. 281–283 °C; (yield 70%), IR: 3327, 3270 (NH) and 1694 (C=O); ^{1}H NMR δ *ppm*: 4.19 (brs, 2H, C$\underline{H}_2$-C=O), 6.85–6.88 (m, 1H, Ar-H), 6.95 (m, 1H, Ar-H), 7.05 (t, 1H, Ar-H, *J* = 7.5 Hz), 7.19 (t, 1H, Ar-$\overline{\text{H, }J}$ = 7.5 Hz), 7.32 (m, 2H, Ar-H), 7.58 (d, 1H, Ar-H, *J* = 7.5 Hz), 8.14 (s, 1H, Ar-H), 10.82 (s, 1H, NH of isatin, D$_2$O exchangeable), 10.96 (s, 1H, NH of hydrazide, D$_2$O exchangeable), 11.23 (s, 1H, NH of indol, D$_2$O exchangeable); ^{13}C NMR δ *ppm*: 28.26 ($\underline{\text{C}}$H$_2$-C=O), 107.26, 107.89, 111.33, 111.42, 113.06, 113.26, 115.58, 115.65, 18.52, 118.63, 121.07, 124.56, 127.28, 136.02, 138.54, 139.95, 156.56, 158.44, 162.57 (C=O of isatin), 164.81 (C=O of hydrazide); MS *m/z* [%]: 356.15 [M$^+$, 100], 157.03 [11.35], 130.12 [25.10]; Anal. Calcd. for C$_{18}$H$_{13}$FN$_4$O$_2$: C, 64.28; H, 3.90; N, 16.66; found C, 63.93; H, 3.94; N, 16.73.

N′-(5-Chloro-2-oxoindolin-3-ylidene)-2-(1H-indol-3-yl)acetohydrazide (**7b**)

Orange powder, m.p. 292–294 °C; (yield 74%), IR: 3450, 3338 (NH) and 1694 (C=O); ^{1}H NMR δ *ppm*: 4.11 (brs, 2H, C$\underline{H}_2$-C=O), 6.87 (d, 1H, Ar-H, *J* = 8.5 Hz), 6.96-6.99 (m, 1H, Ar-H), 7.05 (t, 1H, Ar-H, *J* = 7.5 Hz), 7.31-7.35 (m, 2H, Ar-H), 7.38 (d, 1H, Ar-H, *J* = 8.0 Hz), 7.57 (d, 1H, Ar-H, *J* = 8.0 Hz), 8.33 (s, 1H, Ar-H), 10.92 (s, 1H, NH of isatin, D$_2$O exchangeable), 11.20 (s, 1H, NH of hydrazide, D$_2$O exchangeable), 11.33 (s, 1H, NH of indol, D$_2$O exchangeable); Anal. Calcd. for C$_{18}$H$_{13}$ClN$_4$O$_2$: C, 61.28; H, 3.71; N, 15.88; found C, 60.95; H, 3.66; N, 15.97.

N′-(5-Bromo-2-oxoindolin-3-ylidene)-2-(1H-indol-3-yl)acetohydrazide (**7c**)

Orange powder, m.p. > 300 °C; (yield 85%), IR: 3365, 3219, 3182 (NH) and 1726, 1692 (C=O); ^{1}H NMR δ *ppm*: 4.11 (brs, 2H, C$\underline{H}_2$-C=O), 6.83 (d, 1H, Ar-H, *J* = 8.5 Hz), 6.96–6.99 (m, 1H, Ar-H), 7.04 (t, 1H, Ar-H, *J* = 7.5 Hz), 7.31-7.35 (m, 2H, Ar-H), 7.51 (d, 1H, Ar-H, *J* = 8.0 Hz), 7.57 (d, 1H, Ar-H, *J* = 8.0 Hz), 8.43 (s, 1H, Ar-H), 10.93 (s, 1H, NH of isatin, D$_2$O exchangeable), 10.96 (s, 1H, NH of hydrazide, D$_2$O exchangeable), 11.33 (s, 1H, NH of indol, D$_2$O exchangeable); Anal. Calcd. for C$_{18}$H$_{13}$BrN$_4$O$_2$: C, 54.43; H, 3.30; N, 14.10; found C, 54.85; H, 3.28; N, 14.02.

2-(1H-Indol-3-yl)-N'-(5-methyl-2-oxoindolin-3-ylidene)acetohydrazide (**7d**)

Red powder, m.p. > 300 °C; (yield 78%), IR: 3363, 3309, 3191 (NH) and 1691 (C=O); ^{1}H NMR δ *ppm*: 2.24, 2.50 (2s, 3H, CH$_3$), 3.87, 4.18 (2s, 2H, CH$_2$-C=O), 6.76-6.81 (m, 1H, Ar-H), 6.97 (t, 1H, Ar-H, *J* = 7.5 Hz), 7.07-7.14 (m, 2H, Ar-H), 7.30–7.59 (m, 4H, Ar-H), 10.95 (s, 1H, NH of isatin, D$_2$O exchangeable), 11.08, 11.12 (2s, 1H, NH of hydrazide, D$_2$O exchangeable), 12.51, 12.96 (2s, 1H, NH of indol, D$_2$O exchangeable); ^{13}C NMR δ *ppm*: 18.57, 20.53, 18.14, 32.32, 106.38, 107.03, 110.83, 111.42, 111.52, 118.45, 118.67, 119.87, 120.92, 121.04, 121.29, 124.32, 124.81, 126.99, 127.27, 131.61, 131.79, 140.02, 162.55, 168.32, 173.32; MS *m/z* [%]: 332.11 [M$^+$, 80.08], 157.24 [40.03], 130.19 [100]; Anal. Calcd. for C$_{19}$H$_{16}$N$_4$O$_2$: C, 68.66; H, 4.85; N, 16.86; found C, 68.83; H, 4.80; N, 16.97.

2-(1H-Indol-3-yl)-N'-(5-nitro-2-oxoindolin-3-ylidene)acetohydrazide (**7e**)

Yellow powder, m.p. 289–291 °C; (yield 80%), IR: 3399, 3147 (NH) and 1728, 1686 (C=O); ^{1}H NMR δ *ppm*: 4.13 (brs, 2H, CH$_2$-C=O), 6.95-7.00 (m, 1H, Ar-H), 7.04–7.08 (m, 2H, Ar-H), 7.33–7.35 (m, 2H, Ar-H), 7.58 (d, 1H, Ar-H, *J* = 8.0 Hz), 8.19-8.28 (m, 1H, Ar-H), 9.04 (s, 1H, Ar-H), 10.96 (s, 1H, NH of isatin, D$_2$O exchangeable), 11.50 (s, 1H, NH of hydrazide, D$_2$O exchangeable), 11.81 (s, 1H, NH of indol, D$_2$O exchangeable); ^{13}C NMR δ *ppm*: 18.57 (CH$_2$-C=O), 111.31, 111.40, 111.49, 115.08, 115.62, 118.51, 118.63, 120.70, 121.03, 121.48, 127.21, 136.00, 136.12, 141.99, 142.75, 147.42, 149.13, 162.76 (C=O of isatin), 165.04 (C=O of hydrazide); Anal. Calcd. for C$_{18}$H$_{13}$N$_5$O$_4$: C, 59.50; H, 3.61; N, 19.28; found C, 59.67; H, 3.57; N, 9.34.

N'-(5,7-Dimethyl-2-oxoindolin-3-ylidene)-2-(1H-indol-3-yl)acetohydrazide (**7f**)

Brown powder, m.p. 294–296 °C; (yield 73%), IR: 3432, 3267, 3180 (NH) and 1725, 1692 (C=O); ^{1}H NMR δ *ppm*: 2.13 (s, 3H, CH$_3$ of isatin), 2.20 (s, 3H, CH$_3$ of isatin), 4.14 (brs, 2H, CH$_2$-C=O), 6.96–6.99 (m, 2H, Ar-H), 7.07 (brs, 1H, Ar-H), 7.35 (br s, 2H, Ar-H), 7.58 (d, 1H, Ar-H, *J* = 8.0 Hz), 7.84 (s, 1H, Ar-H), 10.71 (s, 1H, NH of isatin, D$_2$O exchangeable), 10.98 (s, 1H, NH of hydrazide, D$_2$O exchangeable), 11.06 (s, 1H, NH of indol, D$_2$O exchangeable); ^{13}C NMR δ *ppm*: 15.80, 15.99, 20.34, 107.21, 111.40, 111.48, 115.01, 118.32, 118.59, 119.59, 121.05, 121.27, 124.67, 127.28, 130.55, 131.55, 134.05, 136.05, 139.85, 162.20; Anal. Calcd. for C$_{20}$H$_{18}$N$_4$O$_2$: C, 69.35; H, 5.24; N, 16.17; found C, 69.46; H, 5.20; N, 16.11.

N'-(1-Allyl-2-oxoindolin-3-ylidene)-2-(1H-indol-3-yl)acetohydrazide (**9a**)

Orange powder, m.p. 218–220 °C; (yield 75%), IR: 3354, 3238 (NH) and 1704 (C=O); ^{1}H NMR δ *ppm*: 4.12 (brs, 2H, CH$_2$-C=O), 4.36 (s, 2H, *N*-CH$_2$), 5.13-5.16 (m, 2H, CH$_2$=CH-), 5.82-5.87 (m, 1H, *N*-CH$_2$-CH), 6.96-7.09 (m, 4H, Ar-H), 7.34 (s, 2H, Ar-H), 7.39 (t, 1H, Ar-H, *J* = 8.0 Hz), 7.58 (d, 1H, Ar-H, *J* = 8.0 Hz), 8.16 (s, 1H, Ar-H), 10.99 (s, 1H, NH of hydrazide, D$_2$O exchangeable), 11.02 (s, 1H, NH of indol, D$_2$O exchangeable); ^{13}C NMR δ *ppm*: 29.84 (CH$_2$-C=O), 41.48 (*N*-CH$_2$), 107.22, 109.84, 111.45, 114.73, 116.93, 118.64, 121.15, 122.18, 124.62, 125.60, 131.80, 132.27, 136.06, 143.75, 163.09; MS *m/z* [%]: 358.23 [M$^+$, 41.07], 157.15 [34.19], 130.26 [100]; Anal. Calcd. for C$_{21}$H$_{18}$N$_4$O$_2$: C, 70.38; H, 5.06; N, 15.63; found C, 70.21; H, 5.13; N, 15.75.

2-(1H-Indol-3-yl)-N'-(2-oxo-1-propylindolin-3-ylidene)acetohydrazide (**9b**)

Orange powder m.p. 207–208 °C; (yield 77%), IR: 3317, 3282 (NH) and 1705 (C=O); ^{1}H NMR δ *ppm*: 0.85 (brs, 3H, CH$_2$-CH$_3$), 1.59 (brs, 2H, CH$_2$-CH$_3$), 3.68 (t, 2H, *N*-CH$_2$), 3.89, 4.20 (2s, 2H, CH$_2$-C=O), 6.96 (t, 1H, Ar-H, *J* = 7.5 Hz), 7.07–7.19 (m, 2H, Ar-H), 7.30–7.43 (m, 4H, Ar-H), 7.58 (d, 1H, Ar-H, *J* = 7.5 Hz), 7.69 (s, 1H, Ar-H), 10.95, 11.09 (2s, 1H, NH of hydrazide, D$_2$O exchangeable), 12.44, 12.91 (2s, 1H, NH of indol, D$_2$O exchangeable); Anal. Calcd. for C$_{21}$H$_{20}$N$_4$O$_2$: C, 69.98; H, 5.59; N, 15.55; found C, 70.12; H, 5.62; N, 15.43.

N′-(1-(sec-Butyl)-2-oxoindolin-3-ylidene)-2-(1H-indol-3-yl)acetohydrazide (**9c**)

Orange powder, m.p. 199–201 °C; (yield 68%), IR: 3451, 3294 (NH) and 1707 (C=O); ^{1}H NMR δ *ppm*: 0.87 (d, 6H, -CH(CH$_3$)$_2$, J = 2.5 Hz), 2.03 (brs, 1H, -C$\underline{H}$(CH$_3$)$_2$), 3.52 (d, 2H, N-C$\underline{H_2}$-CH(CH$_3$)$_2$, J = 2.5 Hz), 4.13 (brs, 2H, C$\underline{H_2}$-C=O), 6.97–7.07 (m, 3H, Ar-H), 7.12 (d, 1H, Ar-H, J = 8.5 Hz), 7.34 (brs, 2H, Ar-H), 7.39 (t, 1H, Ar-H, J = 8.5 Hz), 7.59 (d, 1H, Ar-H, J = 7.5 Hz), 8.14 (brs, 1H, Ar-H), 10.98 (s, 1H, NH of hydrazide, D$_2$O exchangeable), 11.18 (s, 1H, NH of indol, D$_2$O exchangeable); ^{13}C NMR δ *ppm*: 19.91 (-CH($\underline{C}$H$_3$)$_2$), 26.62 (-$\underline{C}$H(CH$_3$)$_2$), 46.58 (N-CH$_2$), 107.23, 109.74, 111.44, 114.60, 118.66, 121.12, 122.01, 124.61, 125.58, 127.24, 132.31, 132.45, 136.06, 144.38, 163.55; MS *m/z* [%]: 374.23 [M$^+$, 100], 157.10 [18.76], 130.38 [59.05]; Anal. Calcd. for C$_{22}$H$_{22}$N$_4$O$_2$: C, 70.57; H, 5.92; N, 14.96; found C, 70.69; H, 5.87; N, 15.02.

N′-(1-Benzyl-2-oxoindolin-3-ylidene)-2-(1H-indol-3-yl)acetohydrazide (**9d**)

Red powder, m.p. 223–225 °C; (yield 76%), IR: 3308, 3245 (NH) and 1706 (C=O); ^{1}H NMR δ *ppm*: 4.15 (brs, 2H, C$\underline{H_2}$-C=O), 4.96 (s, 2H, benzylic protons), 6.98 (d, 3H, Ar-H, J = 7.5 Hz), 7.06 (t, 1H, Ar-H, J = 7.5 Hz), 7.25–7.35 (m, 8H, Ar-H), 7.60 (d, 1H, Ar-H, J = 8.5 Hz), 8.17 (brs, 1H, Ar-H), 10.99 (s, 1H, NH of hydrazide, D$_2$O exchangeable), 11.23 (s, 1H, NH of indol, D$_2$O exchangeable); ^{13}C NMR δ *ppm*: 42.64, 107.22, 109.86, 111.45, 114.82, 118.66, 121.14, 122.32, 124.64, 125.68, 127.23, 127.51, 128.72, 132.23, 136.08, 136.18, 143.60, 163.54; MS *m/z* [%]: 408.25 [M$^+$, 32.23], 157.11 [100], 130.05 [48.95]; Anal. Calcd. for C$_{25}$H$_{20}$N$_4$O$_2$: C, 73.51; H, 4.94; N, 13.72; found C, 73.63; H, 4.91; N, 13.63.

N′-(1-(4-Fluorobenzyl)-2-oxoindolin-3-ylidene)-2-(1H-indol-3-yl)acetohydrazide (**9e**)

Red powder, m.p. 227–229 °C; (yield 72%), IR: 3353, 3115 (NH) and 1723 (C=O); ^{1}H NMR δ *ppm*: 4.15 (brs, 2H, C$\underline{H_2}$-C=O), 4.94 (s, 2H, benzylic protons), 6.95–7.02 (m, 3H, Ar-H), 7.06 (t, 1H, Ar-H, J = 8.0 Hz), 7.13 (t, 2H, Ar-H, J = 8.0 Hz), 7.34–7.37 (m, 5H, Ar-H), 7.59 (d, 1H, Ar-H, J = 7.5 Hz), 8.16 (brs, 1H, Ar-H), 10.99 (s, 1H, NH of hydrazide, D$_2$O exchangeable), 11.23 (s, 1H, NH of indol, D$_2$O exchangeable); Anal. Calcd. for C$_{25}$H$_{19}$FN$_4$O$_2$ (344.38): C, 70.41; H, 4.49; N, 13.14; found C, 70.73; H, 4.46; N, 13.11.

N′-(1-(4-Cyanobenzyl)-2-oxoindolin-3-ylidene)-2-(1H-indol-3-yl)acetohydrazide (**9f**)

Yellow powder, m.p. 233–234 °C; (yield 80%), IR: 3290, 3247 (NH), 2230 (C≡N) and 1704 (C=O); ^{1}H NMR δ *ppm*: 4.14 (brs, 2H, C$\underline{H_2}$-C=O), 4.95 (s, 2H, benzylic protons), 6.99 (d, 2H, Ar-H, J = 7.5 Hz), 7.06 (t, 2H, Ar-H, J = 7.5 Hz), 7.34 (t, 3H, Ar-H, J = 7.5 Hz), 7.52 (t, 1H, Ar-H, J = 7.5 Hz), 7.60 (d, 1H, Ar-H, J = 7.5 Hz), 7.64 (d, 1H, Ar-H, J = 7.5 Hz), 7.73 (d, 1H, Ar-H, J = 7.5 Hz), 7.85 (s, 1H, Ar-H), 8.14 (brs, 1H, Ar-H), 11.00 (s, 1H, NH of hydrazide, D$_2$O exchangeable), 11.24 (s, 1H, NH of indol, D$_2$O exchangeable); ^{13}C NMR δ *ppm*: 42.01 (CH$_2$), 107.21, 109.68, 111.46, 111.62, 115.02, 118.63, 122.46, 123.28, 124.64, 129.90, 129.98, 130.91, 131.40, 132.10, 137.96, 142.32, 143.32, 160.79 (C=O of isatin), 163.67 (C=O of hydrazide); Anal. Calcd. for C$_{26}$H$_{19}$N$_5$O$_2$: C, 72.04; H, 4.42; N, 16.16; found C, 71.82; H, 4.48; N, 16.28.

N′-(1-Benzyl-5-bromo-2-oxoindolin-3-ylidene)-2-(1H-indol-3-yl)acetohydrazide (**9g**)

Orange powder, m.p. 210–211 °C; (yield 82%), IR: 3367, 3282 (NH) and 1739, 1669 (C=O); ^{1}H NMR δ *ppm*: 4.14 (brs, 2H, C$\underline{H_2}$-C=O), 4.96 (s, 2H, benzylic protons), 6.93 (d, 1H, Ar-H, J = 8.0 Hz), 6.96 (t, 1H, Ar-H, J = 7.5 Hz), 7.05 (t, 1H, Ar-H, J = 7.5 Hz), 7.25–7.36 (m, 7H, Ar-H), 7.53 (d, 1H, Ar-H, J = 8.0 Hz), 7.59 (d, 1H, Ar-H, J = 8.0 Hz), 8.47 (brs, 1H, Ar-H), 10.97 (s, 1H, NH of hydrazide, D$_2$O exchangeable), 11.48 (s, 1H, NH of indol, D$_2$O exchangeable); ^{13}C NMR δ *ppm*: 42.71 (CH$_2$), 107.21, 111.44, 111.62, 114.35, 116.46, 118.53, 118.63, 121.06, 124.54, 127.19, 127.55, 128.03, 1289.74, 134.24, 135.89, 136.01, 142.65, 163.26; Anal. Calcd. for C$_{25}$H$_{19}$BrN$_4$O$_2$: C, 61.61; H, 3.93; N, 11.50; found C, 61.79; H, 3.88; N, 11.57.

N′-(5-Bromo-1-(4-fluorobenzyl)-2-oxoindolin-3-ylidene)-2-(1H-indol-3-yl)acetohydrazide (**9h**)

Brown powder, m.p. 183–185 °C; (yield 80%), IR: 3423, 3343 (NH) and 1730 (C=O); ^{1}H NMR δ *ppm*: 4.13 (s, 2H, CH$_2$-C=O), 4.95 (s, 2H, benzylic protons), 6.97 (d, 2H, Ar-H, J = 8.5 Hz), 7.05 (t, 1H, Ar-H, J = 8.0 Hz), 7.13 (t, 2H, Ar-H, J = 8.5 Hz), 7.33–7.37 (m, 4H, Ar-H), 7.55 (d, 1H, Ar-H, J = 8.0 Hz), 7.58 (d, 1H, Ar-H, J = 7.5 Hz), 8.45 (s, 1H, Ar-H), 10.97 (s, 1H, NH of hydrazide, D$_2$O exchangeable), 11.49 (s, 1H, NH of indol, D$_2$O exchangeable); Anal. Calcd. for C$_{25}$H$_{18}$BrFN$_4$O$_2$: C, 59.42; H, 3.59; N, 11.09; found C, 59.58; H, 3.55; N, 11.16.

N′-(3,4-Dihydronaphthalen-1(2H)-ylidene)-2-(1H-indol-3-yl)acetohydrazide (**11**)

White crystals, m.p. 244–245 °C; (yield 84%), IR: 3315, 3246 (NH) and 1692 (C=O); ^{1}H NMR δ *ppm*: 1.81 (d, 2H, CH$_2$, J = 6.0 Hz), 2.61 (m, 2H, CH$_2$), 2.74 (d, 2H, CH$_2$, J = 5.6 Hz), 3.78, 4.13 (2s, 2H, CH$_2$-C=O), 6.95–7.09 (m, 2H, Ar-H), 7.19–7.27 (m, 4H, Ar-H), 7.34 (t, 1H, Ar-H, J = 7.6 Hz), 7.57, 7.62 (2d, 1H, Ar-H, J = 8.0, 8.0 Hz), 8.00, 8.10 (2d, 1H, Ar-H, J = 7.2, 7.6 Hz), 10.41, 10.43 (s, 1H, NH of hydrazide, D$_2$O exchangeable), 10.86, 10.90 (s, 1H, NH of indol, D$_2$O exchangeable); Anal. Calcd. for C$_{20}$H$_{19}$N$_3$O: C, 75.69; H, 6.03; N, 13.24; found C, 75.86; H, 5.99; N, 13.32.

3.2. Biological Evaluation

The detailed experimental procedures adopted in the different biological assays for target bis-indoles (**7a–f** and **9a–h**); **11** were supplied in the Supplementary Materials.

3.2.1. Cytotoxic Activity against Human Breast Cancer and Non-Tumorigenic Cell Lines

The two examined human breast cancer cell lines (MCF-7 and Breast MDA-MB-231), and non-tumorigenic human breast epithelial cell line (MCF-10A) have been obtained from the American Type Culture Collection (ATCC). Assessment of cytotoxicity for target indole derivatives has been performed following the SRB colorimetric assay procedures [20], as reported earlier [26].

3.2.2. Cell Cycle Analysis

The influence of bis-indoles **7e** and **9a** on cell cycle progression was examined in breast cancer MCF-7 cells, after 24 h of treatment, through DNA flow cytometric assay by the use of BD FACS Caliber flow cytometer, as described previously [27]. The cell cycle distributions were calculated using CellQuest software (Becton Dickinson).

3.2.3. ELISA Immunoassay

Effects of treatment of breast cancer MCF-7 cells with bis-indoles **7e** and **9a** on the expression levels of the pro-apoptotic markers (Bax, caspase-3, and p53) in addition to the anti-apoptotic protein Bcl-2 marker was assessed by the use of ELISA colorimetric kits as per the manufacturer's instructions, as described earlier [28].

3.2.4. Annexin V-FITC/PI Apoptosis Assay

The apoptotic action of bis-indoles **7e** and **9a** was further explored through the investigation of their effect on the phosphatidylserine externalization in breast cancer MCF-7 cells, using Annexin-V-FITC Apoptosis Detection Kit according to manufacturer's protocol, as reported earlier [27].

3.2.5. CDK2 Kinase Inhibitory Activity

The in vitro CDK2 kinase inhibition assay was carried out by Reaction Biology Corp. (Reaction Biology Corp., Chester, PA, USA) Kinase HotSpotSM service (http://www.reactionbiology.com).

4. Conclusions

In summary, the adopted approach of replacing the rigid heterocyclic spacer in the marine natural products Topsentin and Nortopsentin by the flexible hydrazide linker resulted in the discovery of promising marine-inspired bis-indole scaffold with good in vitro antitumor activities toward breast cancer cell lines. All the synthesized bis-indoles (**7a–f** and **9a–h**), and indole **11** were characterized for their antiproliferative action against human breast cancer (MCF-7 and MDA-MB-231) cell lines. All the examined bis-indoles **7a–f** and **9a–h** displayed excellent to low antiproliferative activities against MCF-7 and MDA-MB-231 cells with IC_{50} values in ranges 0.44–47.1 and 0.34–77.30 µM, respectively. Bioisosteric replacement of the oxindole ring with the carbocyclic tetralin ring (compound **11**) resulted in a dramatic worsening of effectiveness against the examined cancer cell lines (IC_{50} = 84.70 ± 4.02 and > 100 µM, respectively) in comparison to bis-indoles **7**, which pointed out the importance of oxindole moiety for the antiproliferative activity. The most potent congeners **7e** and 9a against MCF-7 cells (IC_{50} = 0.44 ± 0.01 and 1.28 ± 0.04 µM, respectively) induced apoptosis in MCF-7 cells (23.7-, and 16.8-fold increase in the total apoptosis percentage) as evident by the externalization of plasma membrane phosphatidylserine detected by AnnexinV-FITC/PI assay. This evidence was supported by the Bax/Bcl-2 ratio augmentation (18.65- and 11.1-fold compared to control) with a concomitant increase in the level of caspase-3 (11.7- and 9.5-fold) and p53 (15.4- and 11.75-fold). Moreover, scrutinizing results of the cell cycle analysis unraveled that both compounds arrest the cell cycle mainly in the G2/M phase. On the other hand, **7e** and **9a** displayed good selectivity toward tumor cells (*S.I.* = 38.7 and 18.3), upon testing of their cytotoxicity toward non-tumorigenic breast MCF-10A cells. Finally, compounds **7a**, **7b**, **7d**, **7e**, and **9a** were examined for their plausible CDK2 inhibitory action. The obtained results (% inhibition range: 16%–58%) unveiled incompetence of the target bis-indoles to inhibit CDK2 significantly. Collectively, these results suggested that herein reported bis-indoles are good lead compounds for further optimization and development as potential efficient anti-breast cancer drugs.

Supplementary Materials: The following are available online at http://www.mdpi.com/1660-3397/18/4/190/s1.

Author Contributions: For research articles with several authors, a short paragraph specifying their individual contributions must be provided. The following statements should be used "Conceptualization, W.M.E., G.S.H. and G.H.A.-A.; methodology, W.M.E., S.T.A.-R. and G.H.A.-A.; software, H.M.A.; validation, S.T.A.-R. and A.A.A.; formal analysis, W.M.E., H.M.A. and G.H.A.-A.; investigation, G.S.H.; resources, S.T.A.-R. and H.M.A.; data curation, A.A.A. and G.H.A.-A.; writing—original draft preparation, W.M.E. and G.H.A.-A.; writing—review and editing, W.M.E. and G.S.H.; visualization, G.S.H. and H.M.A.; supervision, W.M.E. and G.S.H.; project administration, W.M.E.; funding acquisition, S.T.A.-R." Please turn to the CRediT taxonomy for the term explanation. Authorship must be limited to those who have contributed substantially to the work reported. All authors have read and agreed to the published version of the manuscript.

Funding: The authors would like to extend their sincere appreciation to the Deanship of Scientific Research at King Saud University for its funding of this research through the Research Group Project no. RG-1439–65.

Conflicts of Interest: The authors declare no conflict of interest.

References

1. Lunagariya, J.; Bhadja, P.; Zhong, S.; Vekariya, R.; Xu, S. Marine Natural Product Bis-indole Alkaloid Caulerpin: Chemistry and Biology. *Mini Mini-Rev. Med. Chem.* **2019**, *19*, 751–761. [CrossRef] [PubMed]
2. Gupta, L.; Talwar, A.; Chauhan, P.M.S. Bis and Tris Indole Alkaloids from Marine Organisms: New Leads for Drug Discovery. *Front. Med. Chem.* **2012**, *6*, 361–385.
3. Netz, N.; Opatz, T. Marine indole alkaloids. *Mar. Drugs* **2015**, *13*, 4814–4914. [CrossRef] [PubMed]
4. Gul, W.; Hamann, M.T. Indole alkaloid marine natural products: An established source of cancer drug leads with considerable promise for the control of parasitic, neurological and other diseases. *Life Sci.* **2005**, *78*, 442–453. [CrossRef] [PubMed]
5. Edwards, V.; Benkendorff, K.; Young, F. Marine compounds selectively induce apoptosis in female reproductive cancer cells but not in primary-derived human reproductive granulosa cells. *Mar. Drugs* **2012**, *10*, 64–83. [CrossRef] [PubMed]

6. Esmaeelian, B.; Abbott, C.A.; Le Leu, R.K.; Benkendorff, K. 6-bromoisatin found in muricid mollusc extracts inhibits colon cancer cell proliferation and induces apoptosis, preventing early stage tumor formation in a colorectal cancer rodent model. *Mar. Drugs* **2014**, *12*, 17–35. [CrossRef]

7. Shimizu, S.; Yamamoto, Y.; Inagaki, L.; Koshimura, S. Antitumor effect and structure-activity relationship of asterriquinone analogs. *Gann = Gan* **1982**, *73*, 642–648.

8. Kohmoto, S.; Kashman, Y.; McConnell, O.J.; Rinehart, K.L., Jr.; Wright, A.; Koehn, F. Dragmacidin, a new cytotoxic bis(indole)alkaloid from a deep water marine sponge, Dragmacidon sp. *J. Org. Chem.* **1988**, *53*, 3116–3118. [CrossRef]

9. Morris, S.A.; Andersen, R.J. Brominated bis(indole)alkaloids from the marine sponge Hexadella sp. *Tetrahedron* **1990**, *46*, 715–720. [CrossRef]

10. Fahy, E.; Potts, B.C.M.; Faulkner, D.J.; Smith, K. 6-Bromotryptamine derivatives from the Gulf of California tunicate Didemnum candidum. *J. Nat. Prod.* **1991**, *54*, 564–569. [CrossRef]

11. Wright, A.E.; Pomponi, S.A.; Cross, S.S.; McCarthy, P. A new bis-(indole)alkaloids from a deep-water marine sponge of the genus Spongosorites. *J. Org. Chem.* **1992**, *57*, 4772–4775. [CrossRef]

12. Blunt, J.W.; Copp, B.R.; Hu, W.; Munro, M.H.G.; Northcote, P.T.; Prinsep, M.R. Marine natural products. *Nat. Prod. Rep.* **2008**, *36*, 35–94. [CrossRef] [PubMed]

13. Bao, B.; Zhang, P.; Lee, Y.; Hong, J.; Lee, C.; Jung, J.H. Monoindole Alkaloids from a Marine SpongeSpongosorites sp. *Mar. Drugs* **2007**, *5*, 31–39. [CrossRef] [PubMed]

14. Bartik, K.; Braekman, J.; Daloze, D.; Stoller, C. Topsentins, new toxic bis-indole alkaloids from the marinesponge Topsentia genitrix. *Can. J. Chem.* **1987**, *65*, 2118–2121. [CrossRef]

15. Abdel-Aziz, H.A.; Eldehna, W.M.; Ghabbour, H.; Al-Ansary, G.H.; Assaf, A.M.; Al-Dhfyan, A. Synthesis, crystal study, and anti-proliferative activity of some 2-benzimidazolylthioacetophenones towards triple-negative breast cancer MDA-MB-468 cells as apoptosis-inducing agents. *Int. J. Mol. Sci.* **2016**, *17*, 1221. [CrossRef]

16. Eldehna, W.M.; El-Naggar, D.H.; Hamed, A.R.; Ibrahim, H.S.; Ghabbour, H.A.; Abdel-Aziz, H.A.; Ghabbour, H.A. Abdel-Aziz One-pot three-component synthesis of novel spirooxindoles with potential cytotoxic activity against triple-negative breast cancer MDAMB-231 cells. *J. Enzym. Inhib. Med. Chem.* **2018**, *33*, 309–318. [CrossRef]

17. Ismail, R.S.; Abou-Seri, S.M.; Eldehna, W.M.; Ismail, N.S.; Elgazwi, S.M.; Ghabbour, H.A.; Ahmed, M.S.; Halaweish, F.T.; El Ella, D.A.A. Novel series of 6-(2-substitutedacetamido)-4-anilinoquinazolines as EGFR-ERK signal transduction inhibitors in MCF-7 breast cancer cells. *Eur. J. Med. Chem.* **2018**, *155*, 782–796. [CrossRef]

18. Petreni, A.; Bonardi, A.; Lomelino, C.; Osman, S.M.; ALOthman, Z.A.; Eldehna, W.M.; El-Haggar, R.; McKenna, R.; Nocentini, A.; Supuran, C.T. Inclusion of a 5-fluorouracil moiety in nitrogenous bases derivatives as human carbonic anhydrase IX and XII inhibitors produced a targeted action against MDA-MB-231 and T47D breast cancer cells. *Eur. J. Med. Chem.* **2020**, *190*, 112112. [CrossRef]

19. Johnson, H.E.; Crosby, D.G. 3-Indoleacetic Acid. *J. Org. Chem.* **1963**, *28*, 1246–1248. [CrossRef]

20. Skehan, P.; Storeng, R.; Scudiero, D. New colorimetric cytotoxicity assay for anticancer-drug screening. *J. Natl. Cancer Inst.* **1990**, *82*, 1107–1112. [CrossRef]

21. Lopez, J.; Tait, S.W.G. Mitochondrial apoptosis: Killing cancer using the enemy within. *Br. J. Cancer* **2015**, *112*, 957. [CrossRef] [PubMed]

22. Hu, W.; Kavanagh, J.J. Anticancer therapy targeting the apoptotic pathway. *Lancet Oncol.* **2003**, *4*, 721–729. [CrossRef]

23. Delbridge, A.R.; Grabow, S.; Strasser, A.; Vaux, D.L. Thirty years of BCL-2: Translating cell death discoveries into novel cancer therapies. *Nat. Rev. Cancer* **2016**, *16*, 99. [CrossRef]

24. Jiang, H.; Zhao, P.J.; Su, D.; Feng, J.; Ma, S.L. Paris saponin I induces apoptosis via increasing the Bax/Bcl-2 ratio and caspase-3 expression in gefitinib-resistant non-small cell lung cancer in vitro and in vivo. *Mol. Med. Rep.* **2014**, *9*, 2265–2272. [CrossRef] [PubMed]

25. Sen-Gupta, A.k.; Gupta, A.A. Synthesis of some new indolinone derived hydrazones as possible antibacterial agents. *Eur. J. Med. Chem.* **1983**, *18*, 181–184.

26. Eldehna, W.M.; El Kerdawy, A.M.; Al-Ansary, G.H.; Al-Rashood, S.T.; Ali, M.M.; Mahmoud, A.E. Type IIA-Type IIB protein tyrosine kinase inhibitors hybridization as an efficient approach for potent multikinase inhibitor development: Design, synthesis, anti-proliferative activity, multikinase inhibitory activity and molecular modeling of novel indolinone-based ureides and amides. *Eur. J. Med. Chem.* **2019**, *163*, 37–53. [PubMed]

27. El-Naggar, M.; Eldehna, W.M.; Almahli, H.; Elgez, A.; Fares, M.; Elaasser, M.M.; Abdel-Aziz, H.A. Novel thiazolidinone/thiazolo [3, 2-*a*] benzimidazolone-isatin conjugates as apoptotic anti-proliferative agents towards breast cancer: One-pot synthesis and in vitro biological evaluation. *Molecules* **2018**, *23*, 1420. [CrossRef]

28. Eldehna, W.M.; Almahli, H.; Al-Ansary, G.H.; Ghabbour, H.A.; Aly, M.H.; Ismael, O.E.; Al-Dhfyan, A.; Abdel-Aziz, H.A. Synthesis and in vitro antiproliferative activity of some novel isatins conjugated with quinazoline/phthalazine hydrazines against triple-negative breast cancer MDA-MB-231 cells as apoptosis-inducing agents. *J. Enz. Inhib. Med. Chem.* **2017**, *32*, 600–613. [CrossRef]

 *marine drugs* 

Article

Synthesis and Biological Evaluation of Four New Ricinoleic Acid-Derived 1-*O*-alkylglycerols

René Pemha [1,2,*] , Victor Kuete [3,4] , Jean-Marie Pagès [4] , Dieudonné Emmanuel Pegnyemb [5] and Paul Mosset [6,*]

1   Univ Rennes, Ecole Nationale Supérieure de Chimie de Rennes, CNRS, ISCR-UMR 6226, F-35000 Rennes, France
2   AGIR, EA 4294, UFR of Pharmacy, Jules Verne University of Picardie, 80037 Amiens, France
3   University of Dschang, Faculty of Science, Department of Biochemistry, P.O. Box 67 Dschang, Cameroon; kuetevictor@yahoo.fr
4   UMR_MD1, U-1261, Aix-Marseille Univ, INSERM, IRBA. Membranes et Cibles Thérapeutiques, Faculté de Pharmacie, 13385 Marseille cedex 05, France; jean-marie.pages@univ-amu.fr
5   Département de ChimieOrganique, Faculté des Sciences, Université de Yaoundé I, BP 812 Yaoundé, Cameroun; pegnyemb@yahoo.com
6   Univ Rennes, CNRS, ISCR (Institut des Sciences Chimiquesde Rennes), UMR 6226, F-35000 Rennes, France
*   Correspondence: pemharene@gmail.com (R.P.); paul.mosset.1@univ-rennes1.fr (P.M.); Tel.: +33-(0)666-752-553 (R.P.); +33-(0)223-237-336 (P.M.)

Received: 12 January 2020; Accepted: 4 February 2020; Published: 15 February 2020

Abstract: A series of novel substituted 1-*O*-alkylglycerols (AKGs) containing methoxy (**8**), *gem*-difluoro (**9**), azide (**10**) and hydroxy (**11**) group at 12 position in the alkyl chain were synthesized from commercially available ricinoleic acid (**12**). The structures of these new synthesized AKGs were established by NMR experiments as well as from the HRMS and elementary analysis data. The antimicrobial activities of the studied AKGs **8–11** were evaluated, respectively, and all compounds exhibited antimicrobial activity to different extents alone and also when combined with some commonly used antibiotics (gentamicin, tetracycline, ciprofloxacin and ampicillin). AKG **11** was viewed as a lead compound for this series as it exhibited significantly higher antimicrobial activity than compounds **8–10**.

Keywords: alkylglycerol (AKG); ricinoleic acid (RA); antimicrobial activity; structure–activity relationship (SAR) studies; antibiotics (gentamicin; tetracycline; ciprofloxacin and ampicillin)

---

1. Introduction

Natural 1-*O*-alkylglycerols (AKGs) **1** are bioactive ether lipids present in body cells and fluids. They are precursors of ether phospholipids, which participate in structures and functions of membranes in certain cells such as white blood cells or macrophages. AKGs are also found in bone marrow lipids and in milk [1]. Marine sources of AKGs such as the liver oil of certain shark species or rat fish (elasmobranch fishes) contain high levels of these compounds as a mixture of few species varying by length and unsaturation or saturation of the alkyl chain.

The usual composition of alkyl chains in AKGs from Greenland shark (*Centrophorus squamosus*) liver oil (SLO) is as follows: 12:0, 1–2%; 14:0, 1–3%; 16:0, 9–13%; 16:1, n-7, 11–13%; 18:0, 1–5%; 18:1, n-9, 54–68%; 18:1, n-7, 4–6%; and minor species (<1%). Beneficial effects of SLO on health have been recognized in traditional medicine of northern countries involved in fishing such as Japan, Norway and Iceland. In these countries, the ancestral use of SLO was empirical as strengthening or wound healing medication [2].

Experimental studies were performed during the last century, aiming to demonstrate whether AKGs from SLO had biological properties and beneficial effects. Indeed, several studies did observe interesting effects such as hematopoiesis stimulation [3], lowering radiotherapy-induced injuries [4], reducing tumor growth [5] and improving vaccination efficiency [6,7]. However, in most cases, conclusions were mainly impaired by the poor definition of the mixtures used in terms of purity as well as chemical composition. It was then established that the alkyl chain is bound to the glycerol backbone at the *sn*-1 position, thus leading to an *S* configuration at the asymmetric carbon [8] (Figure 1).

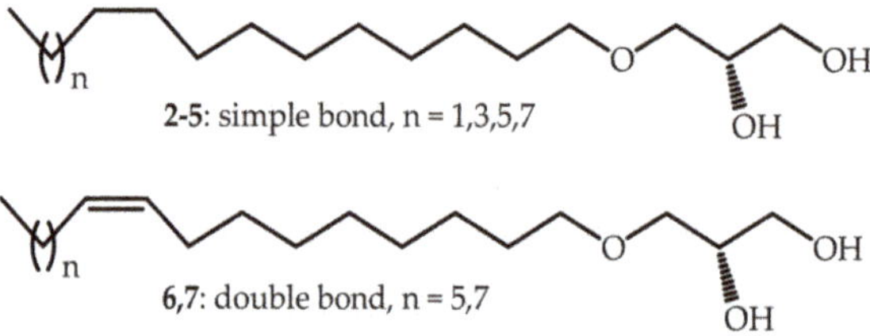

Figure 1. Natural 1-*O*-alkylglycerols (AKGs) (**1**).

To assess the biological activity of each individual AKG, Legrand et al. [9–11] reported the antitumor activity (against lung cancer in mice) of each of the six prominent components **2–7** of the natural mixture. These derivatives were obtained in pure form by total synthesis and it was observed that the biological activity was heavily dependent upon the unsaturation of the alkyl chain. When this chain was saturated, the corresponding 1-*O*-alkylglycerols **2–5** exhibited little or no activity. However, when it was monounsaturated **6–7**, a good antitumor activity was observed, thus indicating that the antitumor activity of the natural SLO mixture was heavily related to its unsaturated components (Figure 2).

Figure 2. Synthesized AKGs **2–7**, prominent components of natural shark liver oil (SLO) mixture.

Currently, resistance to the existing antibiotics and increasing numbers of diseases result in identifying new drug candidates with new forms of activity. Thus, synthesized natural or non-natural AKGs derivatives of natural fatty acids could be one new source of drug delivery systems of antibiotics. Additionally, defined synthetic routes for these targets will facilitate further investigation of biological activities, as natural AKGs were found present in various human cells, but only in trace quantities. Among known natural fatty acids, ricinoleic acid (RA) is one of the major fatty acids occurring in castor oil (almost 90%) [12]. Such a high concentration of this unusual unsaturated fatty acid may be responsible for castor oil's remarkable healing abilities. It is known to be effective in preventing the growth of numerous species of viruses, bacteria, yeasts and molds [13,14]. Due to the many beneficial effects of this fatty acid component, the use of castor oil can be applied topically to treat a wide variety of health complaints [15], and it also has pharmacological effects on the human gastrointestinal tract [16]. An RA-based glycine derivative was reported to exhibit excellent antimicrobial and anti-biofilm activities against the tested Gram-positive bacterial strains and specifically against various *Candida* strains [17], and it is used for the preparation of several bioactive molecules [18–20].

The presence of the hydroxyl group in RA provides a functional group location for performing a variety of chemical reactions including esterification, halogenation, dehydration, alkoxylation and nucleophilic substitution. In this direction, non-natural AKG **8–11** derivatives from ricinoleic acid have been synthesized in a stereo-controlled manner. Taking into account RA and natural alkylglycerols' beneficial effects, herein, we report the synthesis of new non-natural RA-derived methoxy, *gem*-difluorino, azide, and hydroxy-substituted 1-*O*-alkylglycerols **8–11** and their respective antimicrobial activities (Figure 3).

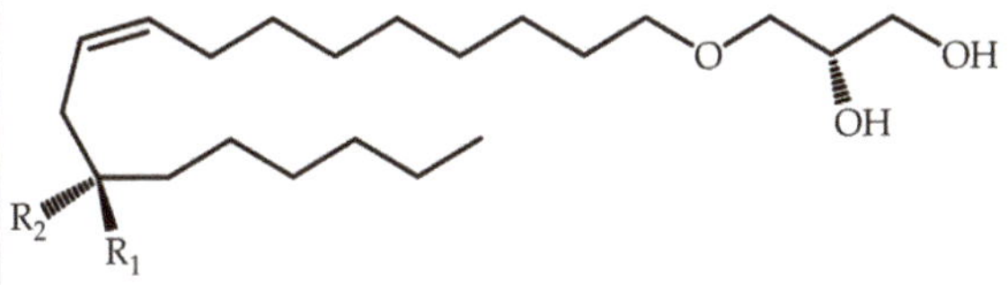

Figure 3. Synthesized AKGs **8–11** from ricinoleic acid.

8: R_1 = OMe, R_2 = H   **10**: R_1 = H, R_2 = N_3

9: R_1 = R_2 = F   **11**: R_1 = OH, R_2 = H

2. Results and Discussion

2.1. Chemistry

Ether lipids bearing a methoxy group at the alkyl chain can be divided into two groups of compounds, namely the methoxylated fatty acids and the methoxy-substituted alkylglycerols [21]. These compounds display interesting biological activities such as antibacterial, antifungal, antitumor and antiviral activities and have been isolated from either bacterial or marine sources, or are mainly of synthetic origin [22]. Hallgren et al. [23] reported that the mixture of methoxylated alkylglycerols made up to 4% of the glyceryl ether content, and in three shark species and three ratfish species they accounted for 0.1% to 0.3% of the total liver lipid content [24].

Methoxy-substituted glyceryl ethers displayed antibacterial and antifungal activities and inhibited some cancer cell lines and metastasis formation in mice [21]. AKGs containing a methoxy group at position 2 in the alkyl chain isolated from the natural SLO mixture of Greenland shark were able to inhibit tumor growth and metastasis formation, and also to stimulate the immunoreactivity in mice [25,26]. Likewise, a synthesized AKG from oleic acid bearing a methoxy group at position 2 in the alkyl chain was reported as an analog of bioactive ether lipid [27].

Therefore, methoxy-substituted AKG **8** was designed as an analog of the AKG **7** with the same alkyl chain length C_{18} and Δ9 unsaturation, but with a methoxy group at position 12 in the alkyl chain to evaluate the beneficial effects of a methoxy group in an other position than 2 in a 1-*O*-alkylglycerol alkyl chain [21,27]. The synthesis of **8** began with the esterification of RA **12** into methyl ricinoleate **13** in 75% yield using boron trifluoride in methanol, along with 4% of a by-product (dimer) **14**, which arises from a subsequent esterification of the secondary alcohol of **13** formed with RA [28]. Compound **14** gave spectroscopic properties (^{1}H and ^{13}C NMR, as well as correlation spectra) in full agreement with its structure. This was confirmed by HRMS (ESI, *m/z*) showing 615.4971 for [M + Na]$^+$ (Scheme 1). Although it is not strictly necessary, the process could be improved by converting this very minor by-product **14** into **13** by transesterification and thus eliminating it by a subsequent treatment of **13** containing **14** by potassium carbonate in methanol (see experimental section). The secondary alcohol functionality of **13** was then methylated in situ with methyl iodide in the presence of sodium hydroxide in DMSO to provide **15** in 72% yield. This method compares favorably to the hitherto reported cobalt-catalyzed etherification of **13** using diazomethane [29]. Then, **15** was reduced to alcohol **16** in 82% yield using as a reducing agent, Red-Al in Et$_2$O at 0 °C. Following this, alcohol **16** underwent mesylation using mesyl chloride in dichloromethane (DCM) with triethylamine, affording **17** in 74% yield, which was in turn alkylated with 2,3-isopropylidene-*sn*-glycerol **18** in the presence of potassium hydroxide and tetra-*n*-butylammonium bromide in DMSO to provide acetonide **19** in 93% yield. Acetonide **19** was hydrolyzed under acidic conditions using a catalytic amount of *p*-toluenesulfonic acid monohydrate in MeOH/H$_2$O (9:1) to afford **8** in 99% yield (Scheme 1). Supplementary Materials of all synthetic compounds for this sequence is attached in a file as Figures S1.1–S1.14.

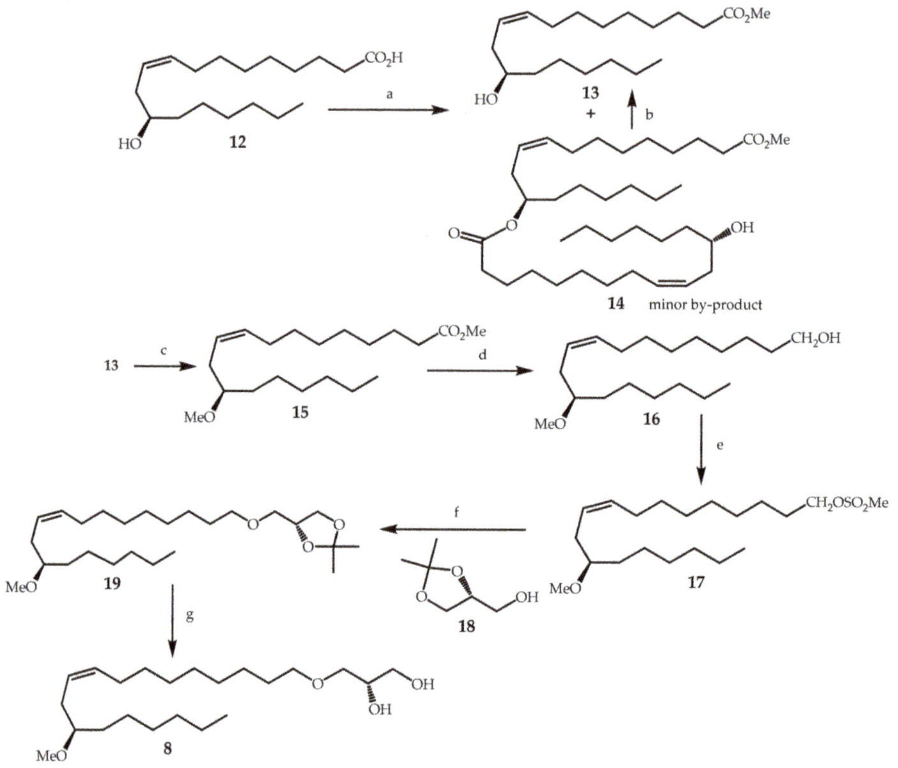

Scheme 1. Synthesis of AKG **8**. Reagents and conditions: (**a**) $BF_3 \cdot 2MeOH$, MeOH, 50 °C, 15 h, 75% of **13** and 4% of **14**; (**b**) K_2CO_3, MeOH, rt, 42 h, quantitative conversion of **14** into **13**; (**c**) MeI, NaOH, *n*-Bu$_4$NBr, DMSO, rt, 18 h, 72%; (**d**) Red-Al, Et$_2$O, 0 °C, 16 h, 82%; (**e**) MsCl, Et$_3$N, DCM, −50 °C, 5 h, 74%; (**f**) KOH, *n*-Bu$_4$NBr, DMSO, 35 °C, 14 h, 93%; (**g**) *p*-TsOH·H$_2$O (0.05 equiv), MeOH/H$_2$O (9:1), 60 °C, 5 h, 99%.

The hydroxyl functionality of RA makes the castor oil a natural polyol providing oxidative stability to the oil and a relatively long shelf life, compared to other oils, by preventing peroxide formation. As a result, this unique functionality allows the castor oil to be used in industrial applications such as paints, coatings, inks and lubricants [30]. With that in mind, the AKG **11** was designed as an analog of **8**, without a methoxy group at the 12 position in the alkyl chain, to study the influence of the hydroxyl group in the AKG's alkyl chain, and the structure–activity relationship (SAR) by comparing their biological activity, respectively. The preparation of the AKG **11** with the hydroxyl group required a protection–deprotection sequence to yield the penultimate intermediate alcohol **21**.

Attempt to protect the hydroxy group of **13** using 2-(bromomethyl)naphthalene in the presence of sodium hydroxide and of tetra-*n*-butylammonium bromide in DMSO for 18 h at rt failed. Thereafter, a trace of **20** was observed when chlorotriisopropylsilane in the presence of DIPEA in dichloromethane was used to protect **13** as silyl ether [31]. When DIPEA was replaced by imidazole as a base and in DMF, **20** was obtained in 61% yield. Afterward, **20** was reduced to the penultimate intermediate alcohol **21** in 75% yield using Red-Al in diethyl ether (Scheme 2). Compound **21**, upon mesylation conditions using mesyl chloride in DCM, and Et$_3$N provided **22** in 76% yield, which was then alkylated under anhydrous conditions with 2,3-isopropylidene-*sn*-glycerol **18** in DMF in the presence of sodium hydride to provide acetonide **23** in 68% yield. Sodium hydride was used as a base instead of KOH to eliminate the by-products formed when reacting **22** with **18**. Following this, silyl ether group was removed using TBAF in THF at rt for 20 h to provide **24** in 92% yield. Easy acetonide cleavage on **24** under acidic conditions (0.05 equiv of *p*-toluenesulfonic acid monohydrate in MeOH/H$_2$O (9:1)) gave **11** in 84% yield (Scheme 2).

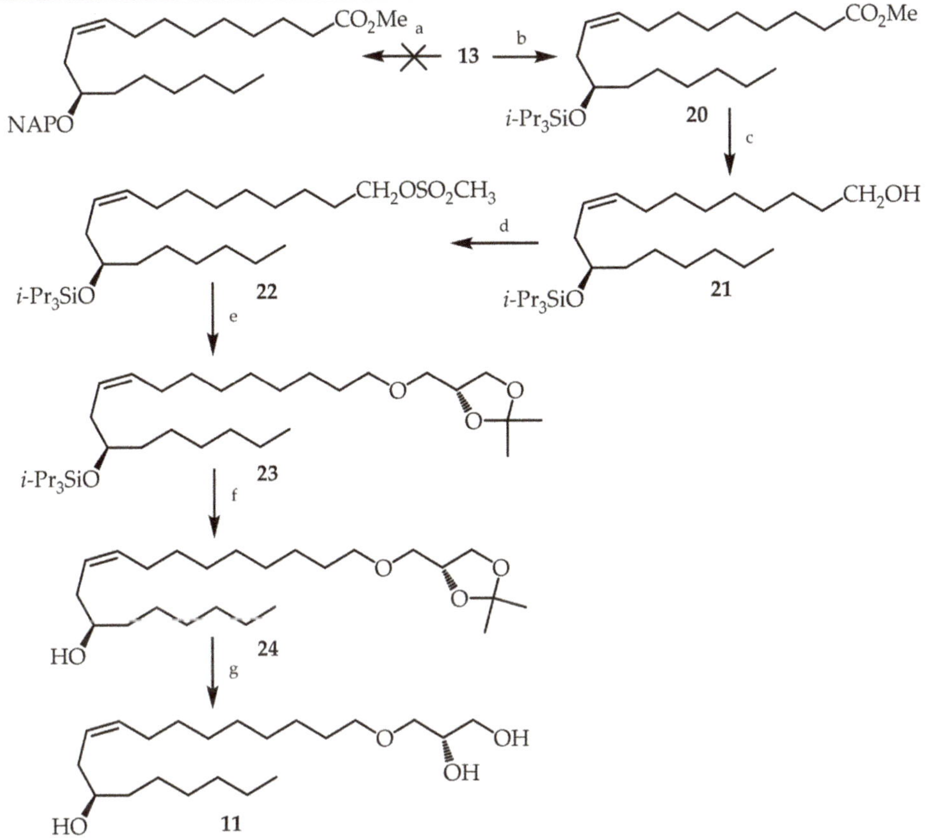

Scheme 2. Synthesis of AKG **11**. Reagents and conditions: (**a**) 2-(bromomethyl)naphthalene, NaOH, *n*-Bu$_4$NBr, DMSO, 18 h, rt; (**b**) ClSi*i*-Pr$_3$, imidazole, DMF, 48 h, rt, 61%; (**c**) Red-Al, Et$_2$O, 0 °C, 5 h, 94%; (**d**) MsCl, Et$_3$N, DCM, −50 °C, 2 h, 76%; (**e**) **18**, NaH, DMF, 15 h, rt, 68%; (**f**) TBAF, THF, rt, 20 h, 92%; (**g**) *p*-TsOH·H$_2$O (0.05 equiv), MeOH/H$_2$O (9:1), 60 °C, 4 h, 84%.

We envisioned that the AKG **9** with a *gem*-difluorinated group in the alkyl chain could exhibit more biological activity than other AKGs studied, as compounds containing a difluoromethylene group were reported to exhibit excellent biological activities [32]. Moreover, introduction of fluorine atoms in molecules heavily modifies their physical, chemical and physiological properties. These fluorinated compounds have found many applications in pharmaceutical and agrochemical fields [33]. Thus, a *gem*-difluorinated **26** key intermediate for the synthesis of **9** was obtained by a classic two-step sequence. Oxidation of **13** by PCC in DCM provided the ketone **25** in 68% yield, which was then subjected to fluorination at rt using (diethylamino)sulfur trifluoride (DAST) in DCM, and the fluorinated product **26** was obtained in 54% yield. Following this, **26** was reduced to alcohol **27** in 79% yield using Red-Al in Et$_2$O, which upon treatment with mesyl chloride in DCM in the presence of Et$_3$N furnished **28** in 68% yield. Alkylation of **28** in DMSO with 2,3-isopropylidene-*sn*-glycerol (**18**) in the presence of 50% aqueous sodium hydroxide and tetra-*n*-butylammonium bromide gave the expected product **29** in 63% yield, along with 6% of a by-product **30** (Scheme 3). Compound **29** was obtained as a pure green oil after column chromatography on florisil gel and showed spectroscopic properties (^{1}H and ^{13}C NMR as well as correlation spectra) in full agreement with its structure, and confirmation was made by HRMS (ESI, *m/z*) that showed 441.3149 for [M + Na]$^+$. Sodium hydroxide solution (50%) in H_2O was used instead of others bases (NaH, KOH) to decrease the formation of the by-product **30**. Acetonide cleavage on compound **29** under acidic conditions using catalytic amount of *p*-toluenesulfonic acid monohydrate in MeOH/H_2O (9:1) provided **9** in 92% yield (Scheme 3).

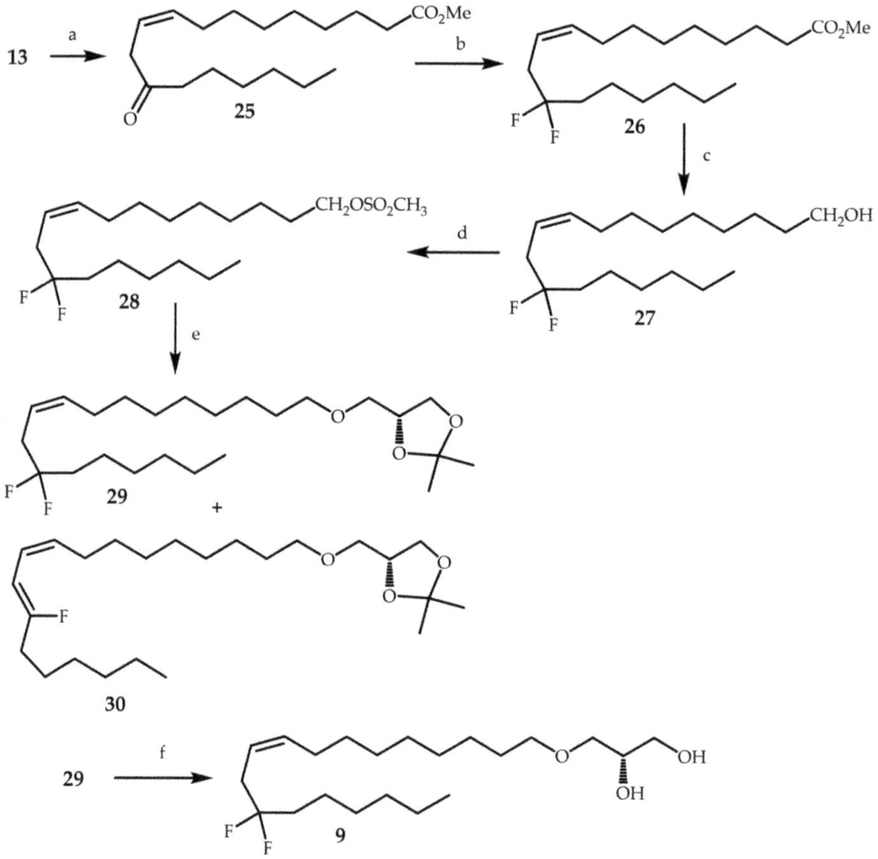

Scheme 3. Synthesis of AKG **9**. Reagents and conditions: (**a**) PCC, DCM, 1 h, rt, 68%; (**b**) DAST, DCM, 21 days, rt, 54%; (**c**) Red-Al, Et$_2$O, 0 °C, 5 h, 77%; (**d**) MsCl, Et$_3$N, DCM, −35 °C to −5 °C, 4–5 h, 69% from **26**; (**e**) **18**, 50% aqueous NaOH, *n*-Bu$_4$NBr, DMSO, 40 °C, 15 h, 63%; (**f**) *p*-TsOH·H_2O (0.05 equiv), MeOH/H_2O (9:1), 60 °C, 5 h, 92%.

At last, the AKG **10** was designed as another analog with an azide group at the same position in the alkyl chain to evaluate the beneficial effects of the azide group on the biological activity, and to estimate the dissimilarity between the studied AKGs. Subsequently, **10** was obtained in a classical seven step sequence. Starting under mesylation conditions of methyl ricinoleate **13** using mesyl chloride and Et$_3$N in DCM, **31** was obtained in 65% yield, which under substitution of the intermediate mesylate group by S$_N$2 substitution reaction using sodium azide in DMSO provided **32** in 80% yield. Dibal in Et$_2$O was used to reduce the ester functionality of **32** into alcohol **34**, but surprisingly an aldehyde **33** was formed in 68% yield instead of the expected alcohol **34**. Dibal was chosen as a reducing agent for its non-interaction with the azide group. Moreover, use of another reducing system such as Zn-AlCl$_3$

was reported to reduce the azide **32** into the amino derivative [17,19]. Thereafter, aldehyde **33** was then reduced into alcohol **34** in 74% yield using $NaBH_4$ in EtOH, which upon reaction with mesyl chloride in DCM in the presence of Et_3N provided **35** in 59% yield (Scheme 4). Following this, alkylation of **35** in DMSO with 2,3-isopropylidene-*sn*-glycerol (**18**) in the presence of 50% aqueous NaOH and tetra-*n*-butylammonium bromide provided a ~1:1 mixture of two compounds: the expected product **36** along with a closer polar by-product **37**, arising from a subsequent elimination of the azide group as indicated in Scheme 4.

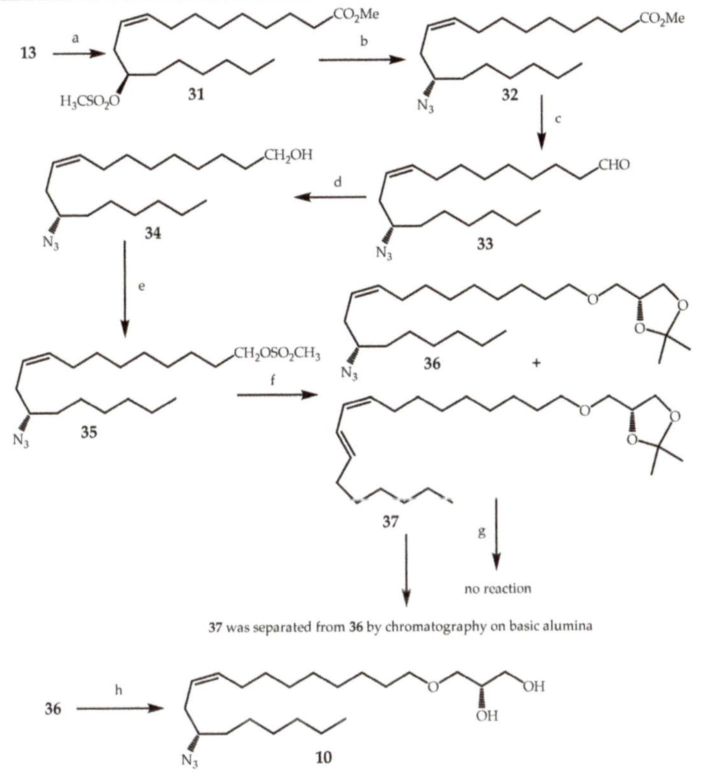

Scheme 4. Synthesis of AKG **10**. Reagents and conditions: (**a**) MsCl, Et_3N, DCM, −10 °C, 4 h, 65%; (**b**) NaN_3, DMSO, 80 °C, 18 h, 80%; (**c**) Dibal, Et_2O, −80 °C, 2 h, 68%; (**d**) $NaBH_4$, EtOH, 0 °C, 1 h, 74%; (**e**) MsCl, Et_3N, DCM, −50 °C, 4 h and then up to −5 °C, 59%; (**f**) **18**, 50% aqueous NaOH, *n*-Bu_4NBr, DMSO, 40 °C, 15 h; (**g**) maleic anhydride, cyclohexane, 72 h, 45 °C; (**h**) *p*-TsOH·H_2O (0.05 equiv), MeOH/H_2O (9:1), 60 °C, 5 h, 97%.

An attempt to separate the mixture of compounds (**36**,**37**) failed when subjected to maleic anhydride in cyclohexane for 72 h at 45 °C with vigorous stirring. Under these conditions, only **37** was supposed to react with maleic anhydride, leading to a polar product other than **36**, which could ease separation via column chromatography. Compound **36** was then obtained without a trace of by-product **37** by column chromatography on basic alumina gel, and visualization of these two compounds (**36** and **37**) was easily followed on TLC plates as they stained differently with an acidic ethanolic solution of *p*-anisaldehyde. Easy acetonide cleavage on **36** under acidic conditions (*p*-toluenesulfonic acid monohydrate in MeOH/H_2O (9:1)) provided **10** in 97% yield (Scheme 4). Supplementary Materials of all synthetic compounds for this sequence is attached in a file as Figures S4.1–S4.14.

The synthesized compounds **8–11** were evaluated for their respective antimicrobial activities.

2.2. Antimicrobial Activities of AKGs **8–11**

The results of the MIC determination presented in Table 1 indicate detectable values recorded for the (*S*)-3-(((*R*,*Z*)-12-hydroxyoctadec-9-en-1-yl)oxy)propane-1,2-diol (**11**) on all the eleven studied organisms including bacteria and fungi. All other AKGs, namely (*S*)-3-(((*R*,*Z*)-12-methoxyoctadec-9-en-1-yl)oxy)propane-1,2-diol (**8**), (*S*,*Z*)-3-((12,12-difluorooctadec-9-en-1-yl)oxy)propane-1,2-diol (**9**), and (*S*)-3-(((*S*,*Z*)-12-azidooctadec-9-en-1-yl)oxy)propane-1,2-diol (**10**) showed selective activity. Compound **10** was active on 5 of the 11 (45.5%) whilst AKGs **8** and **9** were active on 8 of the 11 (72.7%) studied microbial species. The lowest MIC value of 19.53 µg/mL was recorded for compound **8** (52.42 µM) on *E. coli* LMP701 and *S. faecalis*, and compound **11** (54.47 µM) on *E. coli* LMP701, *S. typhi* and *C. glabrata*. This lowest MIC value was in some cases equal to that of gentamicin or nystatin on the corresponding microbial species. It appeared from colony count assay (Figure 4) that AKGs **8** and **11** were able to reduce the bacterial concentration after 480 min when tested at the MIC values. A more pronounced effect was reported at 4 × MIC (Figure 5). No growth was observed after treatment with compounds **9** and **10** at 4 × MIC. These data suggest that compounds **8** and **11** might exhibit a killing effect, whilst compounds **9** and **10** could induce a bacteriostatic effect on susceptible microorganisms. Concerning the structure–activity relationship (SAR) studies of AKGs **8–11**, it was noticed that the substitution of the hydroxy group at position 12 in the alkyl chain of compound **11** by a methoxy, or a *gem*-difluoro or azide group corresponding to AKGs **8–10** respectively, significantly reduced the antimicrobial activity.

AKGs **8–11** were also tested in combination with some commonly used antibiotics (Table 2). The results showed that synergistic effects could be obtained in some cases, especially when the AKG **8** was combined with gentamicin, and compound **11** was combined with ciprofloxacin and ampicillin. More than four-fold increase of the activity of these antibiotics was recorded on the three selected bacteria, suggesting that the study should be emphasized on such combinations. The overall activity could be considered as important, mainly when viewed that most of the organisms used were antibiotic resistant. This study therefore provides supportive data for the potential use of the studied 1-*O*-alkylglycerols, in particular AKG **11**, as well as in combination with some antibiotics for the treatment of microbial infections. However, this is to be confirmed by further toxicological studies.

Table 1. Minimal inhibition concentration (µg/mL and in µM in parenthesis) of AKGs and reference antibiotics.

Microbial Species [a]	Tested Samples				
	8	9	10	11	RE [b]
Cf	-	-	-	156.25 (435.77)	9.76 (2.04)
Ec1	-	-	312.50 (814.71)	39.06 (108.94)	4.88 (15.10)
Ec2	-	-	312.50 (814.71)	156.25 (435.77)	156.25 (483.55)
Ec3	19.53 (52.42)	78.12 (206.38)	312.50 (814.71)	19.53 (54.47)	4.88 (1.02)
Sd	39.06 (104.84)	156.25 (412.77)	-	156.25 (435.77)	9.76 (2.04)
St	39.06 (104.84)	78.12 (206.38)	312.50 (814.71)	19.53 (54.47)	19.53 (4.09)
Sa	312.50 (838.75)	312.50 (825.54)	-	156.25 (435.77)	4.88 (1.02)
Sf	19.53 (52.42)	78.12 (206.38)	78.12 (203.67)	78.12 (217.89)	4.88 (1.02)
Ca	312.50 (838.75)	312.50 (825.54)	-	39.06 (108.94)	19.53 (2.08)
Cg	78.12 (209.67)	312.50 (825.54)	-	19.53 (54.47)	19.53 (2.08)
Ma	39.06 (104.84)	78.12 (206.38)	-	39.06 (108.94)	19.53 (2.08)

[a] Microbial species: Cf: *Citrobacter freundii*, Ec1: *Escherichia coli* ATCC10536, Ec2: *Escherichia coli* AG102, Ec3: *Escherichia coli* LMP701; Sd: *Shigella dysenteriae*, St: *Salmonalla typhi*, Bc: *Bacillus cereus*, Sa: *Staphylococcus aureus*, Sf: *Streptococcus. faecalis*, Ca: *Candida albicans*, Cg: *Candida glabrata*; Ma: *Microsporum audouinii*, [b] RE or reference antibiotics: choramphenicol for Ec1 and Ec2; gentamicin for other bacteria, nystatin for Ca and Ma; (-): not active as MIC was not detected up to 312.50 µg/mL.

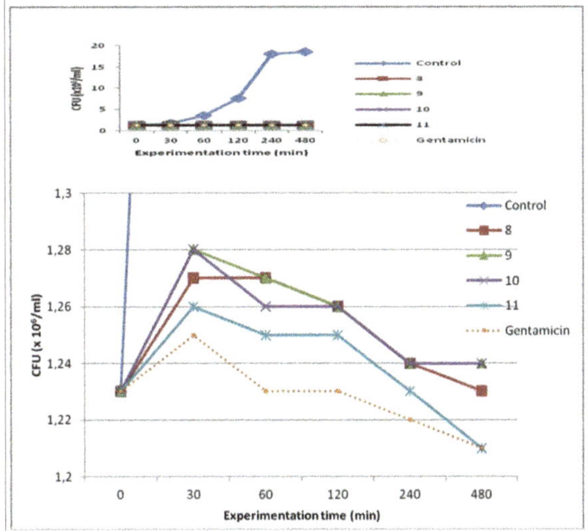

Figure 4. Time-effect of the studied AKGs on the growth of *E. coli* LMP701 when tested with the MIC of the studied samples.

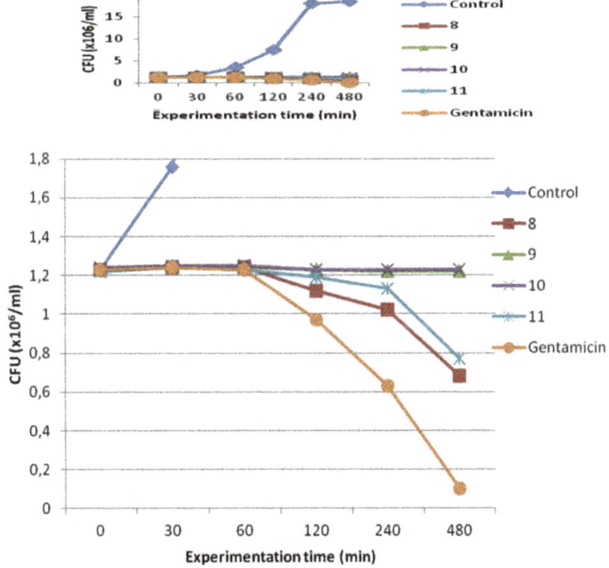

Figure 5. Time-effect of the studied AKGs on the growth of *E. coli* LMP701 when tested with the 4× MIC of the studied samples.

Table 2. Effect of the studied AKGs combined with some commonly used antibiotics on the susceptibility of three bacterial strains to some reference antibiotics.

Samples [a]		Microorganisms, MIC in µg/mL and Fold Restoration [b] of Antibiotic Activity (in Parenthesis)		
		E. coli LMP701	*C. freundii*	*S. aureus*
Gentamicin	alone	4.88	9.76	4.88
	+8	0.61 (8)	1.22 (8)	1.22 (2)
	+9	4.88 (1)	19.53 (-)	4.88 (1)
	+10	4.88 (1)	4.88 (2)	4.88 (1)
	+11	1.22 (2)	9.76 (1)	2.44 (2)
Tetracycline	alone	39.06	78.12	19.53
	+8	19.53 (2)	78.12 (1)	39.06 (1)
	+9	78.12 (-)	78.12 (1)	39.06 (1)
	+10	19.53 (2)	78.12 (1)	39.06 (1)
	+11	9.76 (4)	39.06 (2)	9.76 (2)
Ciprofloxacin	alone	9.76	78.12	39.06 (1)
	+8	4.88 (2)	78.12 (1)	39.06 (1)
	+9	4.88 (2)	78.12 (1)	39.06 (1)
	+10	2.44 (4)	39.06 (2)	39.06 (1)
	+11	2.44 (4)	9.76 (8)	9.76 (4)
Ampicillin	alone	39.06	156.25	78.12
	+8	39.06 (1)	78.12 (2)	19.53 (4)
	+9	39.06 (1)	156.25 (1)	78.12 (1)
	+10	156.25 (-)	78.12 (2)	312.50 (-)
	+11	4.88 (8)	9.76 (16)	9.76 (8)

[a] Samples: Antibiotics and synthesized AKGs **8–11** were tested in combination (1:1). [b] Fold restoration of antibiotic activity was determined according to the initial MIC of the reference antibiotic. (-): no restoration of antibiotic activity.

3. Materials and Methods

3.1. General Experimental Procedures

Moisture sensitive reactions were performed under nitrogen. Anhydrous tetrahydrofuran (THF) and diethyl ether (Et$_2$O) were obtained by percolation through a column of a drying resin. Anhydrous *N,N*-dimethylformamide (DMF) over molecular sieves was used as commercially supplied (Acros). Room temperature (rt) means a temperature generally in the range of 18–20 °C. Column chromatography was performed over silica gel Kielselgel 60 (40–60 µm) or basic alumina (Brockmann activity II; basic; pH 10 ± 0.5). Routine monitoring of reactions was carried out using Merck silica gel 60 F$_{254}$ TLC plates (TLC: thin layer chromatography) purchased from Fluka and visualized by UV light (254 nm) inspection followed by staining with an acidic ethanolic solution of *p*-anisaldehyde, or with a solution of phosphomolybdic acid (5 g in 100 mL 95% ethanol). Infrared spectra were recorded with a Thermo Nicolet Avatar 250 FTIR and were reported using the frequency of absorption (cm^{-1}). ^{1}H NMR spectra (400.13 MHz) and ^{13}C NMR spectra (100.61 MHz) were recorded on an Avance 400 Bruker spectrometer using TMS as an internal standard. Multiplicity was tabulated using standard abbreviations: s for singlet, d for doublet, dd for doublet of doublets, t for triplet, q for quadruplet, dtt for doublet of triplets of triplets and m for multiplet (br means broad). Quite obvious first-order ^{1}H NMR multiplets were analyzed. As a helpful guidance for this analysis, two articles of Hoye et al. appeared in 1994 and 2002 [34,35]. NMR spectra were processed with zero filling (512 k or 1024 k points). Sometimes resolution enhancement in ^{1}H NMR using Traficante facilitated the assignments. Specific rotations were measured on a Perkin Elmer 341 polarimeter, with a cell of 1 dm long and a Na- or Hg-source (Na at 589 nm; Hg at 578 nm, 546 nm, 436 nm and 365 nm), and concentrations were expressed in g/100 mL. High resolution mass spectra (HRMS) were recorded using a MicrO-Tof-Q II spectrometer under electrospray using methanol as solvent. Microanalyses were performed with a CHNS analyzer.

The compound 2,3-isopropylidene-*sn*-glycerol (**18**) (≥95% pure) was purchased from Alfa Aesar (France, article # B23037); and ricinoleic acid (**12**) (~80% pure) was purchased from Fluka (Switzerland, article # 83903). Boron trifluoride dimethanol complex (BF$_3$·2MeOH) was purchased from Acros (Belgium, article # 15890). Products which were used for biological studies were purchased from Maneesh Pharmaceutic PVT (Mumbai, India), Sigma-Aldrich (Johannesburg, South Africa), and Jinling Pharmaceutic Group corp. (Nanjing, China).

3.2. Synthesis of Compounds **8–11,13–17,19–29** and **31–36**

3.2.1. (*R*,*Z*)-Methyl 12-hydroxyoctadec-9-enoate (**13**) (Methyl ricinoleate)

To a solution of ricinoleic acid (**12**) (21 g, technical, ~80% pure, ca. 56 mmol) in methanol (140 mL) with stirring, was added BF$_3$·2MeOH (3.85 mL, 35.5 mmol, 0.63 equiv). Stirring was continued overnight at 50 °C. TLC monitoring showed completion of the reaction mixture after 16 h. Methanol was removed in vacuo, and the resulting oily residue was transferred into a separating funnel with ethyl acetate (100 mL). After washing with brine (3 × 30 mL) and drying over Na$_2$SO$_4$, ethyl acetate was removed under reduced pressure, and the crude product was purified by column chromatography on silica gel (86 g, 0–4% acetone in petroleum ether) to afford methyl ricinoleate **13** as a colorless oil (16.59 g, 75%) along with a small amount (4%) of a slightly less polar by-product **14**. It eluted after a minor amount of a less polar mixture of methyl oleate and methyl linoleate, R$_f$ ca. 0.75 (petroleum ether/acetone 80:20), which was the result of the esterification of the ~15% oleic + linoleic acids, which were contained in technical ricinoleic acid. When potassium carbonate (600 mg) was added to a solution of **13** containing **14** (3.17 g) in methanol (30 mL) with stirring for 42 h (followed by quenching with a solution of citric acid (834 mg) in water (7.5 mL)), the by-product **14** disappeared to afford compound **13** alone, R$_f$ = 0.4 (petroleum ether/acetone 80:20).

IR (KBr) ν 3445 (broad, O–H), 3007, 2928, 2855, 1741 (C=O), 1461, 1436, 1246, 1198, 1173, 725 cm^{-1}.

^{1}H NMR (400 MHz, CDCl$_3$): δ 5.56 (dtt, 1H, *J* = 10.9, 7.3, 1.5 Hz, C$\underline{H}$=CH–CH$_2$–CHOH), 5.40 (dtt, 1H, *J* = 10.9, 7.5, 1.5 Hz, CH=C$\underline{H}$–CH$_2$–CHOH), 3.67 (s, 3H, CO$_2$Me), 3.66–3.56 (m, 1H, C$\underline{H}$OH), 2.30 (dd, 2H, *J* = 7.7, 7.4 Hz, C$\underline{H}_2$CO$_2$Me), 2.24–2.18 (m, 2H, CHOH–C$\underline{H}_2$–CH=CH), 2.05 (br qd, 2H, *J* = 7.2, 1.2 Hz, CH=CH–C$\underline{H}_2$), 1.62 (br tt, 2H, *J* = 7.5, 7.3 Hz, C$\underline{H}_2$CH$_2$CO$_2$Me), 1.58 (br d, 1H, *J* = 0.9 Hz, OH, could be s and overlapped at another δ), 1.51–1.41 (m, 3H, C$\underline{H}_2$CHOH and 1H of CH$_2$CH$_2$CHOH), 1.39–1.23 (m, 15H, 1H of C$\underline{H}_2$CH$_2$CHOH, CH$_3$(C$\underline{H}_2$)$_3$ and (C$\underline{H}_2$)$_4$CH$_2$CH$_2$CO$_2$Me), 0.88 (t (approximately t because not first order due to coupling to a rather close CH$_2$ at ~1.25 ppm), 3H, *J* = 6.9 Hz, CH$_3$).

^{13}C NMR (100 MHz, CDCl$_3$): δ 174.35 (C$_{quat}$, CO), 133.39 ($\underline{C}$H=CH), 125.24 (CH=$\underline{C}$H), 71.52 ($\underline{C}$HOH), 51.46 (CO$_2\underline{C}$H$_3$), 36.87 ($\underline{C}$H$_2$), 35.37 ($\underline{C}$H$_2$), 34.10 ($\underline{C}$H$_2$CO$_2$Me), 31.85 ($\underline{C}$H$_2$CH$_2$CH$_3$), 29.58 ($\underline{C}$H$_2$), 29.37 ($\underline{C}$H$_2$), 29.13 ($\underline{C}$H$_2$), 29.10 (2 $\underline{C}$H$_2$), 27.38 ($\underline{C}$H$_2$), 25.73 ($\underline{C}$H$_2$), 24.93 ($\underline{C}$H$_2$CH$_2$CO$_2$Me), 22.63 ($\underline{C}$H$_2$CH$_3$), 14.10 ($\underline{C}$H$_3$).

[α]$_D^{22}$: +4.2; [α]$_{578}^{22}$: +4.3; [α]$_{546}^{22}$: +4.7; [α]$_{436}^{22}$: +7.0; [α]$_{365}^{22}$: +8.7 (c 6.00, CHCl$_3$),

[α]$_D^{22}$: +5.7; [α]$_{578}^{22}$: +5.9; [α]$_{546}^{22}$: +6.6; [α]$_{436}^{22}$: +10.7; [α]$_{365}^{22}$: +15.4 (c 6.00, acetone).

3.2.2. Physical data for (*R*,*Z*)-(*R*,*Z*)-18-methoxy-18-oxooctadec-9-en-7-yl 12-hydroxyoctadec-9-enoate (**14**)

R$_f$ = 0.61 (petroleum ether/acetone 80:20).

IR (KBr) ν 3465 (O–H), 3010, 2928, 2855, 1737 (C=O), 1466, 1456, 1436, 1245, 1194, 1175, 725 cm^{-1}.

^{1}H NMR (400 MHz, CDCl$_3$): δ 5.56 (dtt, 1H, J = 10.9, 7.3, 1.5 Hz, C<u>H</u>=CH–CH$_2$–CHOH), 5.46 (dtt, 1H, J = 10.9, 7.2, 1.5 Hz, C<u>H</u>=CH–CH$_2$–CHOCO), 5.40 (dtt, 1H, J = 10.9, 7.5, 1.5 Hz, CH=C<u>H</u>–CH$_2$–CHOH), 5.32 (dtt, 1H, J = 10.9, 7.3, 1.5 Hz, CH=C<u>H</u>–CH$_2$–CHOCO), 4.88 (tt, 1H, J = 6.3, 6.3 Hz, C<u>H</u>OCO), 3.67 (s, 3H, CO$_2$Me), 3.61 (br tt, 1H, J = 6.1, 5.7 Hz, C<u>H</u>OH), 2.33–2.24 (m, 6H), 2.23–2.18 (m, 2H), 2.08–1.98 (m, 4H, 2 CH=CH–C<u>H$_2$</u>), 1.66–1.43 (m, ~12H, 2 C<u>H$_2$</u>CH$_2$CO$_2$, C<u>H$_2$</u>CHOCO, C<u>H$_2$</u>CHOH,OH, H$_2$O), 1.38–1.21 (m, 32H, 2 CH$_3$(C<u>H$_2$</u>)$_4$ and 2 (C<u>H$_2$</u>)$_4$CH$_2$CH$_2$CO$_2$), 0.884 (t, 3H, J = 6.9 Hz, CH$_3$), 0.876 (t, 3H, J = 6.9 Hz, CH$_3$).

^{13}C NMR (100 MHz, CDCl$_3$): δ 174.33 (C$_{quat}$, <u>C</u>O$_2$Me), 173.59 (<u>C</u>O$_2$CH), 133.39 (<u>C</u>H=CH), 132.53 (<u>C</u>H=CH), 125.22 (<u>C</u>H=CH), 124.35 (<u>C</u>H=CH), 73.70 (<u>C</u>HOCO), 71.51 (<u>C</u>HOH), 51.46 (CO$_2$<u>C</u>H$_3$), 36.87 (<u>C</u>H$_2$), 35.38 (<u>C</u>H$_2$), 34.68 (<u>C</u>H$_2$), 34.10 (<u>C</u>H$_2$CO$_2$Me), 33.65 (<u>C</u>H$_2$CO$_2$CH), 32.00 (<u>C</u>H$_2$), 31.85 (<u>C</u>H$_2$CH$_2$CH$_3$), 31.76 (<u>C</u>H$_2$), 29.62 (<u>C</u>H$_2$), 29.53 (<u>C</u>H$_2$), 29.36 (<u>C</u>H$_2$), 29.19 (<u>C</u>H$_2$), 29.18 (<u>C</u>H$_2$), 29.16 (<u>C</u>H$_2$), 29.14 (<u>C</u>H$_2$), 29.13 (<u>C</u>H$_2$), 29.12 (2 <u>C</u>H$_2$), 27.40 (<u>C</u>H$_2$), 27.34 (<u>C</u>H$_2$), 25.73 (<u>C</u>H$_2$), 25.37 (<u>C</u>H$_2$), 25.10 (<u>C</u>H$_2$), 24.95 (<u>C</u>H$_2$CH$_2$CO$_2$Me), 22.63 (<u>C</u>H$_2$), 22.59 (<u>C</u>H$_2$), 14.10 (<u>C</u>H$_3$), 14.08 (<u>C</u>H$_3$).

[α]$_D^{21}$: +17.4; [α]$_{578}^{21}$: +18.1; [α]$_{546}^{21}$: +20.7; [α]$_{436}^{21}$: +34.9; [α]$_{365}^{21}$: +54.8 (c 2.56, CHCl$_3$).

HRMS (ESI, *m/z*) calculated for C$_{37}$H$_{68}$O$_5$Na [M + Na]$^+$: 615.4964, found: 615.4971.

3.2.3. (*R,Z*)-Methyl 12-methoxyoctadec-9-enoate (**15**)

In a flask containing a solution of **13** (625 mg, 2.0 mmol), tetra-*n*-butylammonium bromide (709.2 mg, 2.2 mmol, 1.2 equiv) in DMSO (2.0 mL) with stirring, was added finely crushed (with a mortar and pestle) sodium hydroxide (250 mg, 6 mmol, 3 equiv) and methyl iodide (0.63 mL, 10 mmol, 5 equiv). The reaction flask was flushed under nitrogen, tightly stoppered and protected from light by wrapping with an aluminum foil. After stirring overnight for 18 h, TLC monitoring showed that the reaction was mostly done. An aqueous solution of 10% citric acid (10 mL) was added, and extraction was done with petroleum ether/EtOAc (80:20). Organic layers were dried over Na$_2$SO$_4$, and solvent was removed under reduced pressure. Then, the residue was purified by column chromatography on basic alumina (5 g, 0%–0.2% acetone in petroleum ether) to afford **15** as a colorless oil (471 mg, 72%). R_f = 0.6 (petroleum ether/acetone 90:10).

IR (KBr) ν 3465 (small, harmonic of C=O), 3009, 2929, 2855, 1742 (C=O), 1463, 1456, 1436, 1360, 1245, 1195, 1173, 1099 (C–O of OMe), 725 cm^{-1}.

^{1}H NMR (400 MHz, CDCl$_3$): δ 5.50–5.34 (m, 2H: C<u>H</u>=C<u>H</u> partly distorted due to strong coupling at 5.45 and 5.38 ppm (dtt, J = 10.9, 6.9, 1.4 Hz)), 3.67 (s, 3H, CO$_2$Me), 3.34 (s, 3H, CHOC<u>H$_3$</u>), 3.17 (tt, 1H, J = 6.2, 5.5 Hz, C<u>H</u>OMe), 2.30 (dd, 2H, J = 7.7, 7.4 Hz, C<u>H$_2$</u>CO$_2$Me), 2.30–2.17 (m, 2H, CHOMe–C<u>H$_2$</u>–CH=CH), 2.03 (br q, 2H, J = 6.7 Hz, CH=CH–C<u>H$_2$</u>), 1.67–1.57 (m, 2H, C<u>H$_2$</u>CH$_2$CO$_2$Me), 1.49–1.41 (m, 2H, C<u>H$_2$</u>CHOMe), 1.40–1.21 (m, 16H, CH$_3$(C<u>H$_2$</u>)$_4$ and (C<u>H$_2$</u>)$_4$CH$_2$CH$_2$CO$_2$Me), 0.88 (t, 3H, J = 6.9 Hz, CH$_3$).

^{13}C NMR (100 MHz, CDCl$_3$): δ 174.33 (C$_{quat}$, <u>C</u>O), 131.73 (<u>C</u>H=CH), 125.42 (<u>C</u>H=CH), 80.99 (<u>C</u>HOMe), 56.58 (CHO<u>C</u>H$_3$), 51.46 (CO$_2$<u>C</u>H$_3$), 34.11 (<u>C</u>H$_2$CO$_2$Me), 33.57 (<u>C</u>H$_2$), 31.88 (<u>C</u>H$_2$CH$_2$CH$_3$), 31.05 (<u>C</u>H$_2$), 29.56 (<u>C</u>H$_2$), 29.50 (<u>C</u>H$_2$), 29.18 (<u>C</u>H$_2$), 29.15 (<u>C</u>H$_2$), 29.13 (<u>C</u>H$_2$), 27.41 (<u>C</u>H$_2$), 25.36 (<u>C</u>H$_2$), 24.95 (<u>C</u>H$_2$CH$_2$CO$_2$Me), 22.65 (<u>C</u>H$_2$CH$_3$), 14.10 (<u>C</u>H$_3$).

[α]$_D^{17.5}$: +13.6; [α]$_{578}^{17.5}$: +14.1; [α]$_{546}^{17.5}$: +16.1; [α]$_{436}^{17.5}$: +27.5; [α]$_{365}^{17.5}$: +43.2 (c 5.00, CHCl$_3$)

[α]$_D^{17.5}$: +16.2; [α]$_{578}^{17.5}$: +16.9; [α]$_{546}^{17.5}$: +19.2; [α]$_{436}^{17.5}$: +32.6 (neat liquid).

3.2.4. (*R,Z*)-12-Methoxyoctadec-9-en-1-ol (**16**)

Red-Al (0.52 mL, ~3 M in toluene, 1.56 mmol, 1.2 equiv) was added dropwise to a cooled solution of **15** (422 mg, 1.29 mmol) in anhydrous Et_2O (4 mL) at 0 °C with stirring and under nitrogen. After the addition, the stirring was continued overnight for 18 h at 0 °C (use of a Dewar with ice-cooling). TLC monitoring confirmed disappearance of the starting material. A solution of citric acid (400 mg) in distilled water (5 mL) was added to the reaction mixture, which was allowed to stir again for 30 min. Extraction was done with petroleum ether/EtOAc (80:20). Organic layers were dried over Na_2SO_4, and solvent was evaporated under reduced pressure. The crude product was then purified by column chromatography on basic alumina (5 g, 0%–3% acetone in petroleum ether) to afford **16** as a colorless oil (318 mg, 82%). R_f = 0.32 (petroleum ether/acetone 85:15).

IR (KBr) ν 3372 (broad, O–H), 3009, 2927, 2855, 1464, 1456, 1377, 1357, 1099 (C–O of OMe), 1058, 724 cm^{-1}.

^{1}H NMR (400 MHz, $CDCl_3$): δ 5.50–5.34 (m, 2H: C$\underline{H}$=C$\underline{H}$ partly distorted due to strong coupling at 5.46 and 5.38 ppm (dtt, *J* = 10.9, 7.0, 1.4 Hz)), 3.64 (t, 2H, *J* = 6.6 Hz, C$\underline{H}_2$OH), 3.34 (s, 3H, CHOC$\underline{H}_3$), 3.17 (tt, 1H, *J* = 6.2, 5.5 Hz, C$\underline{H}$OMe), 2.31–2.17 (m, 2H, CHOMe–C$\underline{H}_2$–CH=CH), 2.03 (br q, 2H, *J* = 6.7 Hz, CH=CH–C$\underline{H}_2$), 1.66 (br s, 1H, OH), 1.56 (br tt, 2H, *J* = 7.5, 6.6 Hz, C$\underline{H}_2$CH$_2$OH), 1.49–1.41 (m, 2H, C$\underline{H}_2$CHOMe), 1.41–1.23 (m, 18H, CH$_3$(C$\underline{H}_2$)$_4$ and (C$\underline{H}_2$)$_5$CH$_2$CH$_2$OH), 0.88 (t, 3H, *J* = 6.9 Hz, CH$_3$).

^{13}C NMR (100 MHz, $CDCl_3$): δ 131.81 (C$\underline{H}$=CH), 125.38 (C$\underline{H}$=CH), 81.02 (C$\underline{H}$OMe), 63.06 (C$\underline{H}_2$OH), 56.58 (CHOC$\underline{H}_3$), 33.57 (C$\underline{H}_2$), 32.80 (C$\underline{H}_2$CH$_2$OH), 31.88 (C$\underline{H}_2$), 31.07 (C$\underline{H}_2$), 29.61 (C$\underline{H}_2$), 29.50 (2 C$\underline{H}_2$), 29.40 (C$\underline{H}_2$), 29.27 (C$\underline{H}_2$), 27.43 (C$\underline{H}_2$), 25.74 (C$\underline{H}_2$), 25.36 (C$\underline{H}_2$), 22.65 (C$\underline{H}_2$), 14.12 (C$\underline{H}_3$).

$[\alpha]_D^{23.5}$: +12.8; $[\alpha]_{578}^{23.5}$: +13.4; $[\alpha]_{546}^{23.5}$: +15.2; $[\alpha]_{436}^{23.5}$: +25.9; $[\alpha]_{365}^{23.5}$: +40,6 (c 5.02, CHCl$_3$).

$[\alpha]_D^{23.5}$: +17.0; $[\alpha]_{578}^{23.5}$: +17.4; $[\alpha]_{546}^{23.5}$: +19.8; $[\alpha]_{436}^{23.5}$: +33.6; $[\alpha]_{365}^{23.5}$: +53.2 (c 5.02, acetone).

$[\alpha]_D^{23}$: +15.7; $[\alpha]_{578}^{23}$: +16.4; $[\alpha]_{546}^{23}$: +18.6; $[\alpha]_{436}^{23}$: +31.7 (neat liquid).

HRMS (ESI, *m/z*) calculated for $C_{19}H_{38}O_2Na$ [M + Na]$^+$: 321.4892, found: 321.4886.

3.2.5. (*R,Z*)-12-Methoxyoctadec-9-en-1-yl methanesulfonate (**17**)

To a stirred solution of **16** (6.64 g, 22.2 mmol), Et_3N (4.7 mL, 33.4 mmol, 1.5 equiv) in DCM (67 mL) under nitrogen at −30 °C, mesyl chloride (2.2 mL, 28 mmol, 1.25 equiv) in DCM (9 mL) was added dropwise. The addition of mesyl chloride was completed by rinsing with DCM (3 × 0.3 mL). The corresponding mixture was then stirred for 15 h at −30 °C. TLC monitoring (elution with DCM, mesylate showed far bigger R_f than starting alcohol with this eluent) showed completion of the reaction, and distilled water (75 mL) was added to quench the reaction. Extraction was done with DCM. Organic layers were washed with brine and dried over Na_2SO_4. Solvent was removed under reduced pressure to provide the crude product as a light yellow oil. The crude product was then purified by column chromatography on silica gel (10 g, 0%–2% acetone in petroleum ether) to provide **17** as a colorless oil (5.40 g, 74%). R_f = 0.45 (petroleum ether/acetone 80:20).

IR (KBr) ν 3011, 2928, 2855, 1464, 1358, 1177 (S=O), 1098 (C–O of OMe), 974, 945, 831, 724, 529 cm^{-1}.

^{1}H NMR (400 MHz, $CDCl_3$): δ 5.50–5.34 (m, 2H: C$\underline{H}$=C$\underline{H}$ partly distorted due to strong coupling at 5.46 and 5.38 ppm (dtt, *J* = 10.9, 6.9, 1.3 Hz)), 4.22 (t, 2H, *J* = 6.6 Hz, C$\underline{H}_2$OMs), 3.34 (s, 3H, CHOC$\underline{H}_3$), 3.17 (tt, 1H, *J* = 6.1, 5.5 Hz, C$\underline{H}$OMe), 3.00 (s, 3H, SO$_2$C$\underline{H}_3$), 2.31–2.17 (m, 2H, CHOMe–C$\underline{H}_2$–CH=CH), 2.03 (br q, 2H, *J* = 6.7 Hz, CH=CH–C$\underline{H}_2$), 1.75 (br tt, 2H, *J* = 7.5, 6.6 Hz, C$\underline{H}_2$CH$_2$OMs), 1.50–1.22 (m, 20H, CH$_3$(C$\underline{H}_2$)$_5$ and (C$\underline{H}_2$)$_5$CH$_2$CH$_2$OMs), 0.88 (t, 3H, *J* = 6.9 Hz, CH$_3$).

^{13}C NMR (100 MHz, CDCl$_3$): δ 131.70 (C̲H=CH), 125.47 (C̲H=CH), 80.98 (C̲HOMe), 70.16 (C̲H$_2$OMs), 56.59 (CHOC̲H$_3$), 37.38 (SO$_2$C̲H$_3$), 33.57 (C̲H$_2$), 31.88 (C̲H$_2$), 31.08 (C̲H$_2$), 29.57 (C̲H$_2$), 29.49 (C̲H$_2$), 29.34 (C̲H$_2$), 29.21 (C̲H$_2$), 29.13 (C̲H$_2$), 29.02 (C̲H$_2$), 27.41 (C̲H$_2$), 25.43 (C̲H$_2$), 25.36 (C̲H$_2$), 22.64 (C̲H$_2$), 14.10 (C̲H$_3$).

$[\alpha]_D^{23.5}$: +13.7; $[\alpha]_{578}^{23.5}$: +14.1; $[\alpha]_{546}^{23.5}$: +16.0; $[\alpha]_{436}^{23.5}$: +27.5; $[\alpha]_{365}^{23.5}$: +43.4 (c 5.01, acetone).

$[\alpha]_D^{23.5}$: +10.3; $[\alpha]_{578}^{23.5}$: +10.5; $[\alpha]_{546}^{23.5}$: +11.9; $[\alpha]_{436}^{23.5}$: +20.3; $[\alpha]_{365}^{23.5}$: +31.5 (c 2.60, CHCl$_3$).

$[\alpha]_D^{23.5}$: +13.7; $[\alpha]_{578}^{23.5}$: +14.3; $[\alpha]_{546}^{23.5}$: +16.3; $[\alpha]_{436}^{23.5}$: +27.2 (neat liquid).

3.2.6. (*R*)-4-((((*R,Z*)-12-Methoxyoctadec-9-en-1-yl)oxy)methyl)-2,2-dimethyl-1,3-dioxolane (**19**)

In a flask containing a solution of **17** (4.97 g, 13.2 mmol), *n*-Bu$_4$NBr (1.06 g, 3.3 mmol, 0.25 equiv), and finely crushed (with a mortar and pestle) potassium hydroxide (3.48 g, 52.75 mmol, ~85% KOH, 4 equiv) in DMSO (26.4 mL) with stirring and under nitrogen at rt for 10 min, was added (*R*)-solketal (**18**) (2.07 g, ≥95% pure, 14.9 mmol, 1.13 equiv), and the corresponding mixture was stirred overnight for 14 h at 35 °C. TLC monitoring showed completion of the reaction and distilled water (50 mL) was added to the reaction mixture. Extraction was done with petroleum ether/EtOAc (80:20). Organic layers were washed with brine and dried over Na$_2$SO$_4$. Solvent was removed under reduced pressure. The crude product was then purified by column chromatography on basic alumina (25 g, 0%–1% acetone in petroleum ether) to afford **19** as a colorless oil (4.65 g, 93%). R$_f$ = 0.45 (petroleum ether/acetone 95:5).

IR (KBr) ν 2985, 2929, 2856, 2821, 1466, 1456, 1379, 1369, 1256, 1237, 1214, 1118, 1100 (C–O of OMe), 1056, 847, 724, 514 cm^{-1}.

^{1}H NMR (400 MHz, CDCl$_3$): δ 5.50–5.34 (m, 2H: C̲H=C̲H partly distorted due to strong coupling at 5.46 and 5.38 ppm (dtt, *J* = 10.9, 6.9, 1.3 Hz)), 4.27 (dddd (apparent tt), 1H, *J* = 6.4, 6.4, 5.7, 5.6 Hz, C̲H–O in dioxolane), 4.06 (dd, 1H, *J* = 8.2, 6.4 Hz, C̲H$_2$O in dioxolane), 3.73 (dd, 1H, *J* = 8.2, 6.4 Hz, C̲H$_2$O in dioxolane), 3.54–3.39 (m, 4H, C̲H$_2$OC̲H$_2$O with 1H dd at 3.52 ppm, *J* = 9.9, 5.7 Hz and 1H dd at 3.42 ppm, *J* = 9.9, 5.6 Hz), 3.34 (s, 3H, CHOC̲H$_3$), 3.17 (br tt, 1H, *J* = 6.2, 5.4 Hz, C̲HOMe), 2.31–2.17 (m, 2H, CHOMe–C̲H$_2$–CH=CH), 2.03 (br q, 2H, *J* = 6.7 Hz, CH=CH–C̲H$_2$), 1.62–1.52 (m, 2H, C̲H$_2$CH$_2$O), 1.50–1.41 (m, 2H, C̲H$_2$CHOMe), 1.42 (q, 3H, *J* = 0.7 Hz (w coupling), C̲H$_3$), 1.36 (q, 3H, *J* = 0.7 Hz (w coupling), C̲H$_3$), 1.39–1.23 (m, 18H, CH$_3$(C̲H$_2$)$_4$ and (C̲H$_2$)$_3$CH$_2$CH$_2$O), 0.88 (t, 3H, *J* = 6.9 Hz, C̲H$_3$).

^{13}C NMR (100 MHz, CDCl$_3$): δ 131.82 (C̲H=CH), 125.35 (C̲H=CH), 109.37 (C̲Me$_2$), 80.99 (C̲HOMe), 74.76 (C̲H–O), 71.88 and 71.83 (C̲H$_2$OC̲H$_2$(CH$_2$)$_7$), 66.93 (C̲H$_2$OCMe$_2$), 56.58 (CHOC̲H$_3$), 33.57 (C̲H$_2$), 31.88 (C̲H$_2$), 31.04 (C̲H$_2$), 29.63 (C̲H$_2$), 29.56 (C̲H$_2$), 29.51 (C̲H$_2$), 29.50 (C̲H$_2$), 29.45 (C̲H$_2$), 29.30 (C̲H$_2$), 27.44 (C̲H$_2$), 26.78 (C–C̲H$_3$), 26.06 (C̲H$_2$), 25.43 (C–C̲H$_3$), 25.36 (C̲H$_2$), 22.64 (C̲H$_2$), 14.11 (C̲H$_3$).

$[\alpha]_D^{18}$: +4.4; $[\alpha]_{578}^{18}$: +4.1; $[\alpha]_{546}^{18}$: +4.7; $[\alpha]_{436}^{18}$: +7.3; $[\alpha]_{365}^{18}$: +10.7 (c 3.01, acetone).

$[\alpha]_D^{18}$: +2.7; $[\alpha]_{578}^{18}$: +2.8; $[\alpha]_{546}^{18}$: +3.0; $[\alpha]_{436}^{18}$: +4.8; $[\alpha]_{365}^{18}$: +6.3 (c 2.42, CHCl$_3$).

HRMS (ESI, *m/z*) calculated for C$_{25}$H$_{48}$O$_4$Na [M + Na]$^+$: 435.3450, found: 435.3452.

Elementary analysis calculated for C$_{25}$H$_{48}$O$_4$: C, 72.77; H, 11.72, found: C, 73.23; H, 11.94.

3.2.7. (*S*)-3-(((*R,Z*)-12-Methoxyoctadec-9-en-1-yl)oxy)propane-1,2-diol (**8**)

To a solution of acetonide **19** (4.65 g, 11.2 mmol) in methanol (45.1 mL), was added *p*-toluenesulfonic acid monohydrate (107.3 mg, 0.55 mmol, 0.05 equiv) and distilled water (4.5 mL). The flask was then purged with nitrogen, stoppered and dipped in a preheated bath at 60 °C. Stirring was maintained for 5 h at 60 °C. TLC monitoring showed completion of the reaction, and sodium bicarbonate (52.1 mg, 0.62 mmol, 0.055 equiv) was added. Stirring was continued for 1 h at 60 °C. Methanol was then removed under reduced pressure, and the crude product was purified by column chromatography on silica gel (20 g, 0%–5% acetone in petroleum ether and then petroleum ether + 5% acetone + 12% methanol) to afford **8** as a colorless oil (4.16 g, 99%). R_f = 0.03 (petroleum ether/acetone 90:10).

^{1}H NMR (400 MHz, CDCl$_3$): δ 5.50–5.34 (m, 2H: C$\underline{H}$=C$\underline{H}$ partly distorted due to strong coupling at 5.46 and 5.38 ppm (dtt, J = 10.9, 7.0, 1.4 Hz)), 3.90–3.83 (m, 1H, C$\underline{H}$OH), 3.72 (broad ddd, 1H, J = 11.3, 6.7 (with OH), 3.9 Hz, C$\underline{H}_2$OH), 3.65 (broad ddd, 1H, J = 11.3, 5.1 (with OH), 4.9 Hz, C$\underline{H}_2$OH), 3.56–3.42 (m, 4H, C$\underline{H}_2$OC$\underline{H}_2$ with 1H dd at 3.54 ppm, J = 9.7, 4.0 Hz), 3.34 (s, 3H, CHOC$\underline{H}_3$), 3.17 (tt, 1H, J = 6.2, 5.4 Hz, C$\underline{H}$OMe), 2.67 (d, 1H, J = 5.0 Hz, CHO$\underline{H}$), 2.28 (broad dd, 1H, J = 6.7, 5.1 Hz, resolution ω = 1.3 Hz, C$\underline{H}_2$OH), 2.30–2.17 (m, 2H, CHOMe–C$\underline{H}_2$–CH=CH), 2.03 (br q, 2H, J = 6.7 Hz, CH=CH–C$\underline{H}_2$), 1.57 (broad tt, 2H, J = 7.1, 6.7 Hz, C$\underline{H}_2$CH$_2$O), 1.50–1.41 (m, 2H, C$\underline{H}_2$CHOMe), 1.41–1.23 (m, 18H, CH$_3$(C$\underline{H}_2$)$_4$ and (C$\underline{H}_2$)$_5$CH$_2$CH$_2$O), 0.88 (t, 3H, J = 6.9 Hz, CH$_3$).

^{13}C NMR (100 MHz, CDCl$_3$): δ 131.81 ($\underline{C}$H=CH), 125.36 ($\underline{C}$H=CH), 81.00 ($\underline{C}$HOMe), 72.51 and 71.82 ($\underline{C}$H$_2$OCH$_2$(CH$_2$)$_7$), 70.41 ($\underline{C}$HOH), 64.29 ($\underline{C}$H$_2$OH), 56.59 (CHO$\underline{C}$H$_3$), 33.54 ($\underline{C}$H$_2$), 31.88 ($\underline{C}$H$_2$), 31.04 ($\underline{C}$H$_2$), 29.60 ($\underline{C}$H$_2$), 29.57 ($\underline{C}$H$_2$), 29.49 ($\underline{C}$H$_2$), 29.46 ($\underline{C}$H$_2$), 29.40 ($\underline{C}$H$_2$), 29.26 ($\underline{C}$H$_2$), 27.42 ($\underline{C}$H$_2$), 26.06 ($\underline{C}$H$_2$), 25.35 ($\underline{C}$H$_2$), 22.64 ($\underline{C}$H$_2$), 14.11 ($\underline{C}$H$_3$).

$[\alpha]_D^{22.5}$: +9.8; $[\alpha]_{578}^{22.5}$: +9.6; $[\alpha]_{546}^{22.5}$: +11.1; $[\alpha]_{436}^{22.5}$: +19.6; $[\alpha]_{365}^{22.5}$: +31.5 (c 1.10, acetone).

HRMS (ESI, m/z) calculated for C$_{22}$H$_{44}$O$_4$Na [M + Na]$^+$: 395.3137, found: 395.3132.

3.2.8. (*R,Z*)-Methyl 12-((triisopropylsilyl)oxy)octadec-9-enoate (**20**)

To a vigorously stirred solution of **13** (625 mg, 2.0 mmol) and imidazole (334 mg, 4.9 mmol, 2.45 equiv) in DMF (1.6 mL), which was cooled under nitrogen at 0 °C, was added dropwise triisopropylsilyl chloride (0.53 mL, 97% pure, 2.4 mmol, 1.2 equiv). The corresponding mixture was allowed to stir for 48 h at rt. TLC monitoring showed completion of the reaction. Petroleum ether/EtOAc (80:20) was added. After washing with brine and drying over MgSO$_4$, solvent was removed under reduced pressure, and the residue was purified by column chromatography on silica gel (5 g, 0%–0.2% acetone in petroleum ether) to provide **20** as a colorless oil (574.7 mg, 61%). R_f = 0.73 (petroleum ether/acetone 95:5).

IR (KBr) ν 3006, 2929, 2865, 1744, 1464, 1436, 1366, 1096, 883, 678 cm^{-1}.

^{1}H NMR (400 MHz, CDCl$_3$): δ 5.46–5.36 (m, 2H), 3.83 (tt, 1H, J = 5.8, 5.6 Hz), 3.67 (s, 3H), 2.30 (dd, 2H, J = 7.8, 7.4 Hz), 2.27–2.22 (m, 2H), 2.06–1.97 (m, 2H), 1.67–1.57 (m, 2H), 1.54–1.38 (m, 2H), 1.38–1.20 (m, 16H), 1.06 (s, 21H, Si(C$\underline{H}$(C$\underline{H}_3$)$_2$)$_3$, due to the shielding effect on neighboring CH, CH and CH$_3$ of isopropyl groups being very close, so coupling was not seen and these signals were superimposed, HSQC showed CH at 1.061 ppm and CH$_3$ at 1.059 ppm), 0.88 (t, 3H, J = 6.9 Hz, CH$_3$).

^{13}C NMR (100 MHz, CDCl$_3$): δ 174.33 ($\underline{C}$O), 131.31 ($\underline{C}$H=CH), 125.72 ($\underline{C}$H=CH), 72.26 ($\underline{C}$H), 51.46 ($\underline{C}$H$_3$), 36.56 ($\underline{C}$H$_2$), 34.68 ($\underline{C}$H$_2$), 34.11 ($\underline{C}$H$_2$), 31.92 ($\underline{C}$H$_2$), 29.64 (2 $\underline{C}$H$_2$), 29.20 ($\underline{C}$H$_2$), 29.19 ($\underline{C}$H$_2$), 29.15 ($\underline{C}$H$_2$), 27.51 ($\underline{C}$H$_2$), 24.96 ($\underline{C}$H$_2$), 24.81 ($\underline{C}$H$_2$), 22.64 ($\underline{C}$H$_2$), 18.21 (6 $\underline{C}$H$_3$), 14.11 ($\underline{C}$H$_3$), 12.63 (3 $\underline{C}$H).

$[\alpha]_D^{18}$: +12.3; $[\alpha]_{578}^{18}$: +12.5; $[\alpha]_{546}^{18}$: +14.2; $[\alpha]_{436}^{18}$: +24.3; $[\alpha]_{365}^{18}$: +38.6 (c 4.00, acetone).
$[\alpha]_D^{18}$: +10.6; $[\alpha]_{578}^{18}$: +11.0; $[\alpha]_{546}^{18}$: +12.6; $[\alpha]_{436}^{18}$: +21.5; $[\alpha]_{365}^{18}$: +33.7 (c 4.00, CHCl$_3$).

HRMS (ESI, *m/z*) calculated for $C_{26}H_{51}O_3Si$ [M–C_2H_5]$^+$: 439.3607, found: 439.3607, calculated for $C_{25}H_{49}O_3Si$ [M–*i*Pr]$^+$: 425.3451, found: 425.3455, calculated for $C_{24}H_{45}O_2Si$ [M–*i*Pr-MeOH]$^+$: 393.3189, found: 393.3177.

3.2.9. (*R,Z*)-12-((Triisopropylsilyl)oxy)octadec-9-en-1-ol (**21**)

In a flame-dried two-necked flask, a solution of **20** (574.7 mg, 1.22 mmol) in anhydrous Et_2O (3.7 mL) was cooled at 0 °C under nitrogen. A solution of Red-Al (65% in toluene, ~3 M, 0.5 mL, 1.5 mmol, 1.23 equiv) was added dropwise under stirring. Stirring was continued for 5 h at 0 °C. TLC monitoring confirmed completion of the reaction. Citric acid (400 mg) and distilled water (5 mL) were added to the mixture and stirring was continued for 30 min. Extraction was then done with petroleum ether/EtOAc (80:20), and organic layers were dried over Na_2SO_4. Solvent was evaporated under reduced pressure, and the crude product was purified by column chromatography on silica gel (2.5 g, 0%–0.5% acetone in petroleum ether) to afford **21** as a colorless oil (509.6 mg, 94%). $R_f = 0.39$ (petroleum ether/acetone 85:15).

^{1}H NMR (400 MHz, CDCl$_3$): δ 5.47–5.36 (m, 2H), 3.83 (tt, 1H, *J* = 6.0, 5.6 Hz), 3.64 (t, 2H, *J* = 6.6 Hz, C$\underline{H}_2$OH), 2.27–2.22 (m, 2H), 2.02 (br q, 2H, *J* = 6.5 Hz), 1.61–1.52 (m, 3H, CH$_2$ and OH), 1.52–1.40 (m, 2H), 1.40–1.19 (m, 18H), 1.06 (s, 21H), 0.88 (t, 3H, *J* = 6.9 Hz, CH$_3$).

^{13}C NMR (100 MHz, CDCl$_3$): δ 131.38 ($\underline{C}$H=CH), 125.68 ($\underline{C}$H=CH), 72.27 ($\underline{C}$H), 63.10 ($\underline{C}$H$_2$), 36.55 ($\underline{C}$H$_2$), 34.68 ($\underline{C}$H$_2$), 32.81 ($\underline{C}$H$_2$), 31.92 ($\underline{C}$H$_2$), 29.69 ($\underline{C}$H$_2$), 29.64 ($\underline{C}$H$_2$), 29.54 ($\underline{C}$H$_2$), 29.42 ($\underline{C}$H$_2$), 29.32 ($\underline{C}$H$_2$), 27.54 ($\underline{C}$H$_2$), 25.74 ($\underline{C}$H$_2$), 24.81 ($\underline{C}$H$_2$), 22.64 ($\underline{C}$H$_2$), 18.21 (6 $\underline{C}$H$_3$), 14.11 ($\underline{C}$H$_3$), 12.63 (3 $\underline{C}$H).

3.2.10. (*R,Z*)-12-((Triisopropylsilyl)oxy)octadec-9-en-1-yl methanesulfonate (**22**)

To a solution of **21** (509.6 mg, 1.15 mmol) and Et$_3$N (0.245 mL, 1.75 mmol, 1.5 equiv) in DCM (4.6 mL) with stirring and under nitrogen cooled at −50 °C, was added dropwise mesyl chloride (0.112 mL, 1.45 mmol, 1.25 equiv) in DCM (0.6 mL). Transfer of mesyl chloride was completed by rinsing twice with a few drops of DCM. The reaction mixture was then allowed to warm up slowly in 2 h up to −5 °C. TLC monitoring (elution with DCM) showed completion of the reaction, and distilled water (5.8 mL) was added to quench the reaction. Extraction was done with DCM. Organic layers were washed with brine and dried over Na_2SO_4. DCM was removed under reduced pressure, and the crude product was purified by column chromatography on silica gel (4 g, 0%–1% acetone in petroleum ether) to afford **22** as a colorless oil (453 mg, 76%). $R_f = 0.41$ (petroleum ether/acetone 80:20).

^{1}H NMR (400 MHz, CDCl$_3$): δ 5.47–5.37 (m, 2H), 4.22 (t, 2H, *J* = 6.6 Hz, C$\underline{H}_2$OMs), 3.83 (tt, 1H, *J* = 6.0, 5.6 Hz), 3.00 (s, 3H), 2.29–2.19 (m, 2H), 2.08–1.96 (m, 2H), 1.75 (dq, 2H, *J* = 8.2, 6.6 Hz), 1.54–1.20 (m, 20H), 1.06 (s, 21H), 0.88 (t, 3H, *J* = 6.9 Hz, CH$_3$).

^{13}C NMR (100 MHz, CDCl$_3$): δ 131.28 ($\underline{C}$H=CH), 125.75 ($\underline{C}$H=CH), 72.25 ($\underline{C}$H), 70.16 ($\underline{C}$H$_2$), 37.37 ($\underline{C}$H$_3$), 36.56 ($\underline{C}$H$_2$), 34.69 (CH$_2$), 31.91 ($\underline{C}$H$_2$), 29.65 ($\underline{C}$H$_2$), 29.63 ($\underline{C}$H$_2$), 29.37 ($\underline{C}$H$_2$), 29.25 ($\underline{C}$H$_2$), 29.14 ($\underline{C}$H$_2$), 29.04 ($\underline{C}$H$_2$), 27.51 ($\underline{C}$H$_2$), 25.44 ($\underline{C}$H$_2$), 24.81 (CH$_2$), 22.64 ($\underline{C}$H$_2$), 18.22 (6 $\underline{C}$H$_3$), 14.11 ($\underline{C}$H$_3$), 12.64 (3 $\underline{C}$H).

HRMS (ESI, m/z) calculated for $C_{28}H_{58}O_4NaSiS$ [M + Na]$^+$: 541.3723, found: 541.3724.

3.2.11. ((((*R,Z*)-18-(((*R*)-2,2-Dimethyl-1,3-dioxolan-4-yl)methoxy)octadec-9-en-7-yl)oxy)triisopropylsilane (**23**)

A 60% dispersion of sodium hydride in mineral oil (29.5 mg, 0.7 mmol, 2.5 equiv) was washed three times with petroleum ether under nitrogen and with stirring. Anhydrous DMF (0.2 mL) was then added and the mixture was cooled at 0 °C. Following this, a solution of 2,3 isopropylidene-*sn*-glycerol **18** (50.2 mg, ≥95% pure, 0.36 mmol, 1.25 equiv) in DMF (0.2 mL) was added dropwise to the mixture, and the flask containing **18** was rinsed with DMF (2 × 0.1 mL). The corresponding mixture was allowed to stir for 10 min at 0 °C, and a solution of **22** (153 mg, 0.29 mmol) in DMF (0.2 mL) was added to the resulting white suspension. Transfer of **22** was completed by rinsing with DMF (2 × 0.1 mL). The mixture was then vigorously stirred overnight for 16 h at rt. TLC monitoring showed completion of the reaction, and 10% aqueous ammonium acetate was added as a buffer. Extraction was done with petroleum ether. Organic layers were dried over Na_2SO_4. Solvent was removed under reduced pressure, and the crude product was purified by column chromatography on silica gel (1.5 g, 0%–0.5% acetone in petroleum ether) to provide **23** as a colorless oil (110.8 mg, 68%). R_f = 0.71 (petroleum ether/acetone 95:5).

IR (KBr) ν 2930, 2865, 1464, 1379, 1369, 1255, 1097, 883, 849, 678 cm^{-1}.

^{1}H NMR (400 MHz, CDCl$_3$): δ 5.47–5.36 (symmetrical m, 2H), 4.27 (tt, 1H, *J* = 6.4, 5.6 Hz), 4.06 (dd, 1H, *J* = 8.2, 6.4 Hz), 3.83 (broad tt, 1H, *J* = 5.9, 5.5 Hz), 3.73 (dd, 1H, *J* = 8.2, 6.4 Hz), 3.54–3.39 (m, 4H with 1H dd at 3.52 ppm, *J* = 9.9, 5.6 Hz and 1H dd at 3.42 ppm, *J* = 9.9, 5.6 Hz), 2.29–2.20 (m, 2H), 2.07–1.96 (m, 2H), 1.62–1.53 (m, 2H plus signal of water as a singlet at 1.59 ppm), 1.53–1.39 (m, 5H including CH$_3$, q, 3H at 1.42 ppm, *J* = 0.6 Hz), 1.37 (q, 3H, *J* = 0.6 Hz, CH$_3$), 1.36–1.20 (m, 18H), 1.06 (s, 21H), 0.88 (t, 3H, *J* = 6.9 Hz, CH$_3$).

^{13}C NMR (100 MHz, CDCl$_3$): δ 131.40 (<u>C</u>H=CH), 125.66 (CH=<u>C</u>H), 109.37 (<u>C</u>Me$_2$), 74.76 (<u>C</u>H), 72.27 (<u>C</u>H), 71.89 (<u>C</u>H$_2$), 71.83 (<u>C</u>H$_2$), 66.94 (<u>C</u>H$_2$), 36.55 (<u>C</u>H$_2$), 34.68 (<u>C</u>H$_2$), 31.92 (<u>C</u>H$_2$), 29.71 (<u>C</u>H$_2$), 29.64 (<u>C</u>H$_2$), 29.57 (<u>C</u>H$_2$), 29.53 (<u>C</u>H$_2$), 29.47 (<u>C</u>H$_2$), 29.34 (<u>C</u>H$_2$), 27.55 (<u>C</u>H$_2$), 26.78 (<u>C</u>H$_3$), 26.07 (<u>C</u>H$_2$), 25.43 (<u>C</u>H$_3$), 24.81 (<u>C</u>H$_2$), 22.64 (<u>C</u>H$_2$), 18.22 (6 <u>C</u>H$_3$), 14.11 (<u>C</u>H$_3$), 12.64 (3 <u>C</u>H).

$[\alpha]_D^{19}$: +4.1; $[\alpha]_{578}^{19}$: +4.2; $[\alpha]_{546}^{19}$: +4.9; $[\alpha]_{436}^{19}$: +8.2; $[\alpha]_{365}^{19}$: +12.4 (c 3.62, CHCl$_3$).

$[\alpha]_D^{19}$: +4.0; $[\alpha]_{578}^{19}$: +3.9; $[\alpha]_{546}^{19}$: +4.5; $[\alpha]_{436}^{19}$: +7.9; $[\alpha]_{365}^{19}$: +12.2 (c 2.51, acetone).

HRMS (ESI, *m/z*) calculated for C$_{33}$H$_{66}$O$_4$NaSi [M + Na]$^+$: 577.4628, found: 577.4627.

Elementary analysis calculated for C$_{33}$H$_{66}$O$_4$Si: C, 71.42; H, 11.99, found: C, 71.88; H, 12.09.

3.2.12. (*R,Z*)-18-(((*R*)-2,2-Dimethyl-1,3-dioxolan-4-yl)methoxy)octadec-9-en-7-ol (**24**)

To a stirred solution at of **23** (278 mg, 0.5 mmol) in anhydrous THF (1.5 mL) under nitrogen, which was cooled at -20 °C, was added a 1 M solution of TBAF in THF (0.67 mL, 0.67 mmol, 1.3 equiv). The reaction mixture was then left under stirring for 20 h at rt. TLC monitoring showed completion of the reaction, and distilled water (1.5 mL) was added to the mixture. Extraction was done with EtOAc and organic layers were washed with brine and dried over Na_2SO_4. Solvent was removed under reduced pressure, and the crude product was purified by column chromatography on silica gel (2.5 g, 0%–4% acetone in petroleum ether) to provide **24** as a colorless oil (184.2 mg, 92%). R_f = 0.18 (petroleum ether/acetone 95:5).

IR (KBr) ν 3457, 2928, 2856, 1466, 1370, 1256, 1214, 1120, 846, 724 cm^{-1}.

^{1}H NMR (400 MHz, CDCl$_3$): δ 5.56 (dtt, 1H, *J* = 10.9, 7.3, 1.5 Hz), 5.40 (dtt, 1H, *J* = 10.9, 7.3, 1.5 Hz), 4.27 (dddd, 1H, *J* = 6.4, 6.4, 5.7, 5.7 Hz), 4.06 (dd, 1H, *J* = 8.2, 6.4 Hz, ddd with improving the resolution, *J* = 8.2, 6.4, 0.2 Hz), 3.73 (dd, 1H, *J* = 8.2, 6.4 Hz), 3.66–3.56 (m, 1H), 3.54–3.39 (m, 4H with 1H dd at 3.52 ppm, *J* = 9.9, 5.7 Hz (ddd with improving the resolution, *J* = 9.9, 5.7, 0.3 Hz) and 1H dd at 3.42 ppm, *J* = 9.9, 5.6 Hz), 2.24–2.18 (m, 2H), 2.05 (br dtd (apparent qd), 2H, *J* = 7.2, 7.2, 1 Hz), 1.61–1.52 (m, 3H,

CH$_2$ and OH), 1.51–1.43 (m, 2H), 1.43 (q, 3H, J = 0.6 Hz, CH$_3$), 1.37 (q, 3H, J = 0.6 Hz, CH$_3$), 1.37–1.23 (m, 18H), 0.88 (t, 3H, J = 6.9 Hz, CH$_3$).

^{13}C NMR (100 MHz, CDCl$_3$): δ 133.49 ($\underline{C}$H=CH), 125.16 (CH=$\underline{C}$H), 109.37 ($\underline{C}$Me$_2$), 74.76 ($\underline{C}$H), 71.87 ($\underline{C}$H$_2$), 71.83 ($\underline{C}$H$_2$), 71.51 ($\underline{C}$H), 66.94 ($\underline{C}$H$_2$), 36.86 ($\underline{C}$H$_2$), 35.37 ($\underline{C}$H$_2$), 31.85 ($\underline{C}$H$_2$), 29.66 ($\underline{C}$H$_2$), 29.55 ($\underline{C}$H$_2$), 29.46 ($\underline{C}$H$_2$), 29.41 ($\underline{C}$H$_2$), 29.36 ($\underline{C}$H$_2$), 29.25 ($\underline{C}$H$_2$), 27.42 ($\underline{C}$H$_2$), 26.78 ($\underline{C}$H$_3$), 26.04 ($\underline{C}$H$_2$), 25.73 ($\underline{C}$H$_2$), 25.43 ($\underline{C}$H$_3$), 22.63 ($\underline{C}$H$_2$), 14.10 ($\underline{C}$H$_3$).

$[\alpha]_D^{20}$: -5.0; $[\alpha]_{578}^{20}$: −5.4; $[\alpha]_{546}^{20}$: −12.5; $[\alpha]_{436}^{20}$: −11.9; $[\alpha]_{365}^{20}$: −21.2 (c 6.00, CHCl$_3$).

Elementary analysis calculated for C$_{24}$H$_{46}$O$_4$: C, 72.31; H, 11.63, found: C, 71.79; H, 11.66.

3.2.13. (*S*)-3-(((*R,Z*)-12-Hydroxyoctadec-9-en-1-yl)oxy)propane-1,2-diol (**11**)

To a solution of **24** (1.26 g, 3.15 mmol) in methanol (12.6 mL), was added *p*-toluenesulfonic acid monohydrate (30 mg, 0.16 mmol, 0.05 eq) and distilled water (1.26 mL). The flask was then purged with nitrogen, stoppered and dipped in a preheated bath at 60 °C. Stirring was maintained for 4 h at 60 °C. TLC monitoring showed completion of the reaction, and sodium bicarbonate (14.5 mg, 0.173 mmol, 0.055 equiv) was added and stirring was continued for 1 h at 60 °C. Methanol was removed under reduced pressure, and the crude product was purified by column chromatography on silica gel eluting with 0%–5% acetone in petroleum ether to remove non polar impurities and then with petroleum ether + 5% acetone + 12% methanol to give **11** as a colorless oil (945.6 mg, 84%). R$_f$ = 0.02 (petroleum ether/acetone 80:20).

IR (KBr) ν 3383, 2926, 2855, 1654, 1466, 1124, 1045, 858, 724 cm^{-1}.

^{1}H NMR (400 MHz, CDCl$_3$): δ 5.56 (dtt, 1H, J = 10.8, 7.3, 1.4 Hz), 5.40 (dtt, 1H, J = 10.7, 7.4, 1.5 Hz), 3.91–3.81 (m, 1H), 3.72 (very broad dd, 1H, J = 10, 4 Hz (br dd after exchange with D$_2$O, J = 11.3, 3.5 Hz), 1H of C$\underline{H_2}$OH), 3.69–3.58 (m, 2H, 1H of C$\underline{H_2}$OH and H$_{12}$), 3.57–3.41 (m, 4H, C$\underline{H_2}$OC$\underline{H_2}$), 2.80–2.50 (envelope, 1H, CHO$\underline{H}$), 2.42–2.12 (envelope, 1H, CH$_2$O$\underline{H}$), 2.21 (broad ddd, 2H, J = 7.4, 6.2, 1.1 Hz), 2.05 (br dt (apparent q), 2H J = 7.0, 6.9 Hz), 1.90–1.51 (m, 6H with an envelope centered at 1.63 ppm), 1.51–1.41 (m, 3H), 1.40–1.19 (m, 18H), 0.88 (pseudo t, 3H, J = 6.8 Hz, CH$_3$).

^{13}C NMR (100 MHz, CDCl$_3$): δ 133.50 ($\underline{C}$H=CH), 125.18 (CH=$\underline{C}$H), 72.53 ($\underline{C}$H$_2$), 71.79 ($\underline{C}$H$_2$), 71.52 ($\underline{C}$H), 70.40 ($\underline{C}$H), 64.31 ($\underline{C}$H$_2$), 36.83 ($\underline{C}$H$_2$), 35.36 ($\underline{C}$H$_2$), 31.85 ($\underline{C}$H$_2$), 29.61 ($\underline{C}$H$_2$), 29.53 ($\underline{C}$H$_2$), 29.38 ($\underline{C}$H$_2$), 29.36 ($\underline{C}$H$_2$), 29.32 ($\underline{C}$H$_2$), 29.17 ($\underline{C}$H$_2$), 27.39 ($\underline{C}$H$_2$), 26.02 ($\underline{C}$H$_2$), 25.73 ($\underline{C}$H$_2$), 22.63 ($\underline{C}$H$_2$), 14.10 ($\underline{C}$H$_3$).

$[\alpha]_D^{22}$: +1.2; $[\alpha]_{578}^{22}$: +0.5; $[\alpha]_{546}^{22}$: +0.6; $[\alpha]_{436}^{22}$: +1.3; $[\alpha]_{365}^{22}$: +1.6 (c 1.115, acetone).

HRMS (ESI, m/z) calculated for C$_{21}$H$_{42}$O$_4$Na [M + Na]$^+$: 381.2981, found: 381.2982.

3.2.14. Methyl (*Z*)-12-oxooctadec-9-enoate (**25**)

Pyridium chlorochromate (18.06 g, 83.8 mmol, 2.6 equiv) was suspended in DCM (111.7 mL) with stirring for 5 min. Following this, a solution of methyl ricinoleate **13** (10 g, 32 mmol) in DCM (15 mL) was added rapidly to the mixture and the flask containing **13** was rinsed with DCM (3 x 2 mL). Stirring was pursued for 1 h at rt under nitrogen. TLC monitoring showed completion of the reaction, and petroleum ether/EtOAc (90:10) (111.7 mL) was added. The resulting mixture was filtered over a short plug of silica gel with rinsing of silica gel by petroleum ether/EtOAc (90:10). Evaporation of the filtrate under reduced pressure followed by the purification of the crude by column chromatography on silica gel (55 g, 0%–1% acetone in petroleum ether) afforded **25** as a colorless oil (6.73 g, 68%). R$_f$ = 0.36 (petroleum ether/acetone 90:10).

^{1}H NMR (400 MHz, CDCl$_3$): δ 5.62–5.50 (m, 2H), 3.67 (s, 3H), 3.15 (br d, 2H, *J* = 6.0 Hz), 2.43 (dd, 2H, *J* = 7.5, 7.4, Hz), 2.30 (dd, 2H, *J* = 7.7, 7.4 Hz), 2.02 (br dt (apparent q), 2H, *J* = 7.5, 7.0 Hz), 1.67–1.51 (m, 4H), 1.40–1.20 (m, 14H), 0.88 (t, 3H, *J* = 6.8 Hz, CH$_3$).

^{13}C NMR (100 MHz, CDCl$_3$): δ 209.36 (<u>C</u>O), 174.31(<u>C</u>O), 133.53 (<u>C</u>H=CH), 120.99 (<u>C</u>H=CH), 51.45 (<u>C</u>H$_3$), 42.37 (<u>C</u>H$_2$), 41.64 (<u>C</u>H$_2$), 34.04 (<u>C</u>H$_2$), 31.59 (<u>C</u>H$_2$), 29.25 (<u>C</u>H$_2$), 29.11 (<u>C</u>H$_2$), 29.06 (<u>C</u>H$_2$), 29.05 (<u>C</u>H$_2$), 28.88 (<u>C</u>H$_2$), 27.46 (<u>C</u>H$_2$), 24.90 (<u>C</u>H$_2$), 23.77 (<u>C</u>H$_2$), 22.49 (<u>C</u>H$_2$), 14.04 (<u>C</u>H$_3$).

3.2.15. Methyl (Z)-12,12-difluorooctadec-9-enoate (26)

To a solution of **25** (6 g 19.3 mmol) in DCM (22.5 mL) with stirring and under nitrogen at rt, was added dropwise DAST (6.22 mL, 47 mmol, 2.4 equiv). The corresponding mixture was stirred for 21 days at rt. TLC monitoring showed that the reaction was mostly done and saturated aqueous sodium bicarbonate (32 mL) plus water (20 mL) were added to quench unreacted DAST. After partitioning and extraction of the aqueous layer with DCM, combined organic layers were washed with brine and dried over Na$_2$SO$_4$. DCM was removed under reduced pressure, and the crude product was purified by column chromatography on silica gel (20 g). First, elution with 0%–0.5% acetone in petroleum ether) afforded **26** as a colorless oil (3.48 g, 54%). R$_f$ = 0.39 (petroleum ether/acetone 95:5). Then unreacted **25** (2.00 g, 33%) was eluted with 1% acetone in petroleum ether.

IR (KBr) ν 3465, 3021, 2953, 2930, 2856, 1742, 1467, 1436, 1198, 1170, 876, 726 cm^{-1}.

^{1}H NMR (400 MHz, CDCl$_3$): δ 5.64–5.55 (m, 1H), 5.39 (dtt, 1H, *J* = 10.9, 7.3, 1.5 Hz), 3.67 (s, 3H), 2.65–2.52 (m, which could be analyzed as a td (*J$_{HF}$* = 15.9 Hz, *J$_{HH}$* = 7.3 Hz) with further very small couplings, 2H), 2.30 (dd, 2H, *J* = 7.7, 7.4 Hz), 2.03 (br dt (apparent q), 2H, *J* = 7, 7 Hz), 1.87–1.72 (m, 2H), 1.67–1.57 (m, 2H), 1.51–1.41 (m, 2H), 1.40–1.22 (m, 14H), 0.89 (t, 3H, *J* = 6.8 Hz, CH$_3$).

^{13}C NMR (100 MHz, CDCl$_3$): δ 174.32 (<u>C</u>O), 134.52 (<u>C</u>H=CH), 124.87 (<u>C</u>$_{12}$, t, *J* = 241.2 Hz), 120.30 (<u>C</u>H=CH, t, *J* = 5.8 Hz), 51.47 (<u>C</u>H$_3$), 35.98 (<u>C</u>H$_2$, t, *J* = 25.0 Hz), 34.62 (<u>C</u>H$_2$, t, *J* = 26.4 Hz), 34.09 (<u>C</u>H$_2$), 31.60 (<u>C</u>H$_2$), 29.30 (<u>C</u>H$_2$), 29.14 (<u>C</u>H$_2$), 29.09 (<u>C</u>H$_2$), 29.08 (<u>C</u>H$_2$), 29.06 (<u>C</u>H$_2$), 27.40 (<u>C</u>H$_2$), 24.93 (<u>C</u>H$_2$), 22.51 (<u>C</u>H$_2$), 22.16 (<u>C</u>H$_2$, t, *J* =4.6 Hz), 14.04 (<u>C</u>H$_3$).

^{19}F NMR (376 MHz, CDCl$_3$): δ −96.88 (pentuplet, *J* = 16.3 Hz on ^{19}F-undecoupled spectrum).

3.2.16. (Z)-12,12-Difluorooctadec-9-en-1-ol (27)

To a stirred solution of **26** (199.5 mg, 0.6 mmol) in anhydrous Et$_2$O (2 mL) under nitrogen, which was cooled at 0 °C, was added dropwise a solution of Red-Al (65% in toluene, ~3 M, 0.3 mL, 0.9 mmol, 1.5 equiv); then stirring was continued for 5 h at 0 °C. TLC monitoring confirmed completion of the starting material. A solution of citric acid (400 mg) in water (5 mL) was added, and stirring was still continued for 30 min. Extraction was done with petroleum ether/EtOAc (80:20), and organic layers were dried over Na$_2$SO$_4$. Solvent was evaporated under reduced pressure, and the crude product was purified by column chromatography on silica gel (4 g, 0%–2% acetone in petroleum ether) to afford **27** as a colorless oil (141.3 mg, 77%). R$_f$ = 0.21 (petroleum ether/acetone 85:15). This reduction was subsequently performed on a larger scale (15 ×), and the crude alcohol, which was thus obtained, was used as such for the next step.

^{1}H NMR (400 MHz, CDCl$_3$): δ 5.65–5.56 (m, 1H), 5.39 (dtt, 1H, *J* = 10.9, 7.4, 1.6 Hz), 3.64 (t, 2H, *J* = 6.6 Hz), 2.65–2.53 (m, which could be analyzed as a tdd with further very small couplings at 2.59 ppm, 2H, *J$_{HF}$* = 15.9 Hz, *J$_{HH}$* = 7.3, 1.4 Hz), 2.04 (br dt (apparent q), 2H, *J* = 7, 7 Hz), 1.87–1.72 (m, 2H), 1.61–1.52 (m, 2H), 1.50–1.41 (m, 2H), 1.40–1.23 (m, 16H), 0.89 t, 3H, *J* = 6.9 Hz, CH$_3$).

^{13}C NMR (100 MHz, CDCl$_3$): δ 134.59 (<u>C</u>H=CH), 124.89 (<u>C</u>F$_2$, t, J = 241.2 Hz), 120.25 (t, <u>C</u>H=CH, J = 5.8 Hz), 63.07 (<u>C</u>H$_2$), 35.97 (t, <u>C</u>H$_2$, J = 25.0 Hz), 34.62 (t, <u>C</u>H$_2$, J = 26.4 Hz), 32.78 (<u>C</u>H$_2$), 31.60 (<u>C</u>H$_2$), 29.47 (<u>C</u>H$_2$), 29.38 (<u>C</u>H$_2$), 29.36 (<u>C</u>H$_2$), 29.22 (<u>C</u>H$_2$), 29.06 (<u>C</u>H$_2$), 27.42 (<u>C</u>H$_2$), 25.73 (<u>C</u>H$_2$), 22.51 (<u>C</u>H$_2$), 22.16 (t, <u>C</u>H$_2$, J = 4.6 Hz), 14.04 (<u>C</u>H$_3$).

^{19}F NMR (376 MHz, CDCl$_3$): δ −96.85 (pentuplet, J = 16.3 Hz on ^{19}F-undecoupled spectrum).

HRMS (ESI, *m/z*) calculated for C$_{18}$H$_{34}$OF$_2$Na [M + Na]$^+$: 327.2475 found: 327.2478, calculated for C$_{18}$H$_{33}$OFNa [M − HF + Na]$^+$: 307.2413, found: 307.2415.

3.2.17. (Z)-12,12-Difluorooctadec-9-en-1-yl methanesulfonate (28)

To a stirred solution of crude **27** (made from 3.09 g of **26**, 9.3 mmol) and Et$_3$N (1.95 mL, 14.0 mmol, 1.5 equiv) in DCM (28 mL) under nitrogen, which was cooled at −35 °C, was added dropwise mesyl chloride (0.9 mL, 11.6 mmol, 1.25 equiv) in DCM (3.7 mL). Transfer of mesyl chloride was completed by rinsing with DCM (3 × 0.2 mL). The reaction mixture was then allowed to warm up slowly to −5 °C (in 4–5 h) and TLC monitoring (elution with DCM) showed completion of the reaction. Distilled water (96 mL) was added to quench the reaction, and extraction of the aqueous layer was done with DCM. Combined organic layers were washed with brine and dried over Na$_2$SO$_4$. DCM was removed under reduced pressure, and the crude product was purified by column chromatography on silica gel (10 g, 0%–1% acetone in petroleum ether) to afford **28** as a colorless oil (2.44 g, 69% from **26**). R$_f$ = 0.61 (petroleum ether/acetone 80:20).

^{1}H NMR (400 MHz, CDCl$_3$): δ 5.65–5.55 (m, 1H), 5.39 (dtt, 1H, J = 10.9, 7.3, 1.6 Hz), 4.22 (t, 2H, J = 6.6 Hz), 3.00 (s, 3H), 2.65–2.52 (m, which could be analyzed as a tdd with further very small couplings at 2.59 ppm, 2H, J_{HF} = 16.0 Hz, J_{HH} = 7.3, 1.3 Hz), 2.04 (br dt (apparent q), 2H, J = 7, 7 Hz), 1.87–1.70 (m, 4H), 1.58–1.21 (m, 18H), 0.89 (t, 3H, J = 6.9 Hz, CH$_3$).

^{13}C NMR (100 MHz, CDCl$_3$): δ 134.51 (<u>C</u>H=CH), 124.87 (<u>C</u>$_{12}$, t, J = 241.2 Hz), 120.30 (t, <u>C</u>H=CH, J = 5.8 Hz), 70.16 (<u>C</u>H$_2$), 37.36 (<u>C</u>H$_3$), 35.97 (t, <u>C</u>H$_2$, J = 25.0 Hz), 34.61 (t, <u>C</u>H$_2$, J = 26.4 Hz), 31.60 (<u>C</u>H$_2$), 29.31 (<u>C</u>H$_2$), 29.29 (<u>C</u>H$_2$), 29.13 (<u>C</u>H$_2$), 29.12 (<u>C</u>H$_2$), 29.05 (<u>C</u>H$_2$), 28.98, (<u>C</u>H$_2$), 27.39 (<u>C</u>H$_2$), 25.40 (<u>C</u>H$_2$), 22.50 (<u>C</u>H$_2$), 22.16 (t, <u>C</u>H$_2$, J = 4.6 Hz), 14.05 (<u>C</u>H$_3$).

^{19}F NMR (376 MHz, CDCl$_3$): δ −96.86 (pentuplet, J = 16.3 Hz on ^{19}F-undecoupled spectrum).

3.2.18. (R,Z)-4-(((12, 12-Difluorooctadec-9-en-1-yl)oxy)methyl)-2,2-dimethyl-1,3-dioxolane (29)

To a stirred mixture of **28** (115 mg, 0.3 mmol), *n*-Bu$_4$NBr (24.2 mg, 0.075 mmol, 0.25 equiv), DMSO (0.6 mL) and 50% aqueous NaOH (63 µL, 1.2 mmol of NaOH, 4 equiv), was added (*R*)-solketal **18** (49 mg, ≥95% pure, 0.35 mmol, 1.17 equiv). The corresponding mixture was stirred for 5 h at 60 °C. TLC monitoring showed completion of the reaction, and distilled water was added. Extraction was done with petroleum ether/EtOAc (80:20). Organic layers were washed again with brine and dried over Na$_2$SO$_4$. Solvent was removed under reduced pressure, and the crude product was purified by column chromatography on silica gel (5 g, 0%–0.5% acetone in petroleum ether) to provide **29** as a colorless oil (79.3 mg, 63%). R$_f$ = 0.44 (petroleum ether/acetone 95:5).

^{1}H NMR (400 MHz, CDCl$_3$): δ 5.64–5.56 (m, 1H), 5.38 (dtt, 1H, J = 10.9, 7.3, 1.6 Hz), 4.27 (dddd (apparent tt), 1H, J = 6.4, 6.4, 5.7, 5.7 Hz), 4.06 (dd, 1H, J = 8.2, 6.4 Hz), 3.73 (dd, 1H, J = 8.2, 6.4 Hz), 3.54–3.39 (m, 4H), 2.65–2.53 (m, which could be analyzed as a tdd with further very small couplings at 2.59 ppm, 2H, J_{HF} = 15.9 Hz, J_{HH} = 7.4, 1.3 Hz), 2.03 (br dt (apparent q), 2H, J = 7, 7 Hz), 1.87–1.72 (m, 2H), 1.62–1.51 (m, 2H), 1.43 (q, 3H, J = 0.6 Hz), 1.37 (q, 3H, J = 0.6 Hz), 1.40–1.21 (m, 18H), 0.89 (t, 3H, J = 6.8 Hz, CH$_3$).

^{13}C NMR (100 MHz, CDCl$_3$): δ 134.59 (CH=CH), 124.88 (CF$_2$, t, *J* = 241.2 Hz), 120.23 (t, CH=CH, *J* = 5.8 Hz), 109.37 (CMe$_2$), 74.76 (CH), 71.87 (CH$_2$), 71.84 (CH$_2$), 66.93 (CH$_2$), 35.96 (t, CH$_2$, *J* = 25.0 Hz), 34.62 (t, CH$_2$, *J* = 26.4 Hz), 31.60 (CH$_2$), 29.55 (CH$_2$), 29.45 (CH$_2$), 29.41 (CH$_2$), 29.37 (CH$_2$), 29.23 (CH$_2$), 29.06 (CH$_2$), 27.43 (CH$_2$), 26.78 (CH$_3$), 26.05 (CH$_2$), 25.43 (CH$_3$), 22.50 (CH$_2$), 22.15 (t, CH$_2$, *J* = 4.6 Hz), 14.04 (CH$_3$).

^{19}F NMR (376 MHz, CDCl$_3$): δ −96.85 (pentuplet, *J* = 16.3 Hz on ^{19}F-undecoupled spectrum).

HRMS (ESI, *m/z*) calculated for C$_{24}$H$_{44}$O$_3$F$_2$Na [M + Na]$^+$: 441.3156, found: 441.3149, calculated for C$_{24}$H$_{43}$O$_3$FNa [M − HF + Na]$^+$: 421.3094, found: 421.3100, calculated for C$_{24}$H$_{42}$O$_3$Na [M − 2HF + Na]$^+$: 401.3032, found: 401.3044.

3.2.19. (*S*,*Z*)-3-((12,12-Difluorooctadec-9-en-1-yl)oxy)propane-1,2-diol (**9**)

To a solution of acetonide **29** (971.9 mg, 2.32 mmol) in methanol (9.3 mL) and distilled water (0.93 mL) was added *p*-toluenesulfonic acid monohydrate (22.1 mg, 0.116 mmol, 0.05 equiv). The flask was then purged with nitrogen, stoppered and dipped in a preheated bath at 60 °C. Stirring was maintained for 5 h at 60 °C. TLC monitoring showed completion of the reaction, and sodium bicarbonate (10.9 mg, 0.13 mmol, 0.056 equiv) was added to the mixture and the stirring was continued for 1 h at 60 °C. Methanol was removed under reduced pressure, and then the crude product was purified by column chromatography on silica gel (6 g, 0%–5% acetone in petroleum ether and then petroleum ether + 5% acetone + 12% methanol) to provide **9** as a green oil (811.2 mg, 92%). R$_f$ = 0.04 (petroleum ether/acetone 90:10).

^{1}H NMR (400 MHz, CDCl$_3$): δ 5.65–5.55 (m, 1H), 5.39 (dtt, 1H, *J* = 10.9, 7.3, 1.6 Hz), 3.91–3.82 (m (ddt after exchange with D$_2$O, *J* = 6.0, 5.2, 3.9 Hz), 1H), 3.72 (ddd, 1H, *J* = 11.4, 6.9, 3.8 Hz (dd after exchange with D$_2$O, *J* = 11.4, 3.9 Hz)), 3.65 (ddd, 1H, *J* = 11.4, 5.1, 5.0 Hz (dd after exchange with D$_2$O, *J* = 11.4, 5.2 Hz)), 3.56–3.42 (m with 1H dd at 3.54 ppm, *J* = 9.7, 3.9 Hz, 4H), 2.67 (d, which was suppressed after exchange with D$_2$O, 1H, *J* = 5.1, Hz, OH), 2.65–2.53 (m, which could be analyzed as a tdd with further very small couplings at 2.59 ppm, 2H, *J*$_{HF}$ = 15.9 Hz, *J*$_{HH}$ = 7.3, 1.2 Hz, 2H), 2.26 (br dd, 1H, *J* = 6.6, 5.6 Hz), 2.04 (br dt (apparent q), 2H, *J* = 7, 7 Hz), 1.87–1.73 (m, 2H), 1.58 (br tt, 2H, *J* = 7.2, 6.7 Hz), 1.51–1.41 (m, 2H), 1.40–1.22 (m, 16H), 0.89 (t, 3H, *J* = 6.9 Hz, CH$_3$).

^{13}C NMR (100 MHz, CDCl$_3$): δ 134.58 (CH=CH), 124.89 (CF$_2$, t, *J* = 241.2 Hz), 120.25 (t, CH=CH, *J* = 5.8 Hz), 72.53 (CH$_2$), 71.82 (CH$_2$), 70.41 (CH), 64.30 (CH$_2$), 35.96 (t, CH$_2$, *J* = 25.0 Hz), 34.62 (t, CH$_2$, *J* = 26.4 Hz), 31.60 (CH$_2$), 29.57 (CH$_2$), 29.44 (CH$_2$), 29.40 (CH$_2$), 29.36 (CH$_2$), 29.22 (CH$_2$), 29.05 (CH$_2$), 27.42 (CH$_2$), 26.07 (CH$_2$), 22.50 (CH$_2$), 22.15 (t, CH$_2$, *J* = 4.6 Hz), 14.05 (CH$_3$).

^{19}F NMR (376 MHz, CDCl$_3$): δ −96.84 (pentuplet, *J* = 16.3 Hz on ^{19}F-undecoupled spectrum).

[α]$_D$^{24}: −4.6; [α]$_{578}$^{24}: −5.8; [α]$_{546}$^{24}: −6.4; [α]$_{436}$^{24}: −9.9; [α]$_{365}$^{24}: −14.3 (c 0.94, acetone).

HRMS (ESI, *m/z*) calculated for C$_{21}$H$_{40}$O$_3$F$_2$Na [M + Na]$^+$: 401.2843, found: 401.2840, calculated for C$_{21}$H$_{39}$O$_3$FNa [M − HF + Na]$^+$: 381.2781, found: 381.2791, calculated for C$_{21}$H$_{38}$O$_3$Na [M − 2HF + Na]$^+$: 361.2719, found: 361.2726.

3.2.20. Methyl (*R*,*Z*)-12-((methylsulfonyl)oxy)octadec-9-enoate (**31**)

To a stirred solution of **13** (10.0 g, 32.0 mmol) and Et$_3$N (9.15 mL, 65.5 mmol, 2.05 equiv) in DCM (80 mL), which was cooled under nitrogen cooled at −40 °C, mesyl chloride (5.0 mL, 64.0 mmol, 2.0 equiv) in DCM (10 mL) was added dropwise. Transfer of mesyl chloride was completed by rinsing with DCM (2 × 1 mL). The reaction mixture was allowed to warm up slowly to −10 °C and then further stirred for 4 h at −10 °C. TLC monitoring showed completion of the reaction, and distilled water (120 mL) was added. Extraction of the aqueous layer was done with DCM. Combined organic layers were washed with brine and dried over Na_2SO_4. DCM was removed under reduced pressure, and the crude product was purified by column chromatography on silica gel (36 g, 0%–1% acetone in petroleum ether) to afford **31** as a colorless oil (8.06 g, 65%). R_f = 0.26 (petroleum ether/acetone 80:20).

^{1}H NMR (400 MHz, $CDCl_3$): δ 5.55 (dtt, 1H, J = 10.9, 7.3, 1.5 Hz), 5.37 (dtt, 1H, J = 10.9, 7.2, 1.5 Hz), 4.69 (tt, 1H, J = 6.2, 6.1 Hz), 3.67 (s, 3H), 2.99 (s, 3H), 2.55–2.37 (m, 2H), 2.30 (t, 2H, J = 7.5 Hz), 2.03 (br dt (apparent q), 2H, J = 7, 6.5 Hz), 1.72–1.57 (m, 4H), 1.49–1.18 (m, 16H), 0.88 (t, 3H, J = 6.9 Hz, CH_3).

^{13}C NMR (100 MHz, $CDCl_3$): δ 174.32 (C_{quat}, CO), 133.81 ($\underline{C}H=CH$), 123.03 ($\underline{C}H=CH$), 83.66 ($\underline{C}H$), 51.47 ($\underline{C}H_3$), 38.68 ($\underline{C}H_3$), 34.25 ($\underline{C}H_2$), 34.08 ($\underline{C}H_2$), 32.54 ($\underline{C}H_2$), 31.64 ($\underline{C}H_2$), 29.39 ($\underline{C}H_2$), 29.14 ($\underline{C}H_2$), 29.10 ($\underline{C}H_2$), 29.08 ($\underline{C}H_2$), 29.00 ($\underline{C}H_2$), 27.42 ($\underline{C}H_2$), 25.05 ($\underline{C}H_2$), 24.92 ($\underline{C}H_2$), 22.56 ($\underline{C}H_2$), 14.05 ($\underline{C}H_3$).

$[\alpha]_D^{18}$: +12.3; $[\alpha]_{578}^{18}$: +12.5; $[\alpha]_{546}^{18}$: +14.2; $[\alpha]_{436}^{18}$: +24.3; $[\alpha]_{365}^{18}$: +38.6 (c 4.00, acetone).

$[\alpha]_D^{18}$: +10.6; $[\alpha]_{578}^{18}$: +11.0; $[\alpha]_{546}^{18}$: +12.6; $[\alpha]_{436}^{18}$: +21.5; $[\alpha]_{365}^{18}$: +33.7 (c 4.00, $CHCl_3$).

3.2.21. Methyl (S,Z)-12-azidooctadec-9-enoate (32)

To a solution of **31** (9.1 g, 23.3 mmol) in DMSO (34.9 mL), was added NaN_3 (2.6 g, 40.0 mmol, 1.7 equiv). The flask was then purged with nitrogen, stoppered and dipped in a preheated bath at 80 °C. Stirring was maintained for 16 h at the same temperature. TLC monitoring showed completion of the reaction, and it was brought back to rt, then quenched with an aqueous solution of NH_4Cl. Extraction was done with petroleum ether/EtOAc (80:20, 185 mL), and organic layers were washed with brine and dried over Na_2SO_4. Solvent was removed under reduced pressure to obtain the crude product as a light-yellow oil. The crude product was then purified by column chromatography on silica gel (30 g, 0%–0.5% acetone in petroleum ether) to provide **32** as an oil (6.26 g, 80%). R_f = 0.58 (petroleum ether/acetone 80:20).

IR (KBr) ν 3011, 2929, 2855, 2100, 1742, 1456, 1250, 1197, 726 cm^{-1}.

^{1}H NMR (400 MHz, $CDCl_3$): δ 5.52 (dtt, 1H, J = 10.9, 7.3, 1.5 Hz), 5.38 (dtt, 1H, J = 10.9, 7.2, 1.5 Hz), 3.67 (s, 3H), 3.33–3.25 (m, which could be analyzed as a dddd at 3.29 ppm with J= 7.6, 6.5, 6.5, 5.3 Hz, 1H, $C\underline{H}N_3$), 2.33–2.25 (m, 4H with a dd at 2.34 ppm integrating for 2 protons, J = 7.7, 7.4 Hz), 2.04 (br td (apparent q), 2H, J = 7.0, 6.8 Hz), 1.68–1.57 (m, 2H), 1.56–1.40 (m, 3H), 1.40–1.22 (m, 15H), 0.89 (t, 3H, J = 6.9 Hz, CH_3).

^{13}C NMR (100 MHz, $CDCl_3$): δ 174.33 (C_{quat}, $\underline{C}O$), 133.10 ($\underline{C}H=CH$), 124.58 ($\underline{C}H=CH$), 62.95 ($\underline{C}H$), 51.46 ($\underline{C}H_3$), 34.10 ($\underline{C}H_2$), 33.99 ($\underline{C}H_2$), 32.28 ($\underline{C}H_2$), 31.72 ($\underline{C}H_2$), 29.46 ($\underline{C}H_2$), 29.15 ($\underline{C}H_2$), 29.10 (2 $\underline{C}H_2$), 29.09 ($\underline{C}H_2$), 27.40 ($\underline{C}H_2$), 26.13 ($\underline{C}H_2$), 24.93 ($\underline{C}H_2$), 22.59 ($\underline{C}H_2$), 14.07 ($\underline{C}H_3$).

3.2.22. (S,Z)-12-Azidooctadec-9-enal (33)

To a vigorously stirred solution of **32** (5.063 g, 15 mmol) in Et_2O (60 mL), which was cooled at −80 °C under nitrogen, was added dropwise a solution of Dibal (25% in hexane, 28 mL, 34.5 mmol, 2.3 equiv). The resulting mixture was allowed to stir for 2 h at −80 °C. TLC monitoring showed completion of the reaction. A saturated aqueous solution of potassium sodium tartrate (25 mL) was added, and stirring was continued for 10 min. Extraction was then done with petroleum ether/EtOAc (80:20). Organic layers were washed with brine and dried over Na_2SO_4. Solvent was evaporated under reduced pressure, and the crude product was used as such for the next step. For analytical purposes, the crude reaction product, which was initially obtained from 10 times less product, was purified by column chromatography on silica gel (5.2 g, 0%–1% acetone in petroleum ether) to afford **33** as an ochre oil (313.1 mg from 506.3 mg of **32**, 68%). R_f = 0.4 (petroleum ether/acetone 85:15).

IR (KBr) ν 34,340, 2929, 2856, 2716, 2100, 1727, 1466, 1273, 725 cm^{-1}.

^{1}H NMR (400 MHz, CDCl$_3$): δ 9.77 (t, 1H, *J* = 1.9 Hz, C**H**O), 5.52 (dtt, 1H, *J* = 10.9, 7.2, 1.5 Hz, C**H**=CH–CH$_2$–CHN$_3$), 5.39 (dtt, 1H, *J* = 10.9, 7.2, 1.5 Hz, CH=C**H**–CH$_2$–CHN$_3$), 3.33–3.25 (m, which could be analyzed as a dddd at 3.29 ppm with *J* = 7.6, 6.6, 6.5, 5.2 Hz, 1H, C**H**N$_3$), 2.42 (td, 2H, *J* = 7.4, 1.9 Hz, CH$_2$CHO), 2.35–2.20 (m, 2H, CHC**H$_2$**CH=CH), 2.05 (br td (apparent q), 2H, *J* = 7.0, 6.8 Hz, CH=CHC**H$_2$**), 1.68–1.58 (m, 2H, C**H$_2$**CH$_2$CHO), 1.57–1.47 (m, 2H, C**H$_2$**CHN$_3$), 1.47–1.41 (m, 1H, of C**H$_2$**CH$_2$CHN$_3$), 1.41–1.24 (m, 15H, 1H of C**H$_2$**CH$_2$CHN$_3$, CH$_3$(C**H$_2$**)$_3$ and (C**H$_2$**)$_4$CH$_2$CH$_2$CHO), 0.89 (t, 3H, *J* = 6.9 Hz, CH$_3$).

^{13}C NMR (100 MHz, CDCl$_3$): δ 202.92 (**C**HO), 133.05 (**C**H=CH), 124.62 (**C**H=CH), 62.94 (**C**H–N$_3$), 43.91 (**C**H$_2$), 33.99 (**C**H$_2$), 32.29 (**C**H$_2$), 31.72 (**C**H$_2$CH$_2$CH$_3$), 29.44 (**C**H$_2$), 29.25 (**C**H$_2$), 29.11 (**C**H$_2$), 29.08 (**C**H$_2$), 29.07 (**C**H$_2$), 27.39 (**C**H$_2$), 26.13 (**C**H$_2$), 22.58 (**C**H$_2$CH$_3$), 22.05 (**C**H$_2$), 14.07 (**C**H$_3$).

$[\alpha]_D^{18.5}$: −31.5; $[\alpha]_{578}^{18.5}$: −33.4; $[\alpha]_{546}^{18.5}$: −38.2; $[\alpha]_{436}^{18.5}$: −67.5; $[\alpha]_{365}^{18.5}$: −111.0 (c 2.00, acetone).

$[\alpha]_D^{18.5}$: −24.8; $[\alpha]_{578}^{18.5}$: −25.9; $[\alpha]_{546}^{18.5}$: −29.7; $[\alpha]_{436}^{18.5}$: −52.0; $[\alpha]_{365}^{18.5}$: −85.4 (c 2.60, CHCl$_3$).

Elementary analysis calculated for C$_{18}$H$_{33}$N$_3$O: C, 70.31; H, 10.82; N, 13.67 found: C, 70.39; H, 11.01; N, 13.38.

3.2.23. (*S,Z*)-12-Azidooctadec-9-en-1-ol (**34**)

To a vigorous stirred solution of **33** (crude made from 5.063 g, 15 mmol of **32**) in ethanol (15 mL), which was cooled at 0 °C under nitrogen, was added NaBH$_4$ (435 mg, 11.25 mmol, 0.75 equiv). The resulting mixture was allowed to stir for 1 h at 0 °C. TLC monitoring showed completion of the reaction. About 10–20 drops of acetone were added, and ethanol was removed under reduced pressure. After addition of water, extraction with petroleum ether/EtOAc (80:20), organic layers were washed with brine and dried over Na$_2$SO$_4$. Solvent was removed under reduced pressure, and the crude product was purified by two successive column chromatographies on silica gel (15 g, 0%–2% acetone in petroleum ether) to afford **34** as an ochre oil (3.44 g, 74% from **32**). R_f = 0.36 (petroleum ether/acetone 85:15).

^{1}H NMR (400 MHz, CDCl$_3$): δ 5.52 (dtt, 1H, *J* = 10.9, 7.3, 1.5 Hz), 5.38 (dtt, 1H, *J* = 10.9, 7.3, 1.5 Hz), 3.64 (t, 2H, *J* = 6.6 Hz), 3.33–3.25 (m, which could be analyzed as a dddd at 3.29 ppm with *J* = 7.8, 6.8, 6.4, 5.2 Hz, 1H, C**H**N$_3$), 2.35–2.20 (m, 2H), 2.04 (br td (apparent q), 2H, *J* = 7.0, 7.0 Hz), 1.61–1.41 (m, 5H), 1.41–1.23 (m, 17H), 0.89 (t, 3H, *J* = 6.9 Hz, CH$_3$).

^{13}C NMR (100 MHz, CDCl$_3$): δ 133.17 (**C**H=CH), 124.53 (**C**H=CH), 63.07 (**C**H$_2$), 62.96 (**C**H), 33.98 (**C**H$_2$), 32.79 (**C**H$_2$), 32.28 (**C**H$_2$), 31.72 (**C**H$_2$), 29.51 (**C**H$_2$), 29.49 (**C**H$_2$), 29.39 (**C**H$_2$), 29.24 (**C**H$_2$), 29.09 (**C**H$_2$), 27.43 (**C**H$_2$), 26.13 (**C**H$_2$) 25.73 (**C**H$_2$), 22.59 (**C**H$_2$), 14.07 (**C**H$_3$).

3.2.24. (S,Z)-12-Azidooctadec-9-en-1-yl methanesulfonate (**35**)

To a stirred solution of **34** (3.93 g, 12.7 mmol) and Et$_3$N (2.7 mL, 19.1 mmol, 1.5 equiv) in DCM (38 mL), which was cooled at −50 °C under nitrogen, was added dropwise mesyl chloride (1.25 mL, 15.9 mmol, 1.25 equiv) in DCM (5.0 mL). Transfer of mesyl chloride was completed by rinsing with DCM (3 × 0.2 mL). The mixture was then allowed to stir for 4 h at −50 °C and then allowed to warm up slowly to −5 °C. TLC monitoring (elution with DCM) showed completion of the reaction, and distilled water (50 mL) was added to quench the reaction. Extraction was done with DCM, and organic layers were washed with brine and dried over Na_2SO_4. DCM was removed under reduced pressure, and the crude product was purified by two successive column chromatographies on silica gel (15 g and then 40 g, 0%–3% acetone in petroleum ether) to provide **35** as an ochre oil (2.89 g, 59%). R_f = 0.61 (petroleum ether/acetone 80:20).

^{1}H NMR (400 MHz, $CDCl_3$): δ 5.52 (dtt, 1H, *J* = 10.9, 7.3, 1.4 Hz), 5.39 (dtt, 1H, *J* = 10.9, 7.3, 1.5 Hz), 4.22 (t, 2H, *J* = 6.6 Hz), 3.33–3.25 (m, which could be analyzed as a br dddd at 3.29 ppm with *J* = 7.2, 6.8, 6.5, 5.3 Hz, 1H, C$\underline{H}$N$_3$), 3.01 (s, 3H), 2.35–2.22 (m, 2H), 2.05 (br dt (apparent q), 2H, *J* = 7.2, 7.0 Hz), 1.79–1.70 (m, 2H), 1.56–1.22 (m, 20H with 2H at 1.56–1.47 ppm), 0.89 (t, 3H, *J* = 6.9 Hz, CH$_3$).

^{13}C NMR (100 MHz, $CDCl_3$): δ 133.07 ($\underline{C}$H=CH), 124.60 ($\underline{C}$H=CH), 70.17 ($\underline{C}$H$_2$), 62.94 ($\underline{C}$H), 37.37 ($\underline{C}$H$_3$), 33.99 ($\underline{C}$H$_2$), 32.29 ($\underline{C}$H$_2$), 31.72 ($\underline{C}$H$_2$), 29.47 ($\underline{C}$H$_2$), 29.31 ($\underline{C}$H$_2$), 29.16 ($\underline{C}$H$_2$), 29.12 ($\underline{C}$H$_2$), 29.08 ($\underline{C}$H$_2$), 28.99 (CH$_2$), 27.40 ($\underline{C}$H$_2$), 26.13 ($\underline{C}$H$_2$), 25.41 ($\underline{C}$H$_2$), 22.58 ($\underline{C}$H$_2$), 14.07 ($\underline{C}$H$_3$).

3.2.25. (*R*)-4-((((*S*,*Z*)-12-Azidooctadec-9-en-1-yl)oxy)methyl)-2,2-dimethyl-1,3-dioxolane (**36**)

To a stirred mixture at rt of **35** (2.48 g, 6.4 mmol), *n*-Bu$_4$NBr (515.8 mg, 1.6 mmol, 0.25 equiv), DMSO (12.8 mL) and 50% aqueous NaOH (0.67 mL, 2 equiv) was added (*R*)-solketal **18** (1002.8 mg, ≥95% pure, 7.21 mmol, 1.13 equiv). The resulting mixture was then stirred overnight for 15 h at 35 °C. TLC monitoring showed completion of the reaction, and distilled water (40 mL) was added. Extraction was done with petroleum ether/EtOAc (80:20), and organic layers were washed again with brine and dried over Na_2SO_4. Solvent was removed under reduced pressure, and the crude product was purified by two successive column chromatographies on silica gel (11 g and then 30 g, 0%–0.5% acetone in petroleum ether) to afford **36** as an ochre oil (1.41 g, 52%). R_f = 0.56 (petroleum ether/acetone 95:5).

^{1}H NMR (400 MHz, $CDCl_3$): δ 5.52 (dtt, 1H, *J* = 10.9, 7.2, 1.5 Hz), 5.38 (dtt, 1H, *J* = 10.9, 7.2, 1.5 Hz), 4.27 (tt, 1H, *J* = 6.4, 5.7 Hz), 4.06 (dd, 1H, *J* = 8.2, 6.4 Hz), 3.73 (dd, 1H, *J* = 8.2, 6.4 Hz), 3.54–3.39 (m, 4H with 1H dd at 3.52 ppm, *J* = 9.9, 5.7 Hz and 1H dd at 3.42 ppm, *J* = 9.9, 5.6 Hz), 3.33–3.25 (m, 1H, C$\underline{H}$N$_3$), 2.36–2.20 (m, 2H), 2.04 (br td (apparent q), 2H, *J* = 7.1, 7.0 Hz), 1.62–1.40 (m, 7H, 2 CH$_2$ + CH$_3$ q at 1.43 ppm, *J* = 0.6 Hz), 1.40–1.24 (m, 21H, 9 CH$_2$ + CH$_3$ q at 1.37 ppm, *J* = 0.6 Hz), 0.89 (t, 3H, *J* = 6.9 Hz, CH$_3$).

^{13}C NMR (100 MHz, $CDCl_3$): δ 133.18 ($\underline{C}$H=CH), 124.51 ($\underline{C}$H=CH), 109.37 ($\underline{C}$Me$_2$), 74.76 ($\underline{C}$H), 71.88 ($\underline{C}$H$_2$), 71.83 ($\underline{C}$H$_2$), 66.94 ($\underline{C}$H$_2$), 62.95 ($\underline{C}$H), 33.98 ($\underline{C}$H$_2$), 32.28 ($\underline{C}$H$_2$), 31.72 ($\underline{C}$H$_2$), 29.55 ($\underline{C}$H$_2$), 29.52 ($\underline{C}$H$_2$), 29.48 ($\underline{C}$H$_2$), 29.42 ($\underline{C}$H$_2$), 29.26 ($\underline{C}$H$_2$), 29.08 ($\underline{C}$H$_2$), 27.44 ($\underline{C}$H$_2$), 26.78 ($\underline{C}$H$_3$), 26,13 ($\underline{C}$H$_2$), 26.05 ($\underline{C}$H$_2$), 25.43 (CH$_3$), 22.58 (CH$_2$), 14.07 ($\underline{C}$H$_3$).

HRMS (ESI, *m/z*) calculated for $C_{24}H_{45}N_3O_3Na$ [M + Na]$^+$: 446.3358, found: 446.3365.

3.2.26. (*S*)-3-(((*S*,*Z*)-12-Azidooctadec-9-en-1-yl)oxy)propane-1,2-diol (**10**)

To a solution of **36** (610.7 mg, 1.44 mmol) in methanol (5.8 mL), was added *p*-toluenesulfonic acid monohydrate (13.7 mg, 0.072 mmol, 0.05 equiv) and distilled water (0.58 mL). The flask was then purged with nitrogen, stoppered and dipped in a preheated bath at 60 °C. Stirring was maintained for 4 h at 60 °C. TLC monitoring showed completion of the reaction, and sodium bicarbonate (6.7 mg, 0.08 mmol, 0.055 equiv) was added to the mixture, and stirring was continued for 1 h at 60 °C. Methanol was removed under reduced pressure, and the crude product was purified by column chromatography on silica gel (5 g, 0%–5% acetone in petroleum ether and then petroleum ether + 5% acetone + 12% methanol) to provide **10** as an ochre oil (539.1 mg, 97%). R_f = 0.03 (petroleum ether/acetone 80:20).

IR (KBr) ν 3396, 2928, 2855, 2100, 1464, 1254, 1125, 1037 cm^{-1}.

^{1}H NMR (400 MHz, CDCl$_3$): δ 5.52 (dtt, 1H, J = 10.9, 7.3, 1.4 Hz), 5.38 (dtt, 1H, J = 10.9, 7.2, 1.5 Hz), 3.91–3.83 (m, which could be analyzed as a br tt at 3.86 ppm with J = 5.3, 4.1 Hz, exchange with D$_2$O improved a little bit the resolution), 3.77–3.68 (m, 1H (dd at 3.71 ppm after exchange with D$_2$O, J = 11.4, 3.9 Hz)), 3.65 (dd, 1H, J = 11.4, 5.2 Hz, unchanged after exchange with D$_2$O), 3.56–3.42 (m, 4H with 1H dd at 3.54 ppm, J = 9.7, 4.0 Hz), 3.33–3.25 (m, which could be analyzed as a dddd at 3.29 ppm with J = 7.6, 6.6, 6.4, 5.2 Hz, 1H, C$\underline{H}$N$_3$), 2.72 (envelope from 2.85 to 2.59 ppm, 1H, OH), 2.37–2.20 (m, 3H: 1 OH as a partly overlapped envelope, which topped at 2.32 ppm + 1 CH$_2$, which was centered at 2.29 ppm), 2.05 (br dt (apparent q), 2H, J = 7.1, 6.9 Hz), 1.62–1.40 (m, 5H), 1.40–1.23 (m, 17H), 0.89 (t, 3H, J = 6.9 Hz, CH$_3$).

^{13}C NMR (100 MHz, CDCl$_3$): δ 133.16 ($\underline{C}$H=CH), 124.53 (CH=$\underline{C}$H), 72.52 ($\underline{C}$H$_2$), 71.83 ($\underline{C}$H$_2$), 70.43 ($\underline{C}$H), 64.29 ($\underline{C}$H$_2$), 62.95 ($\underline{C}$H), 33.97 ($\underline{C}$H$_2$), 32.28 ($\underline{C}$H$_2$), 31.72 ($\underline{C}$H$_2$), 29.57 ($\underline{C}$H$_2$), 29.51 ($\underline{C}$H$_2$), 29.46 ($\underline{C}$H$_2$), 29.41 ($\underline{C}$H$_2$), 29.24 ($\underline{C}$H$_2$), 29.08 ($\underline{C}$H$_2$), 27.43 ($\underline{C}$H$_2$), 26.13 ($\underline{C}$H$_2$), 26.07 ($\underline{C}$H$_2$), 22.58 ($\underline{C}$H$_2$), 14.07 ($\underline{C}$H$_3$).

$[\alpha]_D^{21}$: -17.7; $[\alpha]_{578}^{21}$: -18.7; $[\alpha]_{546}^{21}$: -21.3; $[\alpha]_{436}^{21}$: -37.9; $[\alpha]_{365}^{21}$: -62.5 (c 2.26, CHCl$_3$).

HRMS (ESI, m/z) calculated for C$_{21}$H$_{41}$N$_3$O$_3$Na [M + Na]$^+$: 406.3046, found: 406.3046, calculated for C$_{21}$H$_{40}$N$_3$O$_3$Na$_2$ [M – H + 2Na]$^+$: 428.2865, found: 428.2882, calculated for C$_{21}$H$_{41}$NO$_3$Na [M – N$_2$ + Na]$^+$: 378.2984, found: 378.2998.

3.3. Chemicals and Culture Media

Nystatin (Maneesh Pharmaceutic PVT) for fungi, chloramphenicol (Sigma) for *E. coli* ATCC 10.536 and AG102, gentamycin (Jinling Pharmaceutic Group corp.), tetracycline, ciprofloxacin and ampicillin (Sigma-Aldrich, South Africa) for bacteria were used as reference antibiotics (RE). Dimethylsulfoxide (DMSO, Sigma) was used to dilute all tested samples. Nutrient agar (NA) was used for bacterial culture. Sabouraud glucose agar was used for the activation of the fungi. The Mueller Hinton broth (MHB) was used to determine the minimal inhibition concentration (MIC) of all samples against the tested microorganisms. Two-fold dilutions were made for MIC determinations, and the results were validated only if at least two of the three replications were similar; standard deviation bars were not appropriate in regards to two-fold dilutions in such study.

3.4. Microbial Strains

The organisms tested included methicillin-resistant *Staphylococcus aureus* LMP805, *Streptococcus faecalis* LMP 806 (Gram-positive bacteria), Gram-negative bacteria, namely β-lactamase positive (βL$^+$) *Escherichia coli* LMP701, *E. coli* ATCC10536, kanamycin and chloramphenicol resistant *E. coli* AG102, ampicillin-resistant *Shigella dysenteriae* LMP820, chloramphenicol-resistant *Salmonella typhi* LMP706, chloramphenicol-resistant *Citrobacter freundii* LMP802 and three fungi, namely *Candida albicans* LMP709U, *Candida glabrata* LMP0413U and *Microsporum audouinii* LMP725D. *E. coli* ATCC10536 and AG102 were provided by UMR-MD1 (Université de la Méditeranée, France). All other microbial species were clinical isolates from the "Centre Pasteur du Cameroun" and provided by the Laboratory of Applied Microbiology and Molecular Pharmacology (LMP) (Faculty of Science, University of Yaoundé I). These were maintained on agar slant at 4 °C and sub-cultured on a fresh appropriate agar plate 24 h prior to any antimicrobial test. The three types of *E. coli* used in the antimicrobial assay were the reference ATCC strain, a wild type and a resistant phenotype.

3.5. Antimicrobial Assays

The antimicrobial assays were conducted using rapid XTT colorimetry and viable count methods. The XTT colorimetric assay was performed according to Pettit et al. [36] as modified by Kuete et al. [37–39]. Concisely, the tested sample (or combined sample with antibiotic) was first of all dissolved in DMSO/MHB. The final concentration of DMSO was lower than 1% and did not affect the microbial growth [37–39]. The solution obtained was then added to MHB and serially diluted two fold (in a 96-well microplate). Then, 100 µl of inoculum 1.5×10^6 CFU/mL was prepared in MHB. The plates were covered with a sterile plate sealer, then agitated to mix the contents of the wells using a plate shaker and incubated at 30 °C for 48 h (*M. audouinii*) or at 37 °C for 24 h (other organisms). The assay was repeated three times. Gentamicin, chloramphenicol (bacteria) and nystatin (fungi) were used as positive controls. Wells containing MHB, 100 µl of inoculum and DMSO to a final concentration of 1% served as negative controls. The MIC of samples was then detected following an addition (40 µL) of 0.2 mg/mL *p*-iodonitrotetrazolium chloride and incubated at 37 °C for 30 min. Viable bacteria reduced the yellow dye to a pink color. MIC was defined as the lowest sample concentration that prevented this change and exhibited complete inhibition of bacterial growth.

Bacterial enumeration was performed on *E. coli* LMP701 as described by Stenger et al. [40]. Cells were treated with samples at their MIC and 4× MIC values as previously determined using XTT assay, and incubated at 37 °C. Viable cells were then determined at 0, 30, 60, 120, 240 and 480 min by performing 10-fold serial dilutions of this suspension in 0.9% saline. Gentamicin was used as reference drug whilst 0.9% saline and DMSO to a final concentration of 1% was used as control. All dilutions were placed on nutrient agar plates that were then incubated at 37 °C for 18 h. Bacterial colonies were then enumerated, and the total CFU/mL at each time was deduced.

4. Conclusions

A series of novel 1-*O*-alkylglycerol compounds **8–11** were synthesized from cheap ricinoleic acid (**12**). The structures of these compounds were characterized by NMR experiments as well as from the HRMS and elementary analysis data. AKGs **8–11** were evaluated for their respective antimicrobial activities. All compounds exhibited antimicrobial activity to different extents alone. Additionally, some beneficial synergistic effects were observed when AKG **8** was combined with gentamicin, and when AKG **11** was combined with ciprofloxacin and ampicillin. AKG **11** was viewed as a lead compound for this series, as it exhibited significant antimicrobial activity alone and when combined with some antibiotics compared with **8–10**. It is now evident that non-natural synthesized 1-*O*-alkylglycerols can be further explored as a new source of drugs, and can be used in diverse preparations of pharmaceutical importance.

Supplementary Materials: The following are available online at http://www.mdpi.com/1660-3397/18/2/113/s1. The copies of ^{1}H and ^{13}C NMR spectra for all synthetic compounds (8–11, 13–17, 19–29 and 31–36) are included in the attached Supplementary Materials as the Figures: S1.1–S1.14; S2.1–S2.12; S3.1–S3.12; and S4.1–S4.14.

Author Contributions: P.M. designed the whole research. R.P. performed the chemical research, analyzed the data and wrote the chemistry of the manuscript. V.K. performed the antimicrobial experiments and wrote the biology of the manuscript under the supervision of J.-M.P., P.M. and D.E.P. were responsible for acquiring the funding of project. All authors have read and agreed to the published version of the manuscript.

Funding: Financial support from Agence Universitaire de la Francophonie (AUF), Région Bretagne, CNRS (UMR 6226) and the Université de Rennes 1 are gratefully acknowledged.

Acknowledgments: We thank André Sasaki of UPJV for going over the manuscript and the CRMPO (Centre Régional de Mesures Physiques de l'Ouest, Rennes, France) for HRMS spectra and microanalyses.

Conflicts of Interest: The authors declare no conflict of interest.

References

1. Hallgren, B.; Larsson, S. The glyceryl ethers in man and cow. *J. Lipids Res.* **1962**, *3*, 39–43.
2. Bordier, C.G.; Sellier, N.; Foucault, A.P.; Le Goffic, F. Purification and characterization of deep sea shark *Centrophorus squamosus* liver oil 1-*O*-aklylglycerol ether lipids. *Lipids* **1996**, *31*, 521–528. [PubMed]
3. Linman, J.W.; Long, M.J.; Korst, D.R.; Bethell, F.H. Studies on the stimulation of hemopoiesis by batyl alcohol. *J. Lab. Clin. Med.* **1959**, *54*, 335–343. [PubMed]
4. Brohult, A.; Brohult, J.; Brohult, S.; Joelsson, I. Effect of alkoxyglycerols on the frequency of injuries following radiation therapy for carcinoma of the uterine cervix. *Acta Obstet. Gynecol. Scand.* **1977**, *56*, 441–448.

5. Brohult, A.; Brohult, J.; Brohult, S. Regression of tumour growth after administration of alkoxyglycerols. *Acta Obstet. Gynecol. Scand.* **1978**, *57*, 79–83.

6. Ngwenya, B.Z.; Foster, D.M. Enhancement of antibody production by lysophosphatidylcholine and alkylglycerol. *Proc. Soc. Exp. Biol. Med.* **1991**, *196*, 69–75.

7. Brohult, A.; Brohult, J.; Brohult, S. Effect of irradiation and alkoxyglycerol treatment on the formation of antibodies after *Salmonella* vaccination. *Experientia* **1972**, *28*, 954–955.

8. Baer, E.; Fisher, H.O.L. Studies on acetone-glyceraldehyde, and optically active glycerides: IX. Configuration of the natural batyl, chimyl, and selachyl alcohols. *J. Biol. Chem.* **1941**, *140*, 397–410.

9. Deniau, A.L.; Mosset, P.; Pedrono, F.; Mitre, R.; Le Bot, D.; Legrand, A.B. Multiple beneficial health effects of natural Alkylglycerols from Shark Liver Oil. *Mar. Drugs* **2010**, *8*, 2175–2184.

10. Deniau, A.L.; Le Bot, D.; Mosset, P.; Legrand, A.B. Activités antitumorale et antimétastasique des alkylglycérols naturels: Relation structure-activité. *OCL* **2010**, *17*, 236–237.

11. Deniau, A.L.; Mosset, P.; Le Bot, D.; Legrand, A.B. Which alkylglycerols from shark liver oil have anti-tumour activities? *Biochimie* **2011**, *93*, 1–3. [PubMed]

12. Hatice, M.; Michael, A.R.M. Castor oil as a renewable resource for the chemical industry. *Eur. J. Lipid Sci. Technol.* **2010**, *112*, 10–30.

13. Merkle, H.P.; Higuchi, W.I. Effects of antibacterial microenvironment on in vitro plaque formation of *Streptococcus mutans* as observed by scanning electron microscopy. *Arzneimittel-Forschung* **1980**, *30*, 1841–1846. [PubMed]

14. Mordenti, J.J.; Lindstrom, R.E.; Tanzer, J.M. Activity of Sodium Ricinoleate Against In Vitro Plaque. *J. Pharm. Sci.* **1982**, *71*, 1419–1421.

15. Gradishar, W.J.; Tjulandin, S.; Davidson, N.; Shaw, H.; Desai, N.; Bhar, P.; Hawkins, M.; O'Shaughnessy, J. Phase III trial of nanoparticle albumin-bound paclitaxel compared with polyethylated castor oil-based paclitaxel in women with breast cancer. *J. Clin. Oncol.* **2005**, *23*, 7794–7803.

16. Burdock, G.A.; Carabin, I.G.; Griffiths, J.C. Toxicology and pharmacology of sodium ricinoleate. *Food Chem. Toxicol.* **2006**, *44*, 1689–1698.

17. Mohini, Y.; Prasad, R.B.N.; Karuna, M.S.L.; Poornachandra, Y.; Ganesh Kumar, C. Synthesis and biological evaluation of ricinoleic acid-based lipoamino acid derivatives. *Bioorg. Med. Chem. Lett.* **2006**, *26*, 5198–5202.

18. Mohini, Y.; Prasad, R.B.N.; Karuna, M.S.L.; Poornachandra, Y.; Ganesh Kumar, C. Synthesis, antimicrobial and anti-biofilm activities of novel Schiff base analogues derived from methyl-12-aminooctadec-9-enoate. *Bioorg. Med. Chem. Lett.* **2014**, *24*, 5224–5227.

19. Mohini, Y.; Shiva, S.K.; Rachapudi, B.N.P.; Karuna, S.L.M.; Poornachandra, Y.; Ganesh Kumar, C. Synthesis of novel (Z)-methyl-12-aminooctadec-9-enoate-based phenolipids as potential antioxidants and chemotherapeutic agents. *Eur. J. Lipid Sci. Technol.* **2016**, *118*, 622–630.

20. Reddy, K.K.; Ravinder, T.; Kanjilal, S. Synthesis and evaluation of antioxidant and antifungal activities of novel ricinoleate-based lipoconjugates of phenolic acids. *Food Chem.* **2012**, *134*, 2201–2207.

21. Magnusson, C.D.; Haraldsson, G.G. Synthesis of enantiomerically pure (Z)-(2′R)-1-O-(2′-methoxyhexadec-4′-enyl)-*sn*-glycerol present in the liver oil of cartilaginous fish. *Tetrahedron Asymmetry* **2010**, *21*, 2841–2847.

22. Carballeira, N.M. New advances in the chemistry of methoxylated lipids. *Prog. Lipid Res.* **2002**, *41*, 437–456. [PubMed]

23. Hallgren, B.; Stallberg, G. Methoxy-substituted Glycerol Ethers Isolated from Greenland Shark Liver Oil. *Acta Chem. Scand.* **1967**, *21*, 1519–1529.

24. Hayashi, K.; Takagi, T. Characteristics of Methoxy-Glyceryl Ethers from Some Cartilaginous Fish Liver Lipids. *Bull. Jpn. Soc. Sci. Fish* **1982**, *48*, 1345–1351.

25. Hallgren, B.; Stallberg, G. Occurrence, synthesis and biological effects of substituted glycerol ethers. *Prog. Chem. Fats Other Lipids* **1978**, *16*, 45–58.

26. Magnusson, C.D.; Haraldsson, G.G. Ether lipids. *Chem. Phys. Lipids* **2011**, *164*, 315–340.

27. Pemha, R.; Pegnyemb, D.E.; Mosset, P. Synthesis of (Z)-(2′R)-1-O-(2′-methoxynonadec-10′-enyl)-sn-glycerol, a new analog of bioactive ether lipids. *Tetrahedron* **2012**, *68*, 2973–2983.

28. Kleiman, R.; Spencer, G.F.; Earle, F.R. Boron trifluoride as catalyst to prepare methyl esters from oils containing unusual acyl groups. *Lipids* **1969**, *4*, 118–122.

29. Davletbakova, A.M.; Maidanova, I.O.; Baibulatova, N.Z.; Dokichev, V.A.; Tomilov, Y.V.; Yunusov, M.S.; Nefedov, O.M. Catalytic cyclopropanation of ricinoleic acid derivatives with diazomethane. *Russ. J. Org. Chem.* **2001**, *37*, 608–611.

30. Patel, V.R.; Dumancas, G.G.; Kasi Viswanath, L.C.; Maples, R.; Subong, B.J.J. Castor Oil: Properties, Uses, and Optimization of Processing Parameters in Commercial Production. *Lipid Insights.* **2016**, *9*, 1–12. [CrossRef]

31. Tsukamoto, T.; Yoshiyama, T.; Kitazume, T. Enantioselective synthesis of β,β-difluoromalic acid via enzymic resolution of furyl substituted derivative. *Tetrahedron Asymmetry* **1991**, *2*, 759–762.

32. Welh, J.T. Tetrahedron report number 221: Advances in the preparation of biologically active organofluorine compounds. *Tetrahedron* **1987**, *43*, 3123–3197.

33. Das, S.; Chandrasekhar, S.; Yadav, J.S.; Grée, R. Ionic liquids as recyclable solvents for diethylaminosulfur trifluoride (DAST) mediated fluorination of alcohols and carbonyl compounds. *Tetrahedron Lett.* **2007**, *48*, 5305–5307.

34. Hoye, T.R.; Hanson, P.R.; Vyvyan, J.R. A Practical Guide to First-Order Multiplet Analysis in ^{1}H NMR Spectroscopy. *J. Org. Chem.* **1994**, *59*, 4096–4103.

35. Hoye, T.R.; Zhao, H. A Method for Easily Determining Coupling Constant Values: An Addendum to "A Practical Guide to First-Order Multiplet Analysis in ^{1}H NMR Spectroscopy". *J. Org. Chem.* **2002**, *67*, 4014–4016.

36. Pettit, R.K.; Weber, C.A.; Kean, M.J.; Hoffmann, H.; Pettit, G.R.; Tan, R.; Franks, K.S.; Horton, M.L. Microplate Alamar Blue Assay for *Staphylococcus epidermidis* Biofilm Susceptibility Testing. *Antimicrob. Agents Chemother.* **2005**, *49*, 2612–2617.

37. Kuete, V.; Mbaveng, T.A.; Tsafack, M.; Beng, P.V.; Etoa, F.X.; Nkengfack, A.E.; Meyer, J.J.; Lall, N. Antitumor, antioxidant and antimicrobial activities of *Bersama engleriana* (Melianthaceae). *J. Ethnopharmacol.* **2008**, *115*, 494–501.

38. Mbaveng, A.T.; Ngameni, B.; Kuete, V.; Simo, I.K.; Ambassa, P.; Roy, R.; Bezabih, M.; Etoa, F.X.; Ngadjui, B.T.; Abegaz, B.M.; et al. Antimicrobial activity of the crude extracts and five flavonoids from the twigs of *Dorstenia barteri* (Moraceae). *J. Ethnopharmacol.* **2008**, *116*, 483–489.

39. Kuete, V.; Azebaze, A.G.B.; Mbaveng, A.T.; Nguemfo, E.L.; Tshikalange, E.T.; Chalard, P.; Nkengfack, A.E. Antioxidant, antitumor and antimicrobial activities of the crude extract and compounds of the root bark of *Allanblackia floribunda*. *Pharm. Biol.* **2011**, *49*, 57–65.

40. Stenger, S.; Hanson, D.A.; Teitelbaum, R.; Dewan, P.; Niazi, K.R.; Froelich, C.J.; Ganz, T.; Thoma-Uszynsky, S.; Melian, A.; Bogdan, C.; et al. An antimicrobial activity of cytolytic T cells mediated by granulysin. *Science* **1998**, *282*, 121–125.

 *marine drugs*

Article

In Vitro Antiproliferative Evaluation of Synthetic Meroterpenes Inspired by Marine Natural Products

Concetta Imperatore [1,†], **Gerardo Della Sala** [2,†], **Marcello Casertano** [1], **Paolo Luciano** [1], **Anna Aiello** [1], **Ilaria Laurenzana** [2], **Claudia Piccoli** [2,3] and **Marialuisa Menna** [1,*]

1   The NeaNat Group, Department of Pharmacy, University of Naples "Federico II", Via D. Montesano 49, 80131 Napoli, Italy; cimperat@unina.it (C.I.); marcello.casertano@unina.it (M.C.); pluciano@unina.it (P.L.); aiello@unina.it (A.A.)

2   Laboratory of Pre-Clinical and Translational Research, IRCCS-CROB, Referral Cancer Center of Basilicata, 85028 Rionero in Vulture, Italy; gerardo.dellasala@crob.it (G.D.S.); ilaria.laurenzana@crob.it (I.L.); claudia.piccoli@unifg.it (C.P.)

3   Department of Clinical and Experimental Medicine, University of Foggia, via L. Pinto c/o OO.RR., 71100 Foggia, Italy

*   Correspondence: mlmenna@unina.it; Tel.: +39-081-678518

†   These authors contributed equally to this work.

Received: 20 November 2019; Accepted: 4 December 2019; Published: 5 December 2019

Abstract: Several marine natural linear prenylquinones/hydroquinones have been identified as anticancer and antimutagenic agents. Structure-activity relationship studies on natural compounds and their synthetic analogs demonstrated that these effects depend on the length of the prenyl side chain and on the type and position of the substituent groups in the quinone moiety. Aiming to broaden the knowledge of the underlying mechanism of the antiproliferative effect of these prenylated compounds, herein we report the synthesis of two quinones **4** and **5** and of their corresponding dioxothiazine fused quinones **6** and **7** inspired to the marine natural product aplidinone A (**1**), a geranylquinone featuring the 1,1-dioxo-1,4-thiazine ring isolated from the ascidian *Aplidium conicum*. The potential effects on viability and proliferation in three different human cancer cell lines, breast adenocarcinoma (MCF-7), pancreas adenocarcinoma (Bx-PC3) and bone osteosarcoma (MG-63), were investigated. The methoxylated geranylquinone **5** exerted the highest antiproliferative effect exhibiting a comparable toxicity in all three cell lines analyzed. Interestingly, a deeper investigation has highlighted a cytostatic effect of quinone **5** referable to a G0/G1 cell-cycle arrest in BxPC-3 cells after 24 h treatment.

Keywords: organic synthesis; meroterpenoids; thiazinoquinones; antiproliferative activity; G0/G1 cell-cycle arrest; cytostatic; solid tumor cell lines

---

1. Introduction

Chemotherapy represents the most applied strategy for cancer treatment; therefore, development of novel and improved antitumor compounds has become mandatory.

The sustainable exploitation of marine natural products as starting leads is a precious and still untapped resource. Many classes of natural molecules have been conceived by nature to play specific roles in cell processes through selective interactions with key cellular targets. In this frame, quinones represent a clinically relevant class of chemotherapeutic agents with antitumor activity already described in several cell lines [1]. The most prominent chemical feature of these compounds, both natural and synthetic derivatives, is the ability to undergo redox cycling to generate reactive oxygen species (ROS), responsible for significant cell damage. In this class of compounds, a heterogeneous group of prenylated structures formed through the mixed terpenoid-polyketide biosynthetic pathway, stands out for their potential anticancer activities [2,3]. This group of molecules, commonly named meroterpenoids, are broadly widespread in both terrestrial plants, insects, fungi, lichens and marine organisms [4], being involved in electron transport processes and into photosynthesis [5]. Among marine ascidians, these compounds have been found almost exclusively in species belonging to the genus *Aplidium*. Since the first report of the presence of geranylhydroquinone in an *Aplidium* sp., a number of structurally diverse meroterpenes have been isolated from several *Aplidium* species. Among them, a wide variety of often very complex molecules, originating from intra- and intermolecular cyclizations and/or rearrangements of the terpene chains to give unique polycyclic or macrocyclic structures, have been discovered [2,5,6]. In the course of our ongoing research program aimed at the search and characterization of new drug candidates of marine origin [7–11], a large group of new meroterpenes with different polycyclic skeletons but all featuring an unusual 1,1-dioxo-1,4-thiazine ring fused with the quinone moiety, i.e., aplidinones A and B, thiaplidiaquinones and conithiaquinones, were isolated from samples of *Aplidium conicum* [12–14]. The valuable antitumor activity shown by these compounds prompted us to further investigate the chemical and pharmacological features of this class of compounds [15]. For this purpose, several synthetic analogs of aplidinone A (**1**), featuring a methoxyl group and a monoprenyl alkyl chain linked at the thiazinoquinones scaffold, have been synthetized, in which the geranyl chain is replaced by other alkyl chains [16–19]. This synthetic chemical library along with the natural metabolite was subjected to cytotoxicity assays and preliminary structure-activity relationships (SAR) studies. This approach allowed us to define that the cytotoxic effects depend on the nature and the length of side chain linked to the benzoquinone ring and, mainly, on its position respect to the dioxothiazine ring.

In order to expand the chemical library and more clearly establish the role of the thiazine ring and of both length and shape of the alkyl side chain on the cytotoxicity, we have synthesized the two prenylated quinones **4** and **5** and we have then converted them into the corresponding thiazinoquinones **6** and **7** (Figure 1). We have then explored their potential effects on viability and proliferation in three different human cancer cell lines, namely MCF-7 (breast adenocarcinoma), Bx-PC3 (pancreas adenocarcinoma), and MG-63 (bone osteosarcoma). We report herein the synthesis, the chemical characterization and the pharmacological profile of compounds **4–7**.

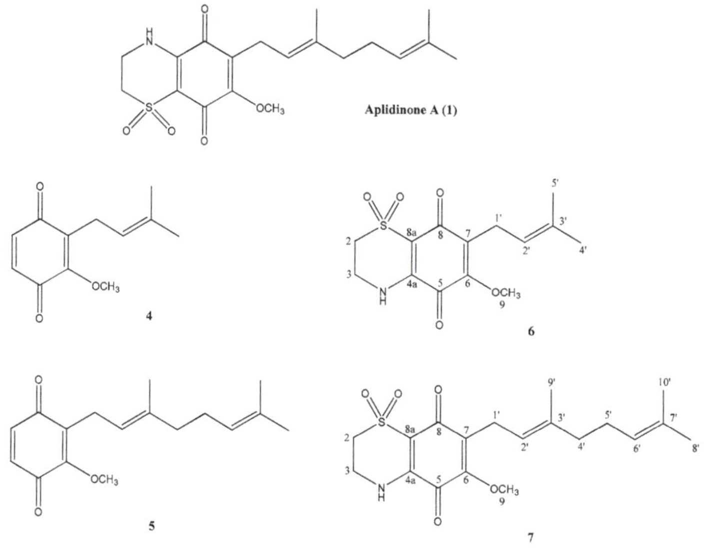

Figure 1. Structures of aplidinone A (**1**) and of the synthetic derivatives **4–7**.

2. Results and Discussion

2.1. Chemistry

The prenylquinones **4** and **5** as well as the relevant thiazinoquinone derivatives **6** and **7** were synthesized using a synthetic protocol previously designed and developed in order to easily produce and enlarge the chemodiversity within the thiazinoquinones library [16–18].

In detail, as reported in Scheme 1, the commercially available 1,2,4-trimethoxybenzene (**2**) has been chosen as the starting material. In the first step, compound **2** was treated with *n*-BuLi in THF solution; 5-bromo-2-methylpent-2-ene or (*E*)-1-bromo-3,7-dimethyl-2,6-octadiene were then added, keeping the relevant mixtures at room temperature overnight. Under these conditions, the monoalkylated products, compounds **3-R$_1$** and **3-R$_2$**, were obtained with very high yield, following that the introduction of the alkyl residue on 1,2,4-trimethoxy benzene resulted highly selective in the 3-position. Subsequently, each compound **3-R$_1$** and **3-R$_2$** was subjected to an oxidation reaction with cerium ammonium nitrate (CAN) providing the quinones **4** (83%) and **5** (76%), respectively. Previously reported attempts to produce selectively 3-monoprenylated quinones afforded mixtures of monoprenylated compounds together with polyprenylated products [20]. Compound **5** itself had been previously obtained through this different synthetic pathway with an overall low yield ($\approx$ 45%). Our procedure afforded monoprenylated quinone **5** with a total yield of 74%, thus resulting in a great improvement over previous synthesis, and the absence of side products, especially of monoalkylated quinones in different position. Afterwards, the thiazinoquinone-bicyclic system of **6** and **7** was built up by condensation of the quinone ring of compounds **4** and **5** with hypotaurine using salcomine as catalyst. Contrary to what previously occurred during the formation of thiazinoquinone scaffold in other unsymmetrical methoxy-quinones [16–19], the addition of hypotaurine to prenylbenzoquinones **4** and **5** led to the formation of only one of the two possible regioisomers (Scheme 1). This highly regioselective outcome could be explained by considering the greater steric hindrance of the prenylated side chains of compounds **4** and **5** when compared to other quinones featuring smaller and/or less flexible side chains [16–19].

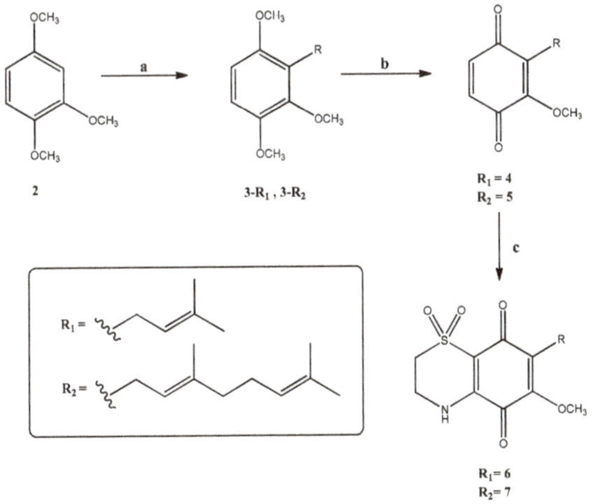

^{a} Reagents and conditions: (a) (1) *n*-BuLi, THF, 0 °C, 1 h; **3-R₁**: 5-bromo-2-methylpent-2-ene, 0 °C → rt, overnight. **3-R₂**: (*E*)-1-bromo-3,7-dimethyl-2,6-octadiene, 0 °C → rt, overnight. (b) CAN, CH₃CN, 0 °C, 45 min. (c) hypotaurine, EtOH/CH₃CN, salcomine, rt, 48 h.

Scheme 1. Synthesis of compounds **4–7**.

The above-reported synthetic strategy is particularly advantageous because it starts from low-cost and commercially available reagents allowing to obtaining several dioxothiazinoquinones in few steps and in very good yields. High purity, more than 99.8%, was easily obtained for quinones **4–5** as well as for thiazinoquinones **6–7** by HPLC; each compound was fully characterized by spectroscopic means (Table 1).

Table 1. ^{1}H (700 MHz) and ^{13}C NMR (125 MHz) spectroscopic data [a] of compounds **6** and **7** in CDCl$_3$.

Pos.	**6** [a] δ_C	**6** [a] δ_H, mult. (*J* in Hz)	**7** [a] δ_C	**7** [a] δ_H, mult. (*J* in Hz)
1	-	-	-	-
2	49.0	3.29, m	49.0	3.28, m
3	39.9	4.04, m	39.7	4.04, m
4	-	-	-	-
4a	142.9	-	143.2	-
5	176.8	-	176.5	-
6	152.7	-	153.1	-
7	136.9	-	137.0	-
8	178.1	-	178.8	-
8a	110.2	-	109.8	-
9	60.8	3.91, s	61.0	3.89, s
1'	22.9	3.20, d (7.3)	22.9	3.20, d (5.5)
2'	119.1	5.04, t (7.6)	119.5	5.04, m [b]
3'	134.5	-	138.5	-
4'	17.8	1.72, s	39.9	1.93, m
5'	25.7	1.65, s	26.5	2.02, m
6'	-	-	124.5	5.04, m [b]
7'	-	-	131.8	-
8'	-	-	17.8	1.56, s
9'	-	-	16.9	1.71, s
10'	-	-	25.3	1.64, s
-NH		6.57, brs	-	6.84, brs

[a] ^{1}H NMR and ^{13}C NMR shifts are referenced to CDCl$_3$ (δ_H = 7.26 ppm and δ_C = 77.0 ppm). [b] Partially overlapped to other resonances.

Structures of compounds **4** and **5** were easily defined similar to the spectroscopic resonances (^{1}H and ^{13}C NMR) of the compounds already reported in literature [20]. HRESI-MS data of compounds **6** and **7** indicated that the two compounds had the molecular formulas C$_{14}$H$_{17}$O$_5$NS and C$_{19}$H$_{25}$O$_5$NS, respectively. Analysis of 1D and 2D NMR spectral data of **6** and **7** (CDCl$_3$) allowed the assignment of all ^{1}H and ^{13}C NMR signals (Table 1) confirming the whole structure of thiazinoquinones **6** and **7**, except for the regiochemistry of the 1,1-dioxo-1,4-thiazine ring. Nevertheless, the close correlation between the experimental NMR signals of bicyclic skeleton of compounds **6** and **7** with ^{13}C chemical shift values previously calculated by theoretical means [14,19] for strictly correlated structures was in agreement with the regiochemistry depicted in Figure 1.

2.2. In Vitro Evaluation of Antiproliferative Activity of Quinones **4–7** in Cancer Cell Lines

Aiming to assess antiproliferative activity and structure-activity relationships of synthetic quinones **4–7** in solid tumor models, potential growth inhibitory effects were evaluated in three different human cancer cell lines, namely MCF-7 (breast adenocarcinoma), Bx-PC3 (pancreas adenocarcinoma), and MG-63 (bone osteosarcoma). The cell viability was monitored by a real-time cell analyzer based upon impedance measurements of cells growing on microelectronic sensors (xCELLigence system-ACEA Biosciences, San Diego, CA, USA). Drug-induced cell growth inhibition prompts alterations of electronic impedance, which are expressed as cell index (CI), a unit-less parameter indicative of cell number and morphology.

Quinones **4–7** were initially tested individually at a single dose exposure (10 μM) for 72 h. Real-time monitoring of cell proliferation (Figure 2) unveiled that a) Bx-PC3 cells were the most sensitive cell line and b) quinones were more effective during the first 24 h, which was selected as time point for our following investigations.

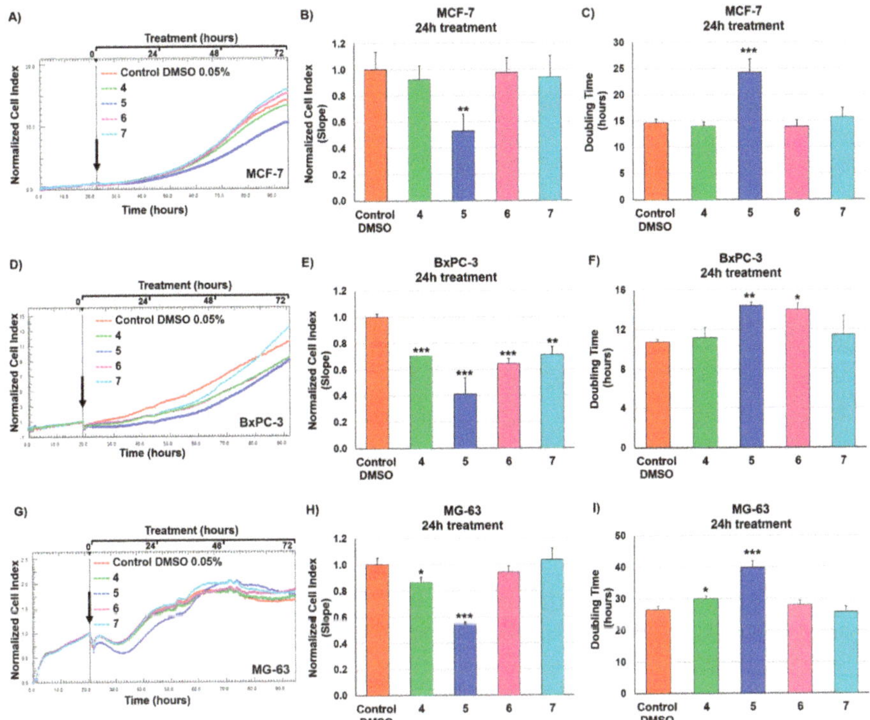

Figure 2. Real time monitoring of cancer cell growth after exposure to quinones **4–7** (10 μM) and DMSO vehicle (0.05%) using the xCELLigence System Real-Time Cell Analyzer. (**A,D,G**) Normalized cell index (NCI) traces of MCF7 (**A**), BxPC-3 (**D**), and MG-63 (**G**) cells exposed to compounds **4–7** and DMSO vehicle for 72 h. Black arrow shows the starting point of drug treatment. Each cell index value was normalized just before treatment. (**B,E,H**) NCI variations of MCF-7 (**B**), BxPC-3 (**E**), and MG-63 (**H**) cells after 24 h exposure to compounds **4–7** (10 μM) and 0.05% DMSO vehicle. Antiproliferative effects are reported as slope of NCI to describe the changing rate of growth curves after drug treatment. NCI slope values are relative to controls treated with DMSO vehicle. (**C,F,I**) Doubling times of NCI of MCF-7 (**C**), BxPC-3 (**F**), and MG-63 (**I**) cells after 24h incubation with 10 μM of quinones **4–7** and 0.05% DMSO. Data are presented as mean ± SD; *n* = 3. Statistical significances are referred to the DMSO control. * *p* < 0.05; ** *p* < 0.01; *** *p* < 0.0001.

Bx-PC3 cells experienced delayed proliferation after exposure to compounds **4**, **6**, and **7**, which were shown to elicit a significant, moderate reduction (approximately 30%) in the slope of the growth curve as compared to the control. On the other hand, MCF-7 and MG-63 cell growth was basically unaffected or slightly delayed after treatment with **4**, **6**, and **7** (Figure 2A–C,G–I). Notably, compound **5** exerted the highest antiproliferative effect and exhibited a comparable toxicity in all three cell lines, as inducing a substantial decrease of cell index (within the range of 50–60%) and a significant increase in cell doubling time.

In the light of these findings, investigation of structure-activity relationships revealed key structural motifs for maintaining growth inhibitory properties of the molecules under examination. The prenyl chain length appears to be crucial to keep a strong bioactivity profile as addition of a second prenyl unit in **5** improves antiproliferative effects in the tested cell lines as compared to **4**. Moreover, the presence of a fused thiazine ring weakens cell viability effects. Indeed, compound **5** resulted to be more active than the relevant thiazinoquinone derivative **7**, where C8a and C4a electrophilic sites are embedded in the bicyclic structure of the molecule and, therefore, are not available for potential covalent binding to specific antitumor targets (e.g., ubiquitin-proteasome pathway) [21].

To evaluate the underlying mechanism of antiproliferative effect of compound **5**, we examined whether it elicited increased apoptotic cell death in BxPC-3 cancer cells by the annexin V-FITC/PI assay (Figure 3). After 24 h incubation with quinone **5** at concentrations of 5, 10 and 20 μM, pancreatic tumor cells exhibited a slight, although significant, increase of early and late apoptotic cells only at the highest dose tested (20 μM), as compared to the control cells treated with DMSO vehicle. The relatively low number of apoptotic cells (approximately 11% at the highest concentration), suggested a cytostatic rather than a cytotoxic effect shown by the compound **5**.

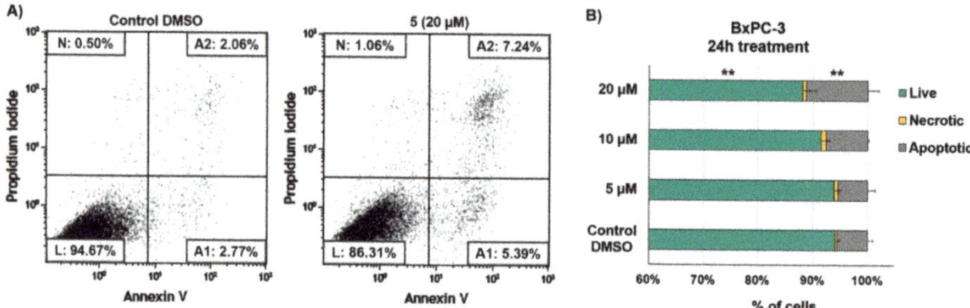

Figure 3. Flow cytometric detection of apoptosis and necrosis with Annexin-V-fluorescein isothiocyanate (FITC) and propidium iodide staining in BxPC3 cells after 24 h exposure to different concentrations of quinone **5** and DMSO vehicle. (**A**) Dot plots show a single representative experiment. Abbreviations: L, live cells; A1, early apoptotic cells; A2, late apoptotic cells; N, necrotic cells. (**B**) Relative amount of live, necrotic, and apoptotic cells after 24 h treatment with different concentrations of **5** (5, 10, and 20 μM) and DMSO vehicle. Percent of apoptotic cells was obtained from the sum of early and late apoptosis. Data are presented as mean ± SD; $n = 3$. Statistical significances are referred to the DMSO control. ** $p < 0.01$.

As drug-mediated growth inhibition observed during real time cell analysis did not correlate with enhanced apoptosis, we next performed cell cycle analysis by flow cytometry based on PI staining (Figure 4). A typical G0/G1 cell-cycle arrest was observed in BxPC-3 cells after 24 h treatment with quinone **5** (20 μM), shown by a significant increase in G0/G1 phase cells shifting from 45.7% to 62.8% ($p < 0.0001$).

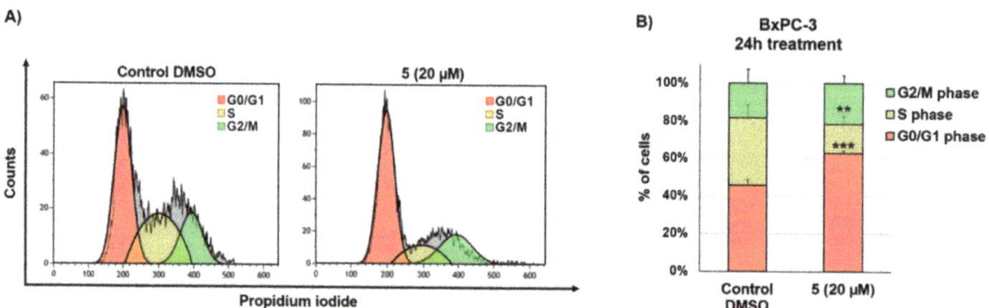

Figure 4. Cell cycle analysis through PI staining and flow cytometry of BxPC-3 cells after 24 h treatment with 20 μM compound **5** and DMSO vehicle. (**A**) Cell cycle histogram plots show a single representative experiment. (**B**) Changes in cell cycle distribution of BxPC-3 cells treated with quinone **5** (20 μM) for 24 h. Data are presented as mean ± SD; $n = 4$. Statistical significances are referred to the DMSO control. ** $p < 0.01$, *** $p < 0.0001$.

3. Materials and Methods

Commercial reagents: Sigma–Aldrich (Saint Louis, MO, USA). Solvents: Carlo Erba (Pomezia, Rome, Italy). TLC: Silica Gel 60 F254 (plates 5 × 20, 0.25 mm) Merck (Kenilworth, NJ, USA). Preparative TLC: Silica Gel 60 F254 plates (20 × 20, 2 mm). Spots revealed by UV lamp then by spraying with 2 N sulfuric acid and heating at 120 °C. Anhydrous solvents: Sigma–Aldrich or prepared by distillation according to standard procedures. High-resolution ESI-MS analyses were performed on a Thermo LTQ Orbitrap XL mass spectrometer (Thermo-Fisher, San Josè, CA, USA). The spectra were recorded by infusion into the ESI (Thermo-Fisher, San Josè, CA, USA) source dissolving the sample in MeOH. ^{1}H (700 MHz and 500 MHz) and ^{13}C (125 MHz) NMR spectra were recorded on a Agilent INOVA spectrometer (Agilent Technology, Cernusco sul Naviglio, Italy) equipped with a ^{13}C enhanced HCN Cold Probe; chemical shifts were referenced to the residual solvent signal (CDCl$_3$: δ_H = 7.26, δ_C = 77.0). For an accurate measurement of the coupling constants, the one-dimensional ^{1}H NMR spectra were transformed at 64 K points (digital resolution: 0.09 Hz). Homonuclear (^{1}H-^{1}H) and heteronuclear (^{1}H-^{13}C) connectivities were determined by COSY and HSQC experiments, respectively. Two and three bond ^{1}H-^{13}C connectivities were determined by gradient 2D HMBC experiments optimized for a $^{2,3}J$ of 8 Hz. $^3J_{\text{H-H}}$ values were extracted from 1D ^{1}H NMR. High performance liquid chromatography (HPLC) separations were achieved on a Shimadzu LC-10AT (Shimadzu, Milan, Italy) apparatus equipped with a Knauer K-2301 (LabService Analytica s.r.l., Anzola dell'Emilia, Italy) refractive index detector.

3.1. Chemistry

3.1.1. Synthesis of 1,2,4-trimethoxy-3-(3-methylbut-2-en-1-yl)benzene (3-R$_1$) and (E)-2-(3,7-dimethylocta-2,6-dien-1-yl)-1,3,4-trimethoxybenzene (3-R$_2$)

A quantity of 500 µL of 1,2,4-trimethoxybenzene (2) (3.4 mmol) was solubilized in 15 mL of THF dry and 2.5 mL of *n*-BuLi (4 mmol) were added to the mixture which was stirred under Argon atmosphere for 1 h at 0 °C. Subsequently, 4 mmol of 3,3-dimethylallyl bromide (470 µL, 4 mmol) for compound 3-R$_1$ and (E)-1-bromo-3,7-dimethyl-2,6-octadiene (800 µL, 4 mmol) for compound 3-R$_2$ were added respectively keeping the relevant mixture under magnetic stirring at room temperature overnight. After 12 h, both obtained mixtures were quenched with an aqueous solution of sodium chloride (30 mL) and extracted two times with diethyl ether (50 mL). The organic layers were dried over anhydrous sodium sulfate and, concentrated in vacuo to afford 3-R$_1$ (787 mg, 98%) and 3-R$_2$ (1 g, 97%) sufficiently pure to the following reaction step.

1,2,4-trimethoxy-3-(3-methylbut-2-en-1-yl)benzene (3-R$_1$): dark yellow oil; HRESIMS *m/z* 259.1307 [M + Na]$^+$ (calcd. for C$_{14}$H$_{20}$O$_3$Na 259.1305). ^{1}H NMR (CDCl$_3$, 500 MHz): δ 6.72 (1H, d, *J* = 8.9 Hz), 6.57 (1H, d, *J* = 8.9 Hz), 5.23 (1H, t), 3.84 (3H, s), 3.83 (3H, s), 3.79 (3H, s), 3.39 (2H, d, *J* = 7.0 Hz), 1.81 (3H, s), 1.69 (3H, s). ^{13}C NMR (CDCl$_3$, 125 MHz): δ 154.6, 150.5, 149.7, 133.8, 127.5, 125.3, 112.4, 108.1, 63.3, 58.8, 58.6, 28.4, 25.8, 20.4. ^{1}H, ^{13}C and HRESIMS spectra are reported in Supporting Information (Figures S1–S3).

(E)-2-(3,7-dimethylocta-2,6-dien-1-yl)-1,3,4-trimethoxybenzene (3-R$_2$): dark yellow oil; HRESIMS *m/z* 327.1942 [M + Na]$^+$ (calcd. for C$_{19}$H$_{28}$O$_3$Na: 327.1931). ^{1}H NMR (CDCl$_3$, 500 MHz): δ 6.73 (1H, d, *J* = 8.9 Hz), 6.58 (1H, d, *J* = 8.9 Hz), 5.24 (1H, t), 5.10 (1H, t), 3.84 (6H, s), 3.80 (3H, s), 3.41 (2H, d, *J* = 6.9 Hz), 2.08 (2H, dd, *J* = 7.5, 6.9 Hz), 2.00 (2H, m), 1.81 (3H, s), 1.67 (3H, s), 1.60 (3H, s). ^{13}C NMR (CDCl$_3$, 125 MHz): δ 154.9, 150.6, 149.9, 137.3, 133.6, 127.8, 127.3, 125.8, 112.3, 108.3, 63.4, 58.9, 58.5, 42.5, 29.4, 28.2, 25.6, 20.4, 18.8. ^{1}H, ^{13}C and HRESIMS spectra are reported in Supporting Information (Figures S4–S6).

3.1.2. Synthesis of 2-methoxy-3-(3-methylbut-2-en-1-yl)cyclohexa-2,5-diene-1,4-dione (4)

A portion of compound 3-R$_1$ (310 mg, 1.3 mmol) was dissolved in 50 mL of acetonitrile (ACN) at 0 °C. A solution of 2.9 g (5.2 mmol) of ammonium cerium nitrate (CAN) in 9 mL of water was prepared and added dropwise stirring the achieved mixture for 45 min at 0 °C. The end of reaction was monitored by TLC with an eluent system chloroform/EtOAc 7:3 before diluting the orange liquid with

100 mL of water and then extracted with diethyl ether (2 × 100 mL). The consequent organic phase was washed with brine, dried over anhydrous Na_2SO_4, filtered and the solvent removed in vacuo. The mixture was chromatographed by HPLC on SiO_2 column (Luna 3 μm, 150 × 4.6 mm, flow rate 1 mL/min) with a mobile phase hexane/EtOAc 9:1 (*v/v*) affording the quinone **4** as a pure compound (221 mg, 83%, t_R 4.9 min).

2-methoxy-3-(3-methylbut-2-en-1-yl)cyclohexa-2,5-diene-1,4-dione (**4**): yellow powder, HRESIMS *m/z* 229.0839 [M + Na]$^+$ (calcd. for $C_{12}H_{14}O_3Na$: 229.0835). ^{1}H NMR (CDCl$_3$, 500 MHz): δ 6.67 (1H, d, *J* = 9.5 Hz), 6.58 (1H, d, *J* = 9.5 Hz), 5.04 (1H, t, *J* = 6.9 Hz), 4.01 (3H, s), 3.13 (2H, d, *J* = 7.3 Hz), 1.73 (3H, s), 1.66 (3H, s). ^{13}C NMR (CDCl$_3$, 125 MHz): δ 187.4, 183.4, 155.0, 136.8, 136.1, 134.4, 131.9, 119.4, 60.7, 25.2, 22.3, 17.5. ^{1}H, ^{13}C and HRESIMS spectra are reported in Supporting Information (Figures S7–S9).

3.1.3. Synthesis of (E)-2-(3,7-dimethylocta-2,6-dien-1-yl)-3-methoxycyclohexa-2,5-diene-1,4-dione (5)

A quantity of 130 mg (0.43 mmol) of compound **3-R$_2$** was dissolved in 20 mL of ACN and an aqueous solution of CAN (938 mg, 1.7 mmol in 5 mL of water) was added dropwise to the mixture in a cold bath at 0 °C. The orange mixture was kept under magnetic stirring for 45 min at the above-mentioned temperature monitoring the reaction progress by TLC (chloroform/EtOAc 7:3). After this time, the solution was diluted with 100 mL of cold water and subjected to an extraction twice with 80 mL of diethyl ether. The collected organic layer was washed with a saturated solution of NaCl, dried over anhydrous Na_2SO_4, filtered and the solvent removal was realized under reduced pressure. The pure quinone **5** (80 mg, 76%) was obtained by HPLC purification of the crude residue on silica gel column (Luna 3 μm, 150 × 4.6mm, flow rate 1 mL/min) and hexane/EtOAc 98:2 (*v/v*) as mobile phase (t_R 20.4 min).

(*E*)-2-(3,7-dimethylocta-2,6-dien-1-yl)-3-methoxycyclohexa-2,5-diene-1,4-dione (**5**): yellow powder, HRESIMS *m/z* 275.1645 [M + H]$^+$ (calcd. for $C_{17}H_{23}O_3$: 275.1642). ^{1}H NMR (CDCl$_3$, 500 MHz): δ 6.67 (1H, d, *J* = 9.5 Hz), 6.59 (1H, d, *J* = 9.5 Hz), 5.05 (2H, m, overlapped), 4.02 (3H, s), 3.16 (2H, d, *J* = 7.3 Hz), 2.05 (2H, m), 1.96 (2H, m), 1.73 (3H, s), 1.65 (3H, s), 1.58 (3H, s). ^{13}C NMR (CDCl$_3$, 125 MHz): δ 187.8, 183.7, 155.2, 137.1, 136.6, 134.4, 132.2, 131.3, 123.9, 119.6, 60.8, 39.5, 26.5, 25.4, 22.4, 17.7, 16.0. ^{1}H, ^{13}C and HRESIMS spectra are reported in Supporting Information (Figures S10–S12).

3.1.4. Synthesis of 6-methoxy-7-(3-methylbut-2-en-1-yl)-3,4-dihydro-2H-benzo[b][1,4]thiazine-5,8-dione-1,1-dioxide (6) and (E)-7-(3,7-dimethylocta-2,6-dien-1-yl)-6-methoxy-3,4-dihydro-2H-benzo[b][1,4]thiazine-5,8-dione-1,1-dioxide (7)

The thiazinoquinone compounds, **6** and **7**, have been synthesized by the coupling of the respective quinone ring, **4** and **5**, with hypotaurine. For this purpose, 137 mg (0.67 mmol) of **4** were solubilized in 15 mL of a mixture ACN/EtOH 1:1 (*v/v*) while 14 mg (0.049 mmol) of **5** were solubilized in 5 mL of the same mixture. A water solution of hypotaurine (66.4 mg, 0.67 mmol in 3 mL for the synthesis of **6** and 6.0 mg, 0.049 mmol in 500 μL for the synthesis of **7**) was added dropwise to relative quinone. Salcomine as catalyst was added in portion and the mixtures were kept under stirring at room temperature for 48 h. After observing a color change from yellow to orange, TLC eluted with a mixture chloroform/EtOAc 7:3 allowed to control the end of the condensation before removing the solvent at rotavapor. The residues were dissolved in water and the mixtures were extracted with diethyl ether (3 × 60 mL). The organic phases were washed with brine, dried, filtered, and concentrated in vacuo. ^{1}H NMR spectra recorded for the two crude residues showed the presence in both cases of a single regioisomer which have been purified by HPLC on silica gel (Luna 3 μm column, 150 × 4.6mm, flow rate 1 mL/min) with a mobile phase hexane/EtOAc 1:1 (*v/v*) giving compounds **6** (190 mg, 92%) and **7** (15,1 mg, 81%), respectively.

6-methoxy-7-(3-methylbut-2-en-1-yl)-3,4-dihydro-2H-benzo[b][1,4]thiazine-5,8-dione-1,1-dioxide (**6**): slight orange powder; HRESIMS *m/z* 312.0909 [M + H]$^+$ (calcd. for $C_{14}H_{18}O_5NS$: 312.0901); *m/z* 334.0728 [M + Na]$^+$ (calcd. for $C_{14}H_{17}O_5NSNa$: 334.0720). ^{1}H and ^{13}C NMR data are reported in Table 1; NMR spectra are reported in Supporting Information (Figures S13–S15). HRESIMS spectrum is reported in Supporting Information (Figure S16).

(*E*)-7-(3,7-dimethylocta-2,6-dien-1-yl)-6-methoxy-3,4-dihydro-2H-benzo[b][1,4]thiazine-5,8-dione-1,1-dioxide (**7**): slight orange powder; HRESIMS *m/z* 380.1519 [M + H]$^+$ (calcd. for $C_{19}H_{26}O_5NS$: 380.1526; HRESIMS *m/z* 402.1337 [M + Na]$^+$ (calcd. for $C_{19}H_{25}O_5NSNa$: 402.1346). ^{1}H and ^{13}C NMR data are reported in Table 1; NMR spectra are reported in Supporting Information (Figures S17–S19). HRESIMS spectrum is reported in Supporting Information (Figure S20).

3.2. In Vitro Evaluation of Antiproliferative Activity of Compounds **4–7** in Cancer Cell Lines

3.2.1. Cell Culture

MCF-7, BxPC-3 and MG-63 cells were purchased from American Type Culture Collection (ATCC, Manassas, VA, USA). MCF-7 and MG-63 cells were cultured in DMEM medium, while BxPC-3 in RPMI medium, at 37 °C in a 5% CO_2 humidified atmosphere. DMEM and RPMI media were supplemented with 10% fetal bovine serum, penicillin–streptomycin (100 U/mL), and 2 mM L-glutamine. Cell morphology was monitored by using an inverted optical microscope. Cells were detached with 0.05% trypsin-EDTA to perform in vitro assays.

3.2.2. xCELLigence Assays

Antiproliferative assays were performed by using the xCELLigence System Real-Time Cell Analyzer (ACEA Biosciences, San Diego, CA, USA), as previously described [22]. MCF-7 cells were seeded at a cell density of 3000 cells/well, BxPC-3 at a cell density of 2500 cells/well, and MG-63 cells at a cell density of 4000 cells/well. Approximately 24 h after seeding, cancer cells were treated with 10 µM of compounds **4–7** and 0.05% DMSO vehicle for 72 h.

For data analysis, cell index (CI) values were normalized just before drug treatment to have normalized cell index (NCI) values. Normalized cell index was calculated as follows: NCI = CI $_{end of treatment}$/CI $_{normalization time}$. Real-time NCI proliferation curves were generated through the Real-Time Cell Analyzer (RTCA)-integrated software (Version 2.0.0.1301, ACEA Biosciences, San Diego, CA, USA). Growth inhibitory effects of compounds **4–7** are expressed either as cell index slopes relative to controls treated with DMSO vehicle or as cell doubling times. Cell index slopes and doubling times were calculated using the RTCA-integrated software within a 24-h time window.

3.2.3. Apoptosis Assay and Cell Cycle Analysis

After treatment with different concentrations (5, 10, and 20 µM) of compound **5**, BxPC-3 cells were stained with annexin-V-fluorescein isothiocyanate (FITC) and propidium iodide, using the FITC Annexin V Apoptosis Detection kit I (Becton Dickinson, BD, Franklin, NJ, USA) for flow cytometric detection of apoptotic and necrotic cells. Samples were prepared according to manufacturer's protocol and three independent experiments were carried out.

For cell cycle analysis of BxPC-3 cells treated with 20 µM of quinone **5** for 24 h, pancreatic cancer cells were permeabilized with 70% cold ethanol for 1 h and stained for 30 min with a solution containing 50 µg/mL propidium iodide (Sigma Aldrich, St. Louis, MO, USA) and 10 µg/mL RNase A (EuroClone S.p.a., Pero, MI, Italy) in calcium and magnesium-free PBS. Four independent experiments were carried out. All samples were acquired by NAVIOS flow cytometer and analysed by Kaluza software (Beckman Coulter). 10,000 events were acquired for each sample.

3.2.4. Statistical Analysis

Data represent the mean (±standard deviation, SD) of at least three independent experiments. The one-way analysis of variance (ANOVA) method was applied to compare means of more than two groups and Dunnett's method was used as post-hoc test to compare multiple groups versus a control group. Two-group comparisons were performed using Student's *t*-test. *p*-values < 0.05 were considered to be statistically significant. Statistical analysis was performed using the GraphPad Prism Software Version 5 (GraphPad Software Inc., San Diego, CA, USA).

4. Conclusions

The synthesis of the prenylated compounds **4–7**, inspired by the marine thiazinoquinone aplidinone A, has been performed through a versatile synthetic protocol, and it represents an example of a successful strategy. In addition, our synthetic procedure resulted in a considerable improvement of the overall yield of **5** with respect to previously reported synthesis [20]. Moreover, the nucleophilic addition of hypotaurine to quinone ring of **4** and **5** to give the thiazinoquinone **6** and **7**, respectively, was completely regioselective with respect to condensation with other quinones featuring smaller and /or less flexible side chains [16–19] indicating a key role of the side chain in the condensation reaction. To assess the anticancer potential and structure-activity relationships of compounds **4–7**, we evaluated their effects on cell viability of MCF-7, Bx-PC3, and MG-63 cell lines. The biological assays demonstrated the quinone **5** as the most relevant compound in the series exhibiting a similar extent of toxicity against three different tumor models. Finally, compound **5** exerted cytostatic activity through induction of cell cycle arrest, resulting in a significant segregation of cells into G0/G1 phase at concentration of 20 μM.

Supplementary Materials: The following are available online at http://www.mdpi.com/1660-3397/17/12/684/s1. Figure S1: ^{1}H NMR spectrum in CDCl$_3$ (500 MHz) of compound 3-**R$_1$**. Figure S2: ^{13}C NMR spectrum in CDCl$_3$ (125 MHz) of compound 3-**R$_1$**. Figure S3: HRESIMS spectrum of compound 3-**R$_1$**. Figure S4: ^{1}H NMR spectrum in CDCl$_3$ (500 MHz) of compound 3-**R$_2$**. Figure S5: ^{13}C NMR spectrum in CDCl$_3$ (125 MHz) of compound 3-**R$_2$**. Figure S6: HRESIMS spectrum of compound 3-**R$_2$**. Figure S7: ^{1}H NMR spectrum in CDCl$_3$ (500 MHz) of compound **4**. Figure S8: ^{13}C NMR spectrum in CDCl$_3$ (125 MHz) of compound **4**. Figure S9: HRESIMS spectrum of compound **4**. Figure S10: ^{1}H NMR spectrum in CDCl$_3$ (500 MHz) of compound **5**. Figure S11: ^{13}C NMR spectrum in CDCl$_3$ (125 MHz) of compound **5**. Figure S12: HRESIMS spectrum of compound **5**. Figure S13: ^{1}H NMR spectrum in CDCl$_3$ (500 MHz) of compound **6**. Figure S14: ^{13}C NMR spectrum in CDCl$_3$ (125 MHz) of compound **6**. Figure S15: ^{1}H-^{13}C HMBC spectrum in CDCl$_3$ (700 MHz) of compound **6**. Figure S16: HRESIMS spectrum of compound **6**. Figure S17: ^{1}H NMR spectrum in CDCl$_3$ (500 MHz) of compound **7**. Figure S18: ^{13}C NMR spectrum in CDCl$_3$ (125 MHz) of compound **7**. Figure S19: ^{1}H-^{13}C HMBC spectrum in CDCl$_3$ (700 MHz) of compound **7**. Figure S20: HRESIMS spectrum of compound **7**. Figure S21: HPLC chromatogram of compound **4**. Figure S22: HPLC chromatogram of compound **5**. Figure S23: HPLC chromatogram of compound **6**. Figure S24: HPLC chromatogram of compound **7**.

Author Contributions: Conceptualization, C.P., C.I. and M.M.; Data curation, G.D.S, M.C., C.I., P.L., A.A. and M.M.; Formal analysis, I.L., G.D.S., M.C. and P.L.; Funding acquisition, M.M.; Investigation, G.D.S., M.C. and C.I.; Methodology, I.L., M.C. and P.L.; Writing—original draft, G.D.S., C.P., C.I. and M.M.; Writing—review & editing, G.D.S., C.P., M.C., C.I., P.L., A.A. and M.M.

Funding: This work was supported by a grant from Regione Campania-POR Campania FESR 2014/2020 "Combattere la resistenza tumorale: piattaforma integrata multidisciplinare per un approccio tecnologico innovativo alle oncoterapie-Campania Oncoterapie" (*Project N.* B61G18000470007). Biological studies were funded by the Italian Ministry of Health, by Current Research Funds to IRCCS-CROB, Rionero in Vulture, Potenza, Italy.

Conflicts of Interest: The authors declare no conflict of interest.

References

1. Polyakov, N.; Leshina, T.; Fedenok, L.; Slepneva, I.; Kirilyuk, I.; Furso, J.; Olchawa, M.; Sarna, T.; Elas, M.; Bilkis, I.; et al. Redox-Active Quinone Chelators: Properties, Mechanisms of Action, Cell Delivery, and Cell Toxicity. *Antioxid. Redox Signal.* **2018**, *28*, 1394–1403. [CrossRef] [PubMed]

2. Menna, M.; Imperatore, C.; D'Aniello, F.; Aiello, A. Meroterpenes from Marine Invertebrates: Structures, Occurrence, and Ecological Implications. *Mar. Drugs* **2013**, *11*, 1602–1643. [CrossRef] [PubMed]

3. Haque, M.A.; Sailo, B.L.; Padmavathi, G.; Kunnumakkara, A.B.; Jana, C.K. Nature-inspired development of unnatural meroterpenoids as the non-toxic anti-colon cancer agent. *Eur. J. Med. Chem.* **2018**, *160*, 256–265. [CrossRef] [PubMed]

4. Li, G.-Y.; Li, B.-G.; Yang, T.; Yin, J.-H.; Qi, H.-Y.; Liu, G.-Y.; Zhang, G.-L. Sesterterpenoids, terretonins A-D, and an alkaloid, asterrelenin, from Aspergillus terreus. *J. Nat. Prod.* **2005**, *68*, 1243–1246. [CrossRef] [PubMed]

5. García, P.A.; Hernández, Á.P.; San Feliciano, A.; Castro, M.Á. Bioactive prenyl- and terpenyl-quinones/hydroquinones of marine origin. *Mar. Drugs* **2018**, *16*, 292. [CrossRef]

6. Zubia, E.; Ortega, M.J.; Salva, J. Natural products chemistry in marine ascidians of the genus Aplidium. *Mini-Rev. Org. Chem.* **2005**, *2*, 389–399. [CrossRef]

7. Menna, M.; Aiello, A.; D'Aniello, F.; Fattorusso, E.; Imperatore, C.; Luciano, P.; Vitalone, R. Further investigation of the mediterranean sponge *Axinella polypoides*: Isolation of a new cyclonucleoside and a new betaine. *Mar. Drugs* **2012**, *10*, 2509–2518. [CrossRef]

8. Imperatore, C.; Luciano, P.; Aiello, A.; Vitalone, R.; Irace, C.; Santamaria, R.; Li, J.; Guo, Y.-W.; Menna, M. Structure and Configuration of Phosphoeleganin, a Protein Tyrosine Phosphatase 1B Inhibitor from the Mediterranean Ascidian *Sidnyum elegans*. *J. Nat. Prod.* **2016**, *79*, 1144–1148. [CrossRef]

9. Imperatore, C.; D'Aniello, F.; Aiello, A.; Fiorucci, S.; D'Amore, C.; Sepe, V.; Menna, M. Phallusiasterols A and B: Two new sulfated sterols from the mediterranean tunicate *Phallusia fumigata* and their effects as modulators of the PXR receptor. *Mar. Drugs* **2014**, *12*, 2066–2078. [CrossRef]

10. Luciano, P.; Imperatore, C.; Senese, M.; Aiello, A.; Casertano, M.; Guo, Y.-W.; Menna, M. Assignment of the Absolute Configuration of Phosphoeleganin via synthesis of Model Compounds. *J. Nat. Prod.* **2017**, *80*, 2118–2123. [CrossRef]

11. Casertano, M.; Imperatore, C.; Luciano, P.; Aiello, A.; Menna, M.; Putra, M.Y.; Gimmelli, R.; Ruberti, G. Chemical Investigation of the Indonesian Tunicate *Polycarpa aurata* and Evaluation of the Effects Against *Schistosoma mansoni* of the Novel Alkaloids Polyaurines A and B. *Mar. Drugs* **2019**, *17*, 278. [CrossRef] [PubMed]

12. Menna, M.; Aiello, A.; D'Aniello, F.; Imperatore, C.; Luciano, P.; Vitalone, R.; Irace, C.; Santamaria, R. Conithiaquinones A and B, Tetracyclic Cytotoxic Meroterpenes from the Mediterranean Ascidian *Aplidium conicum*. *Eur. J. Org. Chem.* **2013**, 3241–3246. [CrossRef]

13. Aiello, A.; Fattorusso, E.; Luciano, P.; Macho, A.; Menna, M.; Munoz, E. Antitumor Effects of Two Novel Naturally Occurring Terpene Quinones Isolated from the Mediterranean Ascidian *Aplidium conicum*. *J. Med. Chem.* **2005**, *48*, 3410–3416. [CrossRef] [PubMed]

14. Aiello, A.; Fattorusso, E.; Luciano, P.; Mangoni, A.; Menna, M. Isolation and structure determination of aplidinones A-C from the Mediterranean ascidian *Aplidium conicum*: A successful regiochemistry assignment by quantum mechanical ^{13}C NMR chemical shift calculations. *Eur. J. Org. Chem.* **2005**, *2005*, 5024–5030. [CrossRef]

15. Imperatore, C.; Cimino, P.; Cebrián-Torrejón, G.; Persico, M.; Aiello, A.; Senese, M.; Fattorusso, C.; Menna, M.; Doménech-Carbó, A. Insight into the mechanism of action of marine cytotoxic thiazinoquinones. *Mar. Drugs* **2017**, *15*, 335. [CrossRef] [PubMed]

16. Imperatore, C.; Persico, M.; Aiello, A.; Luciano, P.; Guiso, M.; Sanasi, M.F.; Taramelli, D.; Parapini, S.; Cebrián-Torrejón, G.; Doménech-Carbó, A.; et al. Marine inspired antiplasmodial thiazinoquinones: Synthesis, computational studies and electrochemical assays. *RSC Adv.* **2015**, *5*, 70689–70702. [CrossRef]

17. Imperatore, C.; Persico, M.; Senese, M.; Aiello, A.; Casertano, M.; Luciano, P.; Basilico, N.; Parapini, S.; Paladino, A.; Fattorusso, C.; et al. Exploring the antimalarial potential of the methoxy-thiazinoquinone scaffold: Identification of a new lead candidate. *Bioorg. Chem.* **2019**, *85*, 240–252. [CrossRef]

18. Gimmelli, R.; Persico, M.; Imperatore, C.; Saccoccia, F.; Guidi, A.; Casertano, M.; Luciano, P.; Pietrantoni, A.; Bertuccini, L.; Paladino, A.; et al. Thiazinoquinones as new promising multi-stage schistosomicidal compounds impacting *Schistosoma mansoni* and egg viability. *ACS Infect. Dis.* **2019**. [CrossRef]

19. Aiello, A.; Fattorusso, E.; Luciano, P.; Menna, M.; Calzado, M.A.; Muñoz, E.; Bonadies, F.; Guiso, M.; Sanasi, M.F.; Cocco, G.; et al. Synthesis of structurally simplified analogues of aplidinone A, a pro-apoptotic marine thiazinoquinone. *Bioorg. Med. Chem.* **2010**, *18*, 719–727. [CrossRef]

20. Fedorov, S.N.; Radchenko, O.S.; Shubina, L.K.; Balaneva, N.N.; Bode, A.M.; Stonik, V.A.; Dong, Z. Evaluation of Cancer-Preventive Activity and Structure–Activity Relationships of 3-Demethylubiquinone Q2, Isolated from the Ascidian *Aplidium glabrum*, and its Synthetic Analogs. *Pharm. Res.* **2006**, *23*, 70–81. [CrossRef]

21. Della Sala, G.; Agriesti, F.; Mazzoccoli, C.; Tataranni, T.; Costantino, V.; Piccoli, C. Clogging the Ubiquitin-Proteasome Machinery with Marine Natural Products: Last Decade Update. *Mar. Drugs* **2018**, *16*, 467. [CrossRef] [PubMed]

22. Teta, R.; Della Sala, G.; Esposito, G.; Via, C.W.; Mazzoccoli, C.; Piccoli, C.; Bertin, M.J.; Costantino, V.; Mangoni, A. A joint molecular networking study of a Smenospongia sponge and a cyanobacterial bloom revealed new antiproliferative chlorinated polyketides. *Org. Chem. Front.* **2019**, *6*, 1762–1774. [CrossRef]

MDPI

St. Alban-Anlage 66

4052 Basel

Switzerland

Tel. +41 61 683 77 34

Fax +41 61 302 89 18

www.mdpi.com

Marine Drugs Editorial Office

E-mail: marinedrugs@mdpi.com

www.mdpi.com/journal/marinedrugs